Philosophy of Agricultural Science

JAPANESE SOCIETY SERIES

General Editor: Yoshio Sugimoto

Lives of Young Koreans in Japan
Yasunori Fukuoka

Globalization and Social Change in Contemporary Japan
J.S. Eades, Tom Gill and Harumi Befu

Coming Out in Japan: The Story of Satoru and Ryuta
Satoru Ito and Ryuta Yanase

Japan and Its Others:
Globalization, Difference and the Critique of Modernity
John Clammer

Hegemony of Homogeneity:
An Anthropological Analysis of Nihonjinron
Harumi Befu

Foreign Migrants in Contemporary Japan
Hiroshi Komai

A Social History of Science and Technology in
Contempory Japan, Volume 1
Shigeru Nakayama

Farewell to Nippon: Japanese Lifestyle Migrants in Australia
Machiko Sato

The Peripheral Centre:
Essays on Japanese History and Civilization
Johann P. Arnason

A Genealogy of 'Japanese' Self-images
Eiji Oguma

Class Structure in Contemporary Japan
Kenji Hashimoto

An Ecological View of History
Tadao Umesao

Nationalism and Gender
Chizuko Ueno

Native Anthropology: The Japanese Challenge
to Western Academic Hegemony
Takami Kuwayama

Youth Deviance in Japan: Class Reproduction of Non-Conformity
Robert Stuart Yoder

Japanese Companies: Theories and Realities
Masami Nomura and Yoshihiko Kamii

From Salvation to Spirituality:
Popular Religious Movements in Modern Japan
Susumu Shimazono

The 'Big Bang' in Japanese Higher Education:
The 2004 Reforms and the Dynamics of Change
J.S. Eades, Roger Goodman and Yumiko Hada

Japanese Politics: An Introduction
Takashi Inoguchi

A Social History of Science and Technology in
Contempory Japan, Volume 2
Shigeru Nakayama

Gender and Japanese Management
Kimiko Kimoto

Philosophy of Agricultural Science: A Japanese Perspective
Osamu Soda

A Social History of Science and Technology in
Contempory Japan, Volume 3
Shigeru Nakayama and Kunio Goto

Japan's Urban Underclass: Day Laborers and the Homeless
Hideo Aoki

Social Stratification and Inequality Series

Inequality amid Affluence: Social Stratification in Japan
Junsuke Hara and Kazuo Seiyama

Intentional Social Change: A Rational Choice Theory
Yoshimichi Sato

Constructing Civil Society in Japan:
Voices of Environmental Movements
Koichi Hasegawa

Deciphering Stratification and Inequality: Japan and beyond
Yoshimichi Sato

Social Justice in Japan: Concepts, Theories and Paradigms
Ken-ichi Ohbuchi

Advanced Social Research Series

A Sociology of Happiness
Kenji Kosaka

MODERNITY AND IDENTITY IN ASIA SERIES

Globalization, Culture and Inequality in Asia
Timothy S. Scrase, Todd Miles Joseph Holden and Scott Baum

Looking for Money:
Capitalism and Modernity in an Orang Asli Village
Alberto Gomes

Governance in Asia:
Developmental State, Asian Values and Third-Wave Democracy
Takashi Inoguchi and Chung-Si Ahn

Philosophy of Agricultural Science

A Japanese Perspective

Osamu Soda

Trans Pacific Press
Melbourne

Nougaku Genron by Osamu Soda © 2000 by Osamu Soda
Originally published in Japanese by Iwanami Shoten, Publishers, Tokyo, 2000.

This English language edition published in 2006 by Trans Pacific Press, Melbourne by arrangement with the author c/o Iwanami Shoten, Publishers, Tokyo. Trans Pacific Press, PO Box 120, Rosanna, Melbourne, Victoria 3084, Australia
Telephone: +61 3 9459 3021 Fax: +61 3 9457 5923
Email: info@transpacificpress.com
Web: http://www.transpacificpress.com

Designed and set by digital environs Melbourne. enquiries@digitalenvirons.com

Printed by BPA Print Group, Burwood, Victoria, Australia

Distributors

Asia and the Pacific
Kinokuniya Company Ltd.

Head office:
38-1 Sakuragaoka 5-chome, Setagaya-ku, Tokyo
156-8691, Japan
Telephone: +81 (0)3 3439 0161
Fax: +81 (0)3 3439 0839
Email: bkimp@kinokuniya.co.jp
Web: www.kinokuniya.co.jp

Asia-Pacific office:
Kinokuniya Book Stores of Singapore Pte., Ltd.
391B Orchard Road #13-06/07/08
Ngee Ann City Tower B
Singapore 238874
Telephone: +65 6276 5558
Fax: +65 6276 5570
Email: SSO@kinokuniya.co.jp

Japan
Kyoto University Press
Kyodai Kaikan
15-9 Yoshida Kawara-cho
Sakyo-ku, Kyoto 606-8305
Telephone: (075) 761-6182
Fax: (075) 761-6190
Email: sales@kyoto-up.gr.jp
Web: http://www.kyoto-up.gr.jp

UK and Europe
Asian Studies Book Services
Franseweg 55B, 3921 DE Elst, Utrecht,
The Netherlands
Telephone: +31 318 470 030
Fax: +31 318 470 073
Email: info@asianstudiesbooks.com
Web: http://www.asianstudiesbooks.com

USA and Canada
International Specialized Book
Services (ISBS)
920 NE 58th Avenue, Suite 300
Portland, Oregon 97213-3786
USA
Telephone: (800) 944-6190
Fax: (503) 280-8832
Email: orders@isbs.com
Web: http://www.isbs.com

Australia and New Zealand
UNIREPS
University of New South Wales
Sydney, NSW 2052
Australia
Telephone: +61(0)2-9664-0999
Fax: +61(0)2-9664-5420
Email: info.press@unsw.edu.au
Web: http://www.unireps.com.au

ISSN 1443–9670 (Japanese Society Series)

ISBN 1-876843-26-8, 978-1-876843-26-7 (Hardcover)
ISBN 1-876843-30-6, 978-1-876843-30-4 (Paperback)

Contents

Figures

Photos

Tables

Foreword

The philosophy of agricultural science was initiated as an academic discipline in Germany in 1919 by Richard Krzymowski. By curious coincidence, the philosophy of forestry science was established by H. Weber in the same year. At the time, the world was in turmoil and faced many problems, and philosophical scholarship was required to conform to the clear guiding principles of academic studies. In this sense, both the philosophy of agricultural science in particular and the philosophy of science in general have had to respond to the crises in academic thinking.

The philosophy of agricultural science was introduced to Japan in 1932 when Krzymowski was translated into Japanese for the first time. In 1952, Kyoto University established a chair in the philosophy of agricultural science in the Faculty of Agriculture – the first in Japan. This was a significant development in the turbulence of postwar Japan, which was confronting serious food shortages as well as a range of issues concerning rural communities. At the time, the accumulation of various agricultural problems made it necessary for the academic discipline to define the nature of agriculture and articulate the issues and methods of the philosophy of agricultural science.

To the best of my knowledge, Kyoto University is the only academic institution in the world with an academic unit composed of a full professor, an associate professor and a research associate, which devotes itself to research in the philosophy of agricultural science. The foundation professor of this unit was Professor Kashiwa Sukekata, who was succeeded by Professor Sakamoto Keiichi. I took up the chair after these two pioneers.

The chapters in this book are based on various research papers that I have published over the past fifteen years. I extensively rewrote and expanded them, and have attempted to systematize them in this study. As I will elaborate in Chapter 3, Professor Kashiwa established the philosophy of the agricultural science of production, while Professor Sakamoto developed the agricultural science of life. In contrast, my own philosophy of agricultural science is *ba*, which emphasizes the placeness and regionality of agriculture and points to the significance and limitations of its globalization.

This book is based on philosophical thinking about the experiences of Japanese agriculture and small-farming areas in Asia. I have detailed these thoughts in the hope that those who inhabit the large-farming areas in other parts of the world might better understand and empathize with the differences. It is my modest hope that this book contributes to the proper and fair development of contemporary agriculture in the world and to the future of agricultural science.

For brevity of expression, I have used 'agriculture' or 'agricultural' in chapter titles and in headings where the terms, in some contexts, refer to the entire primary industry sector including forestry and fisheries. This is simply an attempt to avoid verbosity, and I have tried to articulate the point in the text itself.

Throughout the text, I followed the Japanese convention in which the family name or surname precedes the given or personal name.

I am grateful to Professor Yoshio Sugimoto, Professor of Sociology at La Trobe University and Director of Trans Pacific Press for his encouragement and support and to Mr. Karl Smith for his conscientious and painstaking editorial work. The publication of this book was made possible by the financial support of the Japan Society for the Promotion of Science that provided a 2005–2006 Grant-in-Aid for Publication of Scientific Research Results. My research for this book was supported at various stages by grants-in-aid from the Japan Society for the Promotion of Science, a grant from the Nissei Foundation and a scholarship from Mr. Nagai Kōki, to all of which I am most indebted.

I also wish to express my gratitude to many agriculturalists, agricultural organizations and colleagues of mine for making cooperative contributions to the writing of this book.

Soda Osamu
January 2006
Fukui

Preface to the English Edition

The present work was originally written for Japanese researchers in agricultural science and has now been translated into English. Many English-speaking readers probably do not know the history of or natural conditions surrounding Japanese agriculture, or the situation of Japanese farms and rural villages. I am afraid that this may pose considerable difficulties for these readers in understanding this work. Robert Smith, the American rural sociologist who conducted a long-term investigation of Japanese rural communities, says that 'there are villages in Japan, and America has no villages like those in Japan.' Readers of the English edition of this book may not understand what Smith meant by this comment. To help them understand this work, I have added an introductory chapter to this English version to list several points at issue. In addition, the construction and contents of each chapter are explained in advance in an attempt to help comprehend the entire picture of this work.

Placeness of agriculture and totality of rural regions

Industrial production processes inorganic materials and produces useful products. In such production, economic conditions such as wages, land price, market conditions or energy supply conditions are the key elements in investment decision making. In agricultural production, by contrast, natural conditions such as geographical or climatic conditions have significant influence. For example, in automobile production, the same cars can be manufactured anywhere in the world if the same economic conditions are met. Agriculture, in contrast, is an industry dealing with living organisms, where it is important to 'raise' plants and animals by concerted efforts of people and nature, instead of 'making' products as is done in the manufacturing industry. In agriculture, therefore, elements such as the biological characteristics of the plants or animals in question, the northern or southern limits of their production, whether farmlands are flat or sloping, vast or limited – like the land between mountains – their soil conditions and climatic conditions, are crucially important, independently of economic conditions. Such elements do have some influence on industrial production as well, but their significance in industrial production is certainly different from that in agriculture. This characteristic of agriculture is referred to as 'regionality' or 'placeness' of agriculture in this work and is emphasized throughout the work.

People have developed various characteristic lifestyles under their respective climatic and cultural conditions and spend their days in their respective regional communities. Different regions show different autonomous movements in both economic and living aspects. Their main industry may be the textile industry in some regions and the steel industry in others. Small- or medium-scale local industries may be dominant in some regions, while a large-scale chemical industry invited to the region may be dominant in other regions. Reflecting these industrial structures, each region has developed a unique socio-economic structure, which can be seen in its residents' lives and culture, the relations between towns and rural villages in it, its position in Japan, and its relations with foreign countries.

At the same time, different regions are connected by routes running through mountains, valleys, or a basin, while each region generally constitutes an ecosystem unit. It can thus be said that a region is an independent zone of livelihood, economy and ecology while also being open to the wider world. A region is a total entity consisting of three aspects – economic, living and ecological – combined together. Ideally a region should have no problems or irregularities in any of these aspects. I thus believe that a goal for 21st century societies is that regions pursue comprehensive values, into which the three values – economic, life and ecological – are integrated in a well-balanced manner.

Japanese agriculture as *shōnōken* (small-scale farming zone) agriculture

If we build trade relations by giving top priority to market mechanisms based on production costs under specific natural conditions alone, placeness will be largely ignored and regions with poor conditions will simply be regarded as unfit for agriculture. Their agriculture will thus fall behind the competition and decline. Whether this is right or wrong must be questioned. The situation will be complicated further by considering historical nature in addition to placeness. Let us consider as an example the situation of Japanese and US agriculture. For more than 2,000 years, Japan has developed family-run, self-sufficient agriculture mainly for rice farming, with the average size of a farm being a little more than 1 hectare. In addition, Japan's mountainous terrain and limited flat lands has driven roughly half of her farming activities to areas that are situated along valleys or rivers between mountains and which are typically on relatively steep gradients. With the advent of modern industrial society, Japanese farms transformed into small-scale commercial farms. This was followed by the postwar economic growth, during which most farmers could no longer depend solely on agriculture for their livelihood. This resulted in their division into a large number of part-time farms and a small number of full-time farms.

In addition, Japanese rice farming depends on the rainy climate that is characteristic of Asian monsoon regions. But Japanese farming has not only produced rice; it has also produced other food under conditions of intense rainfall and flooding, by controlling nature and protecting low-lying areas from flood damage. In this context, there are indivisible relationships between Japanese agriculture/rural villages and nature, between towns and rural villages, and between upriver areas and down-river areas. This is typical of agriculture in *shōnōken* (small-scale farming zones), which account for a majority of the agriculture in Asia and Africa. In contrast, what may be called the New World *dainōken* (large-scale farming zones), typical of the United States, Australia and Brazil, are significantly different from *shōnōken.*

From the beginning of its development by immigrants, agriculture in the US developed as an export industry producing products for export to Europe, and has been able to develop overwhelming productivity as a large-scale mechanized farming zone, or a type of large-scale farming zone, producing mainly wheat, soy beans and livestock, generally using vast, flat and fertile farmlands. In 2000 the average farm size in the US was about 200 hectares, approximately 140 times as large as the Japanese average (1.4 hectares). The agriculture of different regions and countries reflects their unique natural conditions and historical course.

'Market failure' and the multifunctionality of agriculture

If you visit an agricultural zone in the US, where large-scale farming is practiced, you will see farm houses located 500 meters or a kilometer away from each other. As Smith says, there are no villages like those in Japan. Rather than forming a village, individual farms are independently connected to nearby towns. In Japan, communal water management has been important, and because of their small size, typically 30 to 50 farm houses are clustered, forming a rural community. In EU countries, so-called *chūnōken* (medium-scale farming zones) have been formed which are different from those of

Photo 1: An agricultural village community on the slope of a mountain area in Japan. Villages of this type maintain 40% of agricultural production in Japan. Courtesy of Ikeda Yoshimi, Miyazaki City.

Japan or the US. In those medium farming zones, the relationship between agricultural villages and nature, between towns and rural villages, and how rural communities should be structured, have always been at issue, even if not as prominently as they have been in Japan. I am afraid that understanding the present work would be difficult without understanding the fact that the state of agriculture and rural villages are essentially different in different regions.

Figure 6.1 in Chapter 6 represents a typical pattern of Japanese communities. Readers are encouraged to begin by reading the explanation of this figure in order to develop an idea of a Japanese rural village and, in turn, the Asian small farming zone. Thirty to 50 farms have formed a cooperative community by jointly controlling water, cultivating rice and worshipping their ancestors at a shrine. There about 120,000 such communities throughout Japan. Across the nation there are numerous clusters of 10 to 20 such communities which have formed municipalities called *shi*, *chō* (also called *machi*) or *son* (also called *mura*), which also contain small or medium urban districts. Municipalities located in upriver areas between mountains serve those municipalities located down river by managing forests and farmlands in such a manner that

Photo 2: Rice cultivation in paddy fields in the plains in Aomori Prefecture, Japan. Settlements comprise 30–50 small-scale farming households with an average farm area of 1.4 hectares. Most households engage in part-time farming and household-members work in nearby towns and cities.

flood and other damage is minimized. Residents of large towns and cities visit upriver areas to relax in pastoral scenery and enjoy traditional festivals. These phenomena are part of the multifunctionality of agriculture. They are also external economies in economic terms. Although there is no economic exchange of such functions, they can be regarded as important functions of agriculture and rural villages.

Although the 'multifunctionality of agriculture' has been frequently raised as an issue at the World Trade Organization's agricultural trade negotiations, it has been difficult to reach agreement. Due to its particular history as described above, Japanese agriculture, particularly rice farming, entails complementary relationships between agriculture and nature, between upriver areas and down river areas, and between rural communities, villages and towns. These relationships have a public nature that goes beyond the economic efficiency of agricultural production. This broadly holds true for all small-scale farming zone in Asia. These relationships are the non-economic functions or multifunctionality of agriculture. The 'multifunctionality' theory was originally proposed by EU countries to emphasize differences between US and European agriculture, recognizes the uniqueness of European agriculture and its rural communities, and therefore seeking certain protective measures which are not required by its manufacturing industry. This proposal had strong resonance in Japan also. If market mechanisms are given top priority, the multifunctionality of agriculture is not taken into account, and the prosperity of decline of agriculture/rural villages will be determined solely on the price differentials of agricultural products. Many of agriculture's non-economic functions will thus be lost. Furthermore, as described in Chapter 4, exclusively prioritizing market mechanisms often encourages environmental destruction. In this regard, some have referred to these negative aspects of market

Photo 3: Rice harvesting in the United States, based on large-scale farming in spacious and flat farmland. Courtesy of Nihon Nōgyō Shimbun.

mechanisms as 'market failure', regarding them as phenomena that must be corrected. The present work shares this perspective.

Global environment and regional environment

I believe that in the 21st century, human activities and science should focus on achieving three main values – economic, ecological and life values – in a harmonious manner *in each region*, rather than allowing the fate of regional communities to be determined by market mechanisms that only recognize economic values.

I have argued for many years that in order to realize and maintain comprehensive values in a harmonious manner, minimum agricultural limits (agri-minimum) must be established to be observed in each country or region and, in conjunction with this, maximum industrial limits (indus-maximum) should be observed by their manufacturing industries.

Globalization does not have to mean accumulating compromises by mutually abandoning the individuality of each country or region, nor need it mean simply expanding the international division of labor.

The term 'global environment' is frequently heard. The more international a place is, the more frequently this term is used. It is important that people make global-scale arguments because humankind has its back against the wall due to environmental problems. These arguments, however, are in the wrong order. The global environment is truly protected when individual regions have been established where human well-being and desirable relationship between humans and nature are sought, and when these regions are linked to each other on a global scale. In other words, the regional environment must come before the global environment.

Agriculture and agricultural science of *ba*

In this book, I aim to establish an agricultural science of *ba* and wish to use this Japanese notion throughout the text. *Ba* is a concept which many Japanese philosophers, including Nishida Kitarō, have attempted to elucidate. Though I refer to local communities, using terms such as place, space and region, I use *ba* as a generic concept that covers all of these and base this study on that notion, which comprises four components.

In the first place, *ba* as used in this book, is logical space. I argue that society is a total entity where three major values – economic, ecological and life values – ought to be pursued in harmony. This is a theoretical framework where *ba* plays a central role. Secondly, *ba* is a dynamic and formative place. A local community is not a static existence but a dynamic place that is continuously shaped and reshaped to solve newly arising problems. Thirdly, however, agriculture frequently faces circumstances beyond human control. Engaged in biological production, it is inevitably entangled with regionality or placeness, subject to natural and geographic conditions. Finally, *ba* is an everyday concrete lifeworld whose specific example is a local agricultural community where face-to-face relationships predominate.

In Asia where small-scale agriculture prevails, I maintain that we should study rural communities comprehensively and organically, using the multiple facets of the *ba* concept.

Structure of the book

In this book, I have described, among others, the present situation, issues, courses and research methodology of agricultural science. Here I will summarize the points of each chapter and describe some prospects for future studies.

In Chapter 1, I describe issues and perspectives of *Landwirtschaftslehre*. The mission of *Landwirtschaftslehre* as the philosophy of agricultural science lies in studies such as: (1) the history of agricultural science and its thought; (2) elucidating the problems and roles of agriculture and the rural community; (3) the establishment of values and aims of agricultural science and possible courses toward their realization, based on the elucidation of (2); and (4) research methodology and system. I argue that such studies must have a strong intention to realize a desired `*ba*` (place) where people can lead a human lifestyle, through the reflection and self-consciousness of modern agricultural science and its creative, constructive efforts to solve problems. Creating a place where people can lead a human lifestyle means harmoniously realizing all three sets of values – economic, ecological, and life values – as the minimum values required by modern society. In addition, I explain that each 'community' as a life-world is the most appropriate place to realize such values.

In Chapter 2, I discuss the relationship between human and nature in agriculture. Humans gradually exited from a long hunting-gathering phase by forming agricultural societies in which they raise and utilize domestic animals and crops. Humans have three different relationships with nature: an interdependent and symbiotic relationship with domestic animals, crops, etc.; a mutually exclusive and competitive relationship with noxious animals and weeds; and a habitat segregative and coexistent relationship with general wild animals and plants. There has been, however, a problematic division between humans and nature in each of these relationships. In response, some have begun to argue, deep ecologists for example, that nature – animals, plants and places – must be accorded the same rights as humans. We must not forget, however, that humans must eat to live. While some (hastily) argue for 'coexistence', I make it clear that such coexistence can only be achieved as a result of human's struggle for their lives and will be able to establish an appropriate mutual relationship with nature, or a 'world of formative equilibrium', which will be imagined based on human reflection and self-consciousness.

Recognizing that modern science is in an important phase, in Chapter 3 I discuss the problem of values in science, the social responsibility of scientists, and the values and aims of modern agricultural science. I considered the relationship between scientific investigations and values as well as the understanding of values in agricultural science. Agricultural science is a practical science, where we actively set values and aims and strive to achieve them. I explain that the main values that have been pursued by agricultural science, or for which there have been demands for agricultural science to pursue, after World War II have been, in the order of occurrence: (1) economic value at the survival level, followed by that at the living level; (2) ecological value for life and environment; (3) life value emphasizing living and society; and (4) total values that harmoniously integrate all three sets of values. It is understood that (1) has been proposed by 'agricultural science of production', and (2) and (3) by 'agricultural science of life', while (4) represents my view of 'agricultural science of "*ba*"'. I argue that the aim of modern agricultural science is to harmoniously realize economic, ecological, and life values, which currently have a trade-off relationship with each other, in the '*ba*' (place) of community.

In Chapters 4, 5 and 6, I consider each of the above values in accordance with the framework outlined in Chapter 3 (see in particular Table 3.1), which represents a condensed scheme of this book, and discuss the present situation and future directions.

In Chapter 4, I discuss the relationship between agriculture, forestry and fishing industries and economic value. I point out that agriculture, forestry and fishing industries have, in modern economic society, evolved based on the market mechanisms accompanied by market failures and government failures, as well as three characteristics that make it difficult for these industries to conform to market mechanisms: a technological and managerial characteristic, a multiple-value characteristic, and a regional characteristic (placeness). I understand that these characteristics demand that the market mechanisms be modified for the sake of stable future economic development as well as for the resolution of environmental problems. I finally argue that it is necessary to reconstruct agricultural economics as the economics of living systems and agricultural science as the comprehensive science of biosystems to realize sustainable agriculture.

In Chapter 5, I discuss the relationship between agriculture, forestry and fishing industries and ecological value. These industries are restricted tightly by their ecological environment. Under such circumstances, man has had to achieve increased food production to feed the population that has been continuously expanding in modern times. Agriculture and agricultural science have realized increased food production by strenuous efforts and technological success, or so-called 'industrialization of agriculture'. However, in the process the problem of burdens that are imposed on the environment by agricultural production activities has been revealed. Meanwhile, the argument that agriculture is a multiple-value industry, with many meanings for environmental preservation and people's lives, has become stronger. I assert that, faced with a current global population of 6 billion, which is expected to eventually reach 9 billion, we should start from the urgent question: 'Can humans protect their environment without starving?', and that if we are to be able to answer 'yes' it is necessary to establish a new view of nature and ethics, to develop sustainable agricultural technologies and management, and to construct a comprehensive social recycling system that connects agriculture with consumers.

In Chapter 6, I consider the relationship of agriculture and the rural community with life value from three different perspectives: the social characteristics of the rural community; the meanings of farm work for people; and the reorganization of cultural activities.

The rural village community based on the fundamental unit of settlement undergoes changes and reorganization with the times. At the same time, however, there lies at the basis of such a community something fundamental that exists over time. The rural village community is a unique regional community, fundamentally different from the town: it is surrounded by primary or secondary (artificial) nature, where 30 to 50 farming families raising animals and crops live in a cluster by sharing the management of water and/or land. What forms the basis of such settlements include this community structure, a strong spirit of cooperation, the collective ownership-like nature of farmland within the rural village community, the importance of family members, families and their continuation, and the family-style, small-scale management approach that pursues welfare for the whole family, aiming at reasonable income.

In addition, farm work has now obtained significant meanings for people in both the rural village and the town due to its long-term social traffic flows, diversity, interaction with nature, and self-creativity (educational value), particularly from the viewpoint of forming a comprehensive mode of humanity, which is lacking in modern society.

Furthermore, as has been frequently pointed out, the decline of traditional events does not mean a decline of the vital energy in rural village communities. I point out, based on specific research results, that we should focus on the fact that new group cultural activities are in fact arising. As long as people live there, they strenuously reorganize the contents of their lives to realize their community as an enjoyable 'place of life'. In the future, the rural village will remain a regional community with a strong centripetal force as well as developing, based on being such a community, an additional character as an information network community with a strong centrifugal force that is always heading outward by using information appliances such as the Internet, and will be reorganized and continue as an 'open society based on regional bond' as a whole.

In Chapter 7, I consider, from the perspective that it is necessary to form a sustainable local region that integrates the economic, ecological and life aspects described in Chapters 4, 5 and 6, the contents of such sustainable local region. In other words, I explain the static structure of '*ba*'.

The rural village community is a place of production, an ecological unit, and a place of living. From a scientific standpoint, if these elements, which are analyzed technically from respective perspectives, are viewed synthetically, a perspective can be formed that leads such a community to a *ba* of comprehensive, human 'life' where people can settle. I believe that the harmonious realization of the three values that are the aims of modern society and modern agricultural science, or the realization of comprehensive values, can best be achieved in the local community as a lifeworld where neighborhood relationships are valued. And importantly, as described in Chapter 8, the sustainable local region is completed by combination with a small or medium town.

In Chapter 8, I offer historical and theoretical considerations of the relationship between the town and the rural village as well as pointing out the problematic nature of Japan's single-nucleus centralization. Although I believe that it is necessary to form a sustainable local region, it is impossible for a single rural village to achieve it by itself. A sustainable local region must always incorporate a small or medium town. With this combination, non-agricultural employment opportunities and various conveniences can be satisfactorily provided. I thus consider various stages of the historical process of formation of the relationship between the town and rural village, by focusing on Europe, particularly Germany, which has taken up this relationship with self-awareness and has incorporated it into national policies, in an attempt to derive suggestions for Japan. This is a longtime subject of my research and constitutes one of the pillars of this book. From such a perspective, I criticize the defects of postwar Japan's land policy and the lack of regional policy that led to excessive concentration in the large cities.

In Chapter 9, I theoretically describe the process of endogenous innovation and the diffusion of agricultural technology in relation to the behavioral types of farmers. The development of agriculture and the rural village is accomplished by endogenous innovations and their diffusion. An innovation in agricultural technology begins with pioneers who break the existing, customary track that has been repeated until then and who thus appear to others as outsiders or deviators. Next comes, however, adopters who judge whether an innovation is truly innovative and has long-term, not temporary, stability, and adopt the innovation as well as making improvements wherever possible. Seeing this, people who adapt to the innovation – imitators – who constitute a majority of farmers, imitiate the new technology, feeling assured and confident. In this fashion, each party involved plays its own role, completing one cycle of development, establishing a new customary track, and laying the ground to proceed to a higher stage.

I believe that the process described above, in agreement with the epistemological process of 'abduction–deduction–induction' described in Chapter 10, explains the behavioral cognition processes emphasized by Miki Kiyoshi and Dewey. Agricultural science and agricultural administration can also play a significant role in the development of agriculture and the rural village by contributing to and supporting the spontaneous dynamic process from scientific and policy points of view. This means unification of experiential knowledge and scientific knowledge.

In addition, although in Chapter 9 the process described above is presented as a process to pursue dynamic profits as an economic value, it also theoretically explains changes in rural village communities that focus on the pursuit of ecological and life values and, moreover, comprehensive values. In other words, the process explains the dynamic formation of '*ba*'.

In Chapter 10, I describe the characteristics, research methodology and system of agricultural science. In order to locate agricultural science in general science, I track the establishment of the modern sciences by describing and characterizing, in order, the science of nature (natural science), the science of man and society (human science), and the science of the relationship between nature and man (practical science), and explore their respective research methodologies. I state, however, that these characteristics represent relative differences only. Scientific methodologies, although they may differ, must be coordinated with each other. Considering the numerous problems facing science today, self-restraint and the ethical nature of science must be upheld.

I show that, on its general foundation, agricultural science, as a field of practical science, can be characterized as the study of the pursuit of practical values, the study of life and living things, the study of local regions, and the study of integration. I describe that research methods in agricultural science include methods of all the various sciences. While centering on imaginative methods as a practical science, because it is the study of bioproduction, analogical inference methods by observation must be particularly emphasized. Research methods in agricultural science must be understood not by simple relative comparisons but as a dynamically deepening process that follows three stages. I argue that agricultural science should aim at embodying and putting into practice its research results in the complex place of nature, which is its original mission, and should develop such results using its imagination and in a manner that integrates elements. When doing so, verification by experiment, examination and field survey, deepened by repeating these procedures (trial and error) are important.

The mission of agricultural science is to stand by agriculture, forestry and fishing industries and to contribute to people's peace and well-being by solving the technological, economical and social problems experienced by these industries. I believe that the aim of modern agricultural science in particular is to pursue the three main values – economic, ecological and life values – that are required by modern society. In other words, to realize total values, in each local region, in each country and ultimately, on a global scale. It also seems that, in realizing them, the simple realization of desires is not important. To restore the modern civilization and culture that seem to be in a serious crisis, some kind of self-restraint or ethical nature – perhaps as expressed in the idea of *shōyoku chisoku* (content with few desires and awareness of satisfaction) – is necessary for both individuals and science.

Additionally, I would like to comment on issues in studies of agricultural science in the future. Japanese studies of agricultural science in the past tended to import trends in European agricultural science and then superimpose Japanese agriculture and agricultural science over them. However, recently, Japanese studies of agricultural science have improved significantly, and have been faced with similar problems to the Europeans. While we need to keep accumulating new knowledge on top of what we have obtained through these studies, it is desirable that we, at the same time, conduct studies that take further consideration of the directions of agriculture, forestry and fishing industries in developing countries – considering the severe conditions in the developing countries of Southeast Asia, Africa and Latin America, for example, whose populations account for approximately 70% of the world population – and establish ideas about agricultural science for such studies.

Many argue that society in the 21st century should be reconstructed based on the assumption that nature and humans must coexist. If this is true, agricultural science, which has explored the lifeworld on land and at sea to the deepest level and with the greatest effort as a comprehensive science of biosystems, must further promote its studies with great zest and an adventurous spirit. The 21st century needs to be imagined as an era of well-founded civilization and culture based on the earth, nature, agriculture and the rural village.

1 What is the Philosophy of Agricultural Science?

Agricultural science is a field of academic study that aims to promote the development of agriculture, forestry and fisheries, the various industries linked to these, and rural communities. In doing so, agricultural science also aims to contribute to the pursuit of happiness of individuals involved in agriculture, forestry, and fisheries and, thereby, to contribute to the broader society. The philosophy of agricultural science is the field of academic study that questions the place of agricultural science in the modern world. In this chapter, I will attempt to clarify the central issues of concern to the philosophy of agricultural science. I have proposed an 'agricultural science of "*ba*"', and will discuss its implications in this chapter.

Genealogy and issues

In 1919, Krzymowski, the author of the world's first published book entitled *Philosophie der Landwirtschaftslehre* (Philosophy of Agricultural Science), argued that the philosophy of agricultural science had not previously been proposed in the field of agricultural studies (1919: 1). Understanding philosophy to be the study of principles, the philosophy of agricultural science would therefore consist of an analysis of the principles and methodology of agricultural science.

In the same year, H. Weber published his *Grundlinien einer neuen Forstwirtschaftsphilosophie* (Ground for a new philosophy of forestry science), in which he argued that the philosophy of forestry science was a 'unified understanding of the justifications, systems, and methods of forestry'. He proceeded to unveil a systematic development of forestry – and thus agricultural – science that surpassed Krzymowski's (Weber 1919: 1–2). The philosophy of agricultural science was thus established and provided with solid foundations by these two authors. In his *Die philosophischen Probleme in der Forstwirtschaftslehre* (The philosophical problems in forestry science), R. Strecker (1938) provided a systematic depiction of the philosophy of forestry science with a focus on its values: 'the system and the methodology of forestry, the values (economic, aesthetic, ethical) to be pursued as the object of forestry, and its philosophical perspective'.

In Japan, the first books to address the philosophy of agricultural science were Nitobe Inazō's *Nōgyō honron* (A thesis on agriculture) (1898) and Yokoi Tokiyoshi's *Nōgyō to nōgaku* (Agriculture and agricultural science) (1917). Hashimoto Denzaemon translated the Krzymowski book mentioned above as *Nōgaku genron* (The philosophy of agricultural science), and published it in 1932, thus introducing it to the Japanese public. A number of works followed this, including Morinaga Toshitarō's *Nōgakukō* (On agricultural science) (1951),[1] Noguchi Yakichi's *Nōgaku gairon* (An introduction to agricultural science) (1950), Kashiwa Sukekata's *Nōgaku genron* (The philosophy of agricultural science) (1962) and *Nōgaku no yukue* (The future of agricultural science) (1980), Hishinuma Tatsuya's *Watashi no nōgaku gairon* (My own introduction to agricultural science) (1973), Matsuo Takane's *Kankyōnōgaku gairon* (An introduction to environmental agricultural science) (1974), Tsuno Yukindo's *Nōgaku no shisō* (The thought of agricultural science) (1975), and Iwakata Isoo's *Nōrinsuisan genron* (The philosophy of agriculture, forestry and fisheries) (1977). Kashiwa Sukekata, who was the first in Japan to systematize the philosophy of agricultural science, defines it as 'a field of academic research that attempts to answer the question about what agricultural science is' (1962: 2). The purpose of the philosophy of agricultural science is to seek for its grounds as science, understand the significance of its thought, and make clear whether or not it has a qualification as a science which guarantees objective thought. It consists of (1) a history of agricultural science, (2) an elucidation of the realities and inherent nature of agriculture, (3) the scientific nature, methodology and system of agricultural science, and finally an endeavor to make a final judgment on (3) by shedding light on (1) and (2) (Kashiwa 1962: 2–11). Thus emerged the basic theory of the philosophy of agricultural science.

In *Shizenkagaku gairon* (An introduction to the natural sciences) (1980), Katō Kunioki claims that texts introducing the sciences, including agricultural studies, have typically dealt with two issues. The first is to provide a clear and comprehensive overview of the methodology and the perspective of nature in modern science, and to examine the general and philosophical issues contained therein (Tanabe Hajime's *Kagaku gairon* (An introduction to science) (1918) is representative of this type). The second is to analyze the social aspects of science (J. D. Bernal's *The Social Function of Science* (1939) is representative of this direction) (Katō 1980: iv–v). Both sets of issues continue to confront science today and require further clarification.

I believe, however, especially in the case of a practical science such as agricultural science, that we need to go further than this, and ask how modern science can resolve the various problems confronting it, and what methods and systems are required to do so. These are particularly urgent questions at present in the agricultural and engineering sciences, both of which have to deal with practical issues.

Various discussants of the philosophy of agricultural science differ in their views of what constitutes this field of enquiry. My own opinion is that the philosophy of agricultural science is the study of (1) the history of agricultural science and its thought, (2) the elucidation of the inherent nature and problems of agriculture and rural communities, (3) value objectives of agriculture and agricultural science, and (4) the formulation of systematic methodologies for resolving the problems confronting agriculture and rural communities today. Since I hope to write further on the history of agricultural science at a later date, I limit myself here to only a handful of suggestions about this topic. I examine the following types of issues.

In Chapters 2 and 3, I attempt to clarify the various problems faced by, and the various roles realized by, the concrete experiences of agriculture and rural communities today. We must begin from actual problems.

In Chapters 4, 5, and 6, I will analyze changes in the values and aims of postwar Japan, and elucidate the key ideas, values and aims of modern agricultural science. The aims, methods, and systems of agricultural science are all affected by their particular social and historical background, and are all constructed in such a way as to address the particular problems faced in each period. While the methods and systems of agricultural science have an inherent and universal aspect unique to it, they also must conform to changing realities (these changes are a product of the historical development of particular problems). After all, the most important basis or *raison d'être* of agricultural science is to provide solutions to pressing agricultural and/or rural community problems, and through this to increase the happiness of farmers and thus the welfare of humankind as a whole. It is important to think in terms of basic principles and essentials from a variety of perspectives (in particular, the three perspectives of the economy, ecology and life) when examining the contents and direction of the values and aims of agricultural science and the problems to be resolved. I will have to leave research into the history of agricultural science for a later book, but note here that such research follows the developmental processes of agricultural science and is useful in determining historical conditions and directions.

In Chapters 7, 8, and 9, I will emphasize the importance of community to lifeworld – the community is the place where the values mentioned above might be harmoniously realized. I will attempt to clarify the sustainability and the dynamic development of this situation.

Finally, in Chapter 10, I will present my account of new methods and systems for agricultural science to resolve its central problems, and elucidate the relationship between agricultural science and agriculture.

The above will suffice to fulfill the mission of the philosophy of agricultural science.

Scientific philosophy of agricultural science

I have stated that the philosophy of agricultural science is an investigation of the basic theories or fundamental ideas about agricultural science. What, then, is philosophy?

According to Nishida Kitarō, philosophy is the inquiry into what human beings should do and how they should live. The philosophy of science is an enquiry into the knowledge that lies at the heart of science, a study of all areas of scientific study in the attempt to reveal the nature of science. Therefore, the self is deeply explored, and 'self-reflection' is pursued, while the self is explicitly 'self-conscious'. Thus philosophy is self-reflective and self-conscious study (Nishida 1953).

In his *Kagaku gairon* (An introduction to science), Tanabe Hajime follows Kant and the neo-Kantians in claiming that the philosophy of science examines the basis for and significance of scientific cognition, and asks about its *raison d'être*. This is known as *Kritik*. The philosophy of science, from this perspective, is the 'reflection of scientists about their scientific pursuits' and is a self-critique of science (Tanabe 1918: 20). Kashiwa Sukekata, borrowing from Tanabe, claims that 'the philosophy of agricultural science is the area of study that questions the basis of agricultural science, and therefore is a critical study: the philosophy of agricultural science is the self-criticism of agricultural science, and is a self-reflection on science'. Furthermore, he states that it is self-defining and self-evaluating, or in other words a science of reflection and self-consciousness (Kashiwa 1962: 3). Omodaka Hisayuki argues similarly that the philosophy of medical science is a science of reflection, self-consciousness, and self-criticism that emerges when medical science examines itself (Omodaka 1987: 17–19; cf. 1945a; 1945b; 1959; 1989/1964).

Similarly, others define philosophy as an 'unrelenting reflection on human life' (Noda Matao), an 'endless self-reflection' (Takeichi Akihiro), and 'prudence and reflectiveness, or the internal acts of the self' (Jaspers). In each case, it appears that philosophy deeply examines itself: it is a study of reflection and self-consciousness (e.g., Noda 1963; Takeichi *et al.* 1976; Jaspers 1976).

Here, however, it is important to move beyond the formalities of scholarship, and consider what the philosophy of science must reflect upon, and what are the aims of science itself. Miki Kiyoshi moves a step beyond self-reflection and self-consciousness. He says 'the meaning of self-consciousness, or the self knowing the self, must be the self moving past the self', thus suggesting a path to self-formation and self-creation (Miki 1967 [7]: 60). Bollnow argues that as a result of 'the methodological self-reflection of human sciences', the *raison d'être* of philosophy lies in the increasing richness of 'the dynamism of life itself' and 'the wealth of human life' (1958: 134, 139). In other words, drawing on Bergson's insights concerning 'the creation of the self by the self itself' and Nietzsche's observations about 'the reversal of values – that is the reversal of the creator' and the accompanying claim that 'the only happiness is to create', Bollnow emphasizes the 'creative function' in continuing to create the self through the scientific way of thinking (Bollnow 1958: 34, 138).

Today, although the innumerable achievements of science are indisputable, the various problems produced by science and the culture that has grown with it – e.g., environmental devastation, global warming, etc.; problems that put a question mark over the continuing existence of humankind itself – have become increasingly clear, and raise urgent practical issues. Science, and especially agricultural studies, must take as its starting point the immediate problems that confront us. It must devise new aims, methods and systems for resolving these problems, rethinking itself from its most basic fundamentals. This seems to me to be the fundamental task of the philosophy of agricultural science. Kuhn's theory of a scientific 'paradigm shift' – a comprehensive revision of the mission, essence, and systems of a scientific discipline – explains how science responds to this kind of crisis in society (Kuhn 1962). Kashiwa Sukekata expresses hopes that agricultural science will move beyond introspection to self-formation and self-ordering (Kashiwa 1962: 3).

Bollnow argues that, compared to all the anxieties and despair that members of our generation experienced from 1914, the optimistic relationship of trust that people felt was supported and protected by their comprehensive, root existence appears to have been a childish illusion. Existentialist philosophers have elucidated some of the darker aspects of life, such as 'the hopeless insecurity of human existence'. In contrast, many *lebens* philosophers have explored the more positive aspects of life. However, for us to unfold a new, energized 'creative life', it is no longer possible to oppose fear, anxiety, and despair with trust, hope, and gratitude. Rather, they all have to be recognized as phenomena of the same metaphysical order. In facing this new age of crisis, we must overcome the polarization of the *lebens* philosophies and existentialism, which at first glance appears to be unbridgeable, and move beyond despair and optimism. Here Bollnow sees the light of hope, which forms the basis of what he calls the philosophy of hope (1958: 2).

Likewise, the philosophy of agricultural science must begin from the various problems of modern agriculture, forestry and fisheries, and attempt to become a study of self-formation, creation, and hope in searching for solutions to these problems. In so doing, reflection, self-awareness, and self-critique reach their fullest fruition.

The philosophy of agricultural science starts from the real problems of agriculture and agricultural communities. As a scientific philosophy, it does not focus on particular (i.e., partial) problems, but rather focuses on comprehending the whole. Philosophy, as mentioned, rests on the reflective method, discussing itself in a fundamental, essential, systematic and synthetic way, while at the same time, as is the case with science, without taking any premises for granted. In this book, the 'self' that discusses itself includes both the researchers in agricultural science and their research. Philosophy 'must place itself in a wretched position', which may at first appear to be contradictory, torn between despair and hope, criticism and creation. This is, however, also 'the glory of philosophy' (Hayashi *et al.* 1971/1954: 975).

According to K. Jaspers, the best philosophy of the modern era does not lie with the people who have been most directly entrusted with philosophical doctrine. Philosophy exists among other sciences, prevents them from becoming involved in that which is not worth pursuing, and thus becoming distracted, and provides scientific research with life. This philosophy is a concrete philosophy that is realized within the entirety of a specific science (Jaspers 1976). This view locates living philosophy not in abstract philosophical doctrines, but in synthetic inquiries into specific and individual problems. Such a view provides comfort to agricultural science as a special and particular science. Jaspers' words 'In this modern age, I want to shed light on the modern age itself while also developing, together with my students, an eye for that which is eternal' encapsulate the stance that must be taken when teaching basic principles at university.

Perspective of this work: agricultural science of *ba*

As described in the Preface, in this book I aim to establish an agricultural science of *ba* and wish to use this Japanese notion throughout the text. Here, '*ba*' means the place in which we pursue 'human life' (Sakamoto 1989: 2),[2] including its internal situations and circumstances. As I will discuss further in Chapters 3 and 7, modern society, modern agricultural science and modern rural communities require the harmonious realization of three main values – economic, ecological, and life values. At present, these appear to be contradictory demands whose mutual realization requires some trade-off. As I discuss in detail in Chapter 3, values mean 'the criterion of the selection act necessary for the existence of the subject' (Kurita and Ueyama 1968: 295). With respect to economic values, for example, they do not point to such tangible and specific items as profit, interest and labor costs but to general social orientations that choose the priorities of economic activities in the context of overall human activities. I posit economic, ecological and life values as the mainstay of these orientations.

Although there are many difficulties, overcoming these contradictions in order to realize these comprehensive values is essential.

Contradictory circumstances emerge and are perceived only when contradictions are placed in the same field, in the same theoretical space. Here, questions, imaginative creations and efforts are formulated to resolve the contradictions. The 'place of life' – including animals and plants – is always the 'contradictory place', the 'questioning place', the 'place in which questions are raised', and the 'place in which problems are perceived'. Being a 'place of questions' is the first step towards resolving problems. It must then become a 'place of imagination' in a movement towards problem resolution. Then it becomes a conduct-based 'place of creation' or 'place of formation'. Finally, it reaches completion as a 'place where problems have been resolved', or Bollnow's 'place of hope'. Bollnow's place of hope is a place of life suitable for human beings. In this fashion, place is a 'dynamic' 'place of processes' which is continuously developed and formed. As Miki Kiyoshi says, 'human beings form themselves by forming the environment' (Miki 1967 [7]: 10). Therefore, the human being exists within a place, creates its place, and constructs the situations of its place (Miki 1967 [18]: 245). The place for human beings 'is not given, but is formed. Self-formation is the very font of human happiness' (Miki 1967 [7]: 10; Meguri 1968: 72). Thus *ba* is a total and practical place for creation and formation in which humans live in solving their highly contradictory lives.

As I argue in Chapter 3, the community is the concrete place for this. The local community is the place where the human activities of production and living take place, a geographical space which enjoys a certain degree of economic, social, and natural autonomy and individuality. *The community is a place where face-to-face relations can be realized.* I believe that it is the 'local community as a concrete *ba*', a place in which small and medium-sized cities have merged with rural villages and become *a single unit or daily-life sphere*, or in other words the so-called lifeworld, wherein lies the best potential for solving various problems. The key to resolving many of the problems of today depends on concrete and practical actions in which the residents of each local community are aware of the problems and then, by unifying the knowledge derived from their daily experiences and that derived from science and technology, create new schemes, while joining together in a fraternal manner to recreate the local community. The place I depict is a concrete local community that is a product of these dynamic formative circumstances. It is in this *ba* that the values mentioned above have the greatest possibility of being realized.

However, it can be anticipated that the realization of these values will require acknowledging (understanding/accepting) that we cannot continue the attempt to satisfy endless desires; that is, it will be necessary for *ba* to be a place of self-control, a place 'where one realizes that one has enough', a place where ethics and education are emphasized. This is the 'place of hope', where trust, hope, and gratitude bridge the gap between despair and optimism, and where people are able to form themselves (cf., Bollnow 1958).

The 'community' of which I speak is this type of creative, autonomous, and individualistic place, and at the same time is open to the wider world. I call this perspective *global regionalism* (Soda 1989: 119).

In this book, I aim to examine agriculture, the rural community, and agricultural science from this basic perspective.[3]

2 Human Beings and Nature in Agriculture

In this chapter I discuss the types of relationships that have been established, and the problems that have emerged, between human beings and nature – plants (fruits, nuts and vegetables) and animals – over the long history of agricultural civilizations. First, I trace the steps by which mankind gave birth to agriculture. Second, I analyze the types of relationships humankind created with nature as agriculture emerged. Third, I clarify the problems of modern industrialized agriculture. Fourth, I note that the deep ecology philosophy – a response to the explosion of environmental problems – has produced a notion of plant and animal rights that cannot be ignored. Fifth, I outline my prediction that the direction that agriculture and agricultural science has taken will lead to a new world of human life and responsibility, which I call 'formative equilibrium'.

The birth of agriculture

Hunting, gathering, and fishing stages

Human beings are living creatures that must eat to survive. The earliest humans, in order to acquire foodstuffs, relied on hunting (eating animals), gathering (eating products of the plant world, i.e., fruits, nuts and vegetables), and/or fishing. This was the so-called 'hunting and gathering stage', which preceded agricultural development.

According to Harlan, in general, the diet of the average member of a hunter and gatherer society consisted of 60–70% plant products and 30–40% meat/fish products (with exceptions such as the far northern Inuit, who were compelled by environmental conditions to rely almost entirely on a meat diet). Regional differences in the types of plant products consumed range from about 500 types in the case of the Australian Aborigines to approximately 1,100 types for the Native Americans. The average lifestyle consisted of two days spent gathering and one day resting. It has been estimated that for the Bushmen of southern Africa, 2.46 days a week at six hours a day, i.e., 15 hours per week, was sufficient for food gathering. Children and the elderly were not required to work. Compared to us moderns, the hunter-gatherer enjoyed a surprisingly rich diet, was blessed with good teeth and few cavities, had few chronic diseases, and was a skilled 'practical botanist' with sound knowledge of poisonous and non-poisonous plants, how to counter poisons, and the medical uses of a wide array of plants. Moreover, according to Harlan, although there was no cultivation or domestication per se, there are confirmed cases of hunter-gatherers sowing seeds and propagating the plants used as foodstuffs (Harlan 1984, 1971).

Hunter-gatherer societies were thus relatively healthy societies, without the so-called modern diseases, and are generally considered to have been relaxed and healthy in mind. At the same time, they were clearly also fragile societies that could quickly experience crises when forced to compete with other human groups or large animals for food due to climatic changes, such as floods or droughts, for example.

The origins and spread of agriculture

How, and when, did hunter-gatherer societies become agricultural societies? Presently, three methods are considered useful in the study of the origins of agriculture. These comprise (1) the method that, to the greatest extent possible, makes logical speculation based on imagination and reasoning, (2) the method that, utilizing data derived from ethnological research, makes comparisons based on historical facts, and (3) the archaeological method (Smith 1992: 5–6). Based on these three methods, most researchers locate the beginnings of agriculture approximately 10,000 years before the present era (Hoshikawa 1987: 6). Furthermore, Harlan argues that by 7000 BCE, agriculture was practiced in the areas currently known as the Near East, South America, New Guinea, Southeast Asia, and probably also in Africa, Central America and many other parts of Asia (Harlan 1992: 12).

Various proposals have been advanced to explain the origins of agriculture. The two basic features that define agriculture are (1) the domestication of animals – i.e., raising livestock – and (2) the domestication of plants – i.e., cultivating crops. Entering the agricultural stage, humans no longer had to rely on hunting and gathering for subsistence. In their efforts to secure raw materials for foodstuffs and clothing, they began to raise and consume plants and animals within man-made environments.

Several theories have been proposed to explain the emergence of domestication and its underlying causes. Prominent among these are: Sauer's pet origin theory, Harlan's religious origin theory (animals as objects of sacrifice), and numerous economic origin theories (animals as sources of meat, milk and fur) (Singer 1954: Chapter 8; Watanabe 1977: 68). Childe observes that domesticated animals are 'a living food storage and a walking clothes drawer' (1943: 49). It is easy to imagine how the assumed forefather of the modern dog, the jackal (or wild dog), was, through participation in hunting activities, converted into a pet and accommodated as a family member, and then gradually trained to become a hunting partner, watchdog and shepherd. In most cases, arguably the process of domestication began when people captured infants of various species and proceeded to care for them. It has been confirmed that the dog was domesticated as early as the late Paleolithic era, making it possibly the first species to undergo this process (Watanabe 1977: 68). Soon thereafter, nomadic and settled agricultural lifestyles developed alongside the domestication of other species such as chicken, cat, sheep, cow, horse, pig and reindeer.

Presumably, the domestication of crops developed from an economic desire to acquire sufficient foodstuffs regularly and consistently. Furthermore, the evidence suggests that the possibility of intentionally cultivating crops was first realized when hunter-gatherers noticed that the refuse of fruits, peas, grains and potatoes, which lay scattered around their (hunting and gathering) bases, began to sprout and take root. In the past, there was a place in the garden of every farming house that was designated as a rubbish heap where various useful plants that the family would utilize as foodstuffs grew naturally. We can assume that people had similar experiences in hunting and gathering domiciles. It was this kind of initial accidental process that eventually made possible the growing of easy-to-eat fruits, crops of grains, storable food plants and rhizomes that could be harvested year-round. The most widely accepted theory holds that gradually, with the introduction of artificial and technical elements into the cultivation process, crops were born.

Arguably, then, through a slow transformation from unintentional to intentional processes, from primitive trial-and-error-based technologies to highly knowledge-based ones, spanning a period of thousands of years, the domestication of animals and cultivation of crops gradually took root and began to diversify (Harlan 1984: 12). Accordingly, it is incorrect to hastily attribute the development of agriculture to simple necessity. Sauer, for example, argues that it was the fishermen who had settled in areas with mild climates and year-round access to fresh water fisheries and other foods who first initiated agriculture. He also suggests that no region other than Southeast Asia satisfied the pre-conditions for the emergence of such a fishing-based agrarian culture (Sauer 1952; Watanabe 1977: 70; Sauer/Takeuchi 1960: 56). From this perspective, and considering the surprising richness and stability of the hunter-gatherer societies, it seems clear that agriculture was not a direct result of human need. Smith categorizes the theories that attempt to trace the factors that 'switched on' agricultural development into three groups: (1) those focusing on external factors such as environmental changes, population pressure, technological innovation (the so-called materialist and ecological approaches); (2) those emphasizing internal factors such as social class divisions or hierarchies, religion, ideology and symbols (the dominant approach of anthropology, history and religious studies); and (3) those that consider both external and internal factors together, which can be called 'eclectic theories' (the approach adapted by the overwhelming majority of archaeologists) (Smith 1992: 7–18). Rather than explaining the birth of agriculture per se, though, it is probably better to think of these theories as attempts to highlight the factors that accelerated the formation and development of societies that had already taken the first steps in creating agriculture.

Opinions remain divided over where agricultural development first took place. Some argue that agriculture was born in a single place (the so-called 'unitary theory'), others contend that this happened not in a single but a small number of locations roughly simultaneously, while yet others propose that agriculture emerged in a large number of places around the world in a scattered fashion. The British scholars who inherited Babiloff's theories – the 'small number of locations' view – argued that the origins of agriculture lay in the so-called Fertile Crescent of Tigris and Euphrates where wheat, barley and peas were grown, effectively proposing another 'single location' or 'unitary theory' of agricultural origins. Similarly, Sauer and others have advanced a single location theory, albeit tracing the place of origin to Southeast Asia (cited in Nakao 1966: 15).

Nakao Sasuke has built a grand 'multi-sources' theory to explain the spread and consolidation of agriculture. According to Nakao's scheme, Southeast Asia, with banana and potato as its staple crops, is home to what he calls 'root crop agrarian culture'. Meanwhile, the arrowroot and tea-growing Himalayas, southern China and Japan are host to 'glossy leaved evergreen forest culture' (not shown clearly in Figure 2.1). The African countries like Ethiopia, that primarily grow

Figure 2.1: The formation and propagation of farming cultures

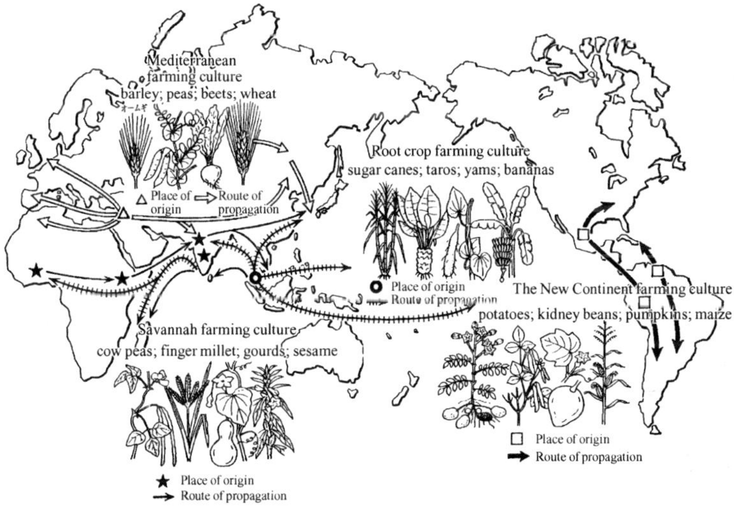

Source: Nakao 1966: vi.

various cereals and beans, comprise the 'savannah agrarian culture'. Furthermore, the Mediterranean region where wheat, barley, rye, oat and peas are cultivated form 'Mediterranean agrarian culture'. Finally, the American continents, where potato and corn are the main crops, form the so-called 'New World agrarian culture'. Similarly, Harlan concludes that 'according to archaeological information, the origins of agriculture sprang up in scattered locations; agriculture did not emerge and spread from a single nor a very small number of locations', putting forth a clear 'multiple scattered sources theory' (Harlan 1979: 12).

Since the intellectual capacities of human beings evolved without a marked gap between the inhabitants of different regions, it is plausible to propose that the seeds of agriculture emerged gradually here and there, primarily catalysed by daily accidental incidents. Nevertheless, with regard to the extent and utility of domestication and the level of cultivation techniques, differences inevitably emerged between developed and underdeveloped regions. Furthermore, it is clear that the typical core agrarian cultures particular to each region evolved reflecting closely the natural climates of their respective regions and only later began spreading elsewhere.

In the study of prehistory, questions about the origin or origins of agriculture and issues such as the relationship between permanent settlement, nomadic life and agriculture, the context and interrelationship of nomadism and agriculture, and the role of women in the initiation of agriculture, are becoming increasingly common themes and are therefore likely to be further clarified through future research (Smith 1992).

The relationship between humans and nature in agriculture

The relationship between humans and nature, or the relationship between humans and plants and animals, is extremely complicated. I am able to discern three sets of relationships. First, there is an 'interdependent symbiotic relationship' between humans and domesticated animals and crops. Second, a 'mutually exclusive and competitive relationship' can be observed between humans and vermin and weeds that invade lands meant for agricultural production. Third, the 'habitat segregative and coexistent relationship' between humans and wild plants and animals that yield no special benefits or damage to either. Furthermore, the natural environment – including geographical, weather, water and soil conditions – must be considered for a comprehensive theory of agricultural production, but these are beyond the scope of this chapter.

Domestic animals and crops: interdependent symbiosis

As previously explained, as agriculture gradually developed and took root, the relationship between humans and plants and animals became inseparable. In agriculture, humans raise and utilize plants and animals to suit their own purposes. The particular plants and animals that humans have chosen for agricultural purposes from the wide variety of species available on Earth are referred to as livestock and crops.

Animal species that have been domesticated include horse, cow, sheep, goat, pig, dog, donkey, camel, elephant, reindeer, rabbit, chicken, duck and turkey. These are utilized for foodstuffs, ploughing, transportation, and as raw materials for clothes and leather goods. A prime example of the ways that animals are put to a variety of uses by humans can be seen by a quick look at 'man's best friend', the dog, that is petted and played with, and also employed as a controller and protector of livestock, or a guardian of property. Domestic animals are linked to everyday human behaviour (that aims to secure livelihood and survival) through activities such as ploughing, carrying, raising, taming, caring, playing, guarding, and then slaughtering, eating and wearing.

As for crops, if classified according to use, the following exist: (1) rice, wheat, barley, rye and oat, corn, bean, potato and other main edible crops, (2) tangerine, apple, fig, banana, pineapple, mango and other fruit, (3) Chinese cabbage, cabbage, leek, carrot, spinach and other vegetables, (4) chrysanthemum, carnation, lily, peony, tulip, rose, orchid, hydrangea, nandin, camellia, plum, maple, pine and other flowers and trees (2, 3 and 4 are referred to collectively as horticultural crops), (5) rapeseed, tobacco, tea, sugar beet, hop, rush grass, sugarcane, various medical herbs and other industrial crops that serve as raw ingredients for processed goods (6) grass for animal feed, lotus flower that works as fertilizer, clover and other feed and fertilizer crops. Similarly to domestic animals, these crops are intertwined in – constituted by and constitutive of – daily human behavior through activities such as planting, growing, harvesting, processing and eating.

Unlike those 'natural' plants and animals that have developed independently of human interference, domestic animals and crops have developed over a long period of time through processes of crossbreeding and adaptation to serve human purposes. Domestic plants and animals cannot survive without human support. Left on their own, they would either perish or slowly revert to their nature after some generations. In short, we can say that humans and domestic plants and animals live in a mutually interdependent and symbiotic relationship.

Pests and weeds: mutually exclusive, competitive relationship

While there are innumerable plants and animals that are beneficial to agriculture, there are also many that are detrimental to it. When unwanted plants and animals invade fields, greenhouses and other spaces of agricultural production and hinder the efficient breeding and cultivation of livestock and crops, it becomes necessary for the farmer to eliminate them from these locations.

For instance, monkeys, wild boars, raccoons, dogs, bears, deer and others eat crops in the field and ravage freshly planted trees. Wolves, foxes and lions attack grazing livestock. Birds feed on sowed seeds and ripening crops. Combined or separately, bird, animal and insect damage and disease can deal a fatal blow to agricultural production. Throughout history such threats have frequently endangered the survival of human populations. Today, in market societies, vermin damage continues to threaten livelihoods by making agricultural products useless for sale. During fieldwork in Tanzania, I observed that establishing scarecrows is one of the more important tasks in African rice farming. Without it, it is not uncommon to have one's harvest decimated by flocks of queleaquelea and other birds (Soda 1996: 11).

Plants can also affect harvest volumes and production efficiency by competing with farm crops for nutrients in the soil they share. Unwanted grasses that sprout up in the spaces between crops are generally regarded as weeds. Some of the plants that grow in pastures can be poisonous to horses, cows and other livestock. There are also plenty of creeping and climbing weeds that can strangle crops and hamper livestock movement. In forestry, a distinction is made between artificial and natural forests. In artificial forests, undergrowth and creeping plants that climb up and attach to tree trunks are classified as weeds.

Since vermin and weeds are harmful to agricultural production, the response is typically to attempt to exterminate them or remove them from production areas. The plants and animals that do not encroach on these areas are simply classified as wild animals and grasses.

We must also acknowledge, of course, a range of insects and birds that are defined neither as livestock or domesticated, but which are nevertheless necessary and beneficial to agricultural production. These include, for example, the dragonfly, which eats the larva of leafhoppers, which are injurious to rice plants, and birds that eat the grasshoppers that can destroy entire crops. Similarly, alongside countless bacteria and microbes that are deleterious to the health of crops and livestock, there are as many organisms that either counter these negative effects or provide other benefits. Yeast, for example, is indispensable in the fermentation of miso (soybean paste), soy sauce, bread, alcohol and other products.

Harmful birds, animals and weeds are entwined in human behavior through activities such as killing, driving away, and removing. In this regard, humans stand in a mutually exclusive and competitive relationship with harmful plants and animals.

Wild animals and plants: habitat segregation and coexistence

We classify as wild (ordinary) animals and plants those outside direct agricultural interests, namely, those animals and plants that are neither beneficial nor harmful to agriculture, or those that live in non-agricultural areas even if they may be called harmful animals or weeds once they are in agricultural space. These wild (ordinary) animals may possibly be linked indirectly with agricultural interests through food chains. Yet most animals and plants live, drawing a line between humans and themselves, without interfering with the space of agricultural production and living. In human life, the nature that these animals and plants inhabit is the object of appreciation and literature and that of human action, such as loving, relaxing, drawing and composing. These ordinary animals and plants are in the habitat segregative, coexistent relationship in which they coexist simultaneously without having mutually antagonistic interests.

Humans and nature in modern agricultural science

Industrialization and life/environmental problems

The spread of the Industrial Revolution from the latter half of the 18th century onwards and the population explosion that accompanied it completely transformed the nature of agriculture. The features of the so-called 'modern civil society' that emerged can be summarized by keywords such as nation building, capitalism or the emergence of a market economy, the development of scientific technologies, industrialization, urbanization, population explosion and agricultural revolution. All this greatly expanded the 'wealth of nations' anticipated by Adam Smith.

Over the two hundred years that followed, agriculture that hitherto could have been described as 'organic' or 'ecological' passed through drastic changes in every country. This change was caused on the one hand by the population explosion, and on the other by the simultaneous outflow of rural inhabitants to urban areas and the subsequent decrease in the relative share of agricultural laborers. In the so-called developed nations, this share eventually fell to 3 to 5 percent of the total work force. Considering these figures and the fact that agriculture encompasses the production of not only foodstuffs, but also wool, cotton and other industrial crops, we see that agricultural productivity rose so remarkably that it became possible for one farm worker to produce enough food to feed thirty people.

Due to experimental on-site improvement of agricultural techniques and the progress of agricultural science benefiting from the innovations of modern science, agriculture was transformed on many fronts. The invention and use of chemical fertilizers, and of new agricultural machinery and materials, the discovery of the laws of heredity and the subsequent improvement in cattle and crop quality, the introduction of agricultural chemicals, the development of new preservation and processing methods of agricultural products, the rationalization of distribution systems, the progress in land and sea transport, the betterment of crop planting practices and improvement in other management methods were among the factors that not only greatly expanded agricultural productivity but also drastically upgraded the capacity of societies to support bigger populations.

By definition, agriculture is an activity that transforms parts of the primary natural environment into artificial natural environments and therefore embodies an aspect of environmental destruction. However, in the past when natural areas remained vast (and mostly untouched), the capacity of nature to purify pollution was also great. Eventually though, triggered by population expansion and industrial development, the use of natural resources and the excessive sprawl of artificial environments increased. Without much awareness of 'environmental problems', the revolution in agricultural methods progressed with the levelling of forests for use as farmland and the growing use of water for irrigation and other purposes. And then, after the Second World War, modern agriculture in developed countries finally reached a stage referred to as the 'industrialization of agriculture'.

In order to respond to the mass consumption needs of the world population (approximately six billion in 1999), techniques to enable more efficient and mass-scale production were developed. Approaches such as specialization, large-scale agriculture, intensive livestock farming, single crop farming (monoculturization), continuous cropping (planting the same crop in the same land for many years in succession), chemicalization (agricultural chemicals and chemical fertilizers as well as materials such as vinyl), mechanization, and the use and expansion of agricultural facilities (the green house, irrigation facilities) were promoted.

This type of agriculture spread gradually after the onset of the Industrial Revolution but soon thereafter set on a more rapid track of expansion. Regardless of the worries of Malthus (who warned that humankind would face severe food shortages and therefore social problems due to an expanding gap caused by the fact that populations grew geometrically [1, 2, 4, 8…] while food production could only grow arithmetically [1, 2, 3, 4,…]), as far as has been observed, the growth of food production worldwide has outpaced that of population growth. This was made possible by the use of nature for developmental purposes, advances in modern science and the development of agricultural science and related industries. These results are monumental achievements reached under great pressure to support the life of expanding populations and to facilitate improving diets.

However, at the same time several problems were exposed. Intensive livestock farming, large-scale monoculturization, the practice of planting the same crop in the same field from year to year, and the increase in the use of chemical fertilizers had a weakening effect on the ability of plants and animals to survive. This accelerated the proliferation of disease-carrying vermin and subsequently increased the mixing of antibiotics into animal feed, and also pushed up the quantities of agricultural chemicals used in crop cultivation. This vicious cycle of modern agriculture placed considerable stress on the ecological environment, caused a decrease in the number of surviving plant and animal species and cast anxiety over food safety. Furthermore, as for energy efficiency, at present, American agriculture uses five or six chemical fuel calories per food calorie that is produced; for British farmers the corresponding figure is three. Therefore, present-day agriculture can be characterized as a great 'energy drain' (Dethier 1976: Chapter 8). In this fashion, from both within and without, modern agricultural practices that ultimately prioritize the production of economic value and efficiency above all other values are now forcing the very people that developed them to 'live a sluggish death'.

Livestock, crops and humans

Behind the modern science that brought about this situation lies a Cartesian-Newtonian view of nature. Francis Bacon, criticizing medieval Scholasticism, strived to establish empirical methods of modern science. Bacon, who lived at the dawn of English capitalism, had a strong interest in improving man's material conditions through thoroughly conquering and remodelling nature. Holding that 'human knowledges and human power meet in one' so to speak science, he promoted man's dominance over nature, founding this stance on 'the idol of man's rule over nature' that strived to build a 'kingdom of humans'. As implied by Descartes' famous words 'I think, therefore I am', this school of thought recognized human agency and spirit independently from the material world. In effect, this separated humans and nature, objectifying the latter. Descartes saw nature as consisting of objects and their movements and when these were based on mathematically verifiable facts, declared them 'scientific truths', thereby adopting a 'mechanical view of nature'. Plants and animals alike were thus considered machines. Furthermore, Newton laid down the concrete fundamentals of modern science by establishing physics around dynamic systems. It was this so-called Cartesian-Newtonian paradigm that really underlay the beginning of modern science.

These scientific perspectives and their implications for agriculture were compiled by A.D. Thaer in his 1812 book *Theory of Rational Agriculture*. Keywords that imply 'rationality' include calculation, deliberation, reason, essence and planning (Haushofer 1963: 30). In agricultural production, economic rationality was given highest priority. Industrialization of agriculture meant the birth of a system that was evaluated by indices such as livestock feed efficiency, energy efficiency in the use of machinery, and ultimately capital efficiency or economic efficiency.

Under industrialized modern agriculture, animals are turned into mere 'production machines'. For instance, chickens are raised in small cages measuring 15 by 30 centimetres that allow for only minimal bodily movement, and in huge economic units of ten or twenty thousand birds at a time, as if they were nothing but 'egg production machines' at egg production plants. The provision of feed and water to hens and the collection, sorting by size, washing and packaging of eggs is all carried out automatically. At the time of writing in 1995, the number of poultry in Japan stood at 260 million, of which 140 million were egg producing hens and 120 million for sale as broilers. This was double the human population of Japan. When production conditions become this badly congested, the physical health of the poultry begins to deteriorate. Furthermore, infectious diseases, in the event of an outbreak, spread rapidly and cause great losses under such large-scale, concentrated farming conditions. To consistently prevent the outbreak of virulent diseases, therefore, it becomes necessary to mix medicine into animal feed. This applies also to the growing of fattened and dairy cows, among others. In addition, in order to improve economic productivity, it is not uncommon to give livestock growth-inducing chemicals. The effects of these measures on the human body are a subject of concern as well as a source of trade friction.[1]

Matters are similar in the case of crops. Due to practices such as large-scale agriculture, monoculture and continuous cropping, the so-called 'sick soil phenomenon' (a condition in which crop growth is hindered and diseases emerge more easily due to planting of the same crop in the same land from year to year) breaks out and disease carrying insects increase. And since consumers often value appearances over quality when purchasing products, practices that completely prevent worm-eating and other damage must be developed in order to preserve product value. Hence, on one hand, disease-resistant crop qualities have been successfully developed, while on the other hand, agricultural chemicals that counter disease-carrying insects are used in great quantities.

Within a market economy and under this production system that has gradually increased its emphasis on economic efficiency, the relationship between humans and livestock and crops, and thereby between humans and nature, became increasingly divided.

It used to be the privilege of those engaged in agricultural production to experience feelings of amazement, sweetness and kindness through the care of living things like cows, horses and others, and exchange emotions with animals and derive excitement and joy from them. This aspect of agricultural production is in danger of being lost. Cows, pigs and chickens are shipped to the slaughterhouse and reduced to pieces of meat upon growing up. The everyday events that take place at slaughtering facilities are usually seen neither by producers nor consumers. Consumers express an interest only for matters such as whether the beef, pork or chicken on the dining table was raised domestically or overseas, whether it was expensive or cheap, and whether it tastes good or bad. As exemplified by the chain 'production–collection–transportation–slaughtering–distribution and processing–wholesale and retail–consumption', under a socio-economic system of extensive division of labor and professionalization, each process is made independent and separate from the other, because such a scheme is the most economically efficient one. In sum, then, the more civilized societies develop, the more the process leading from production to consumption becomes concealed due to exceeding differentiation and systematization.

In the abovementioned process, there is far too large a gap, on the level of human action and emotion, between the stages of raising livestock with hard work and care, slaughtering, and finally eating. *This is a hard-to-neglect, grim state of affairs that ignores questions of what is right or wrong for the cause of human survival and livelihood.* Modern agriculture that prioritizes profitability and economic efficiency does not give much regard to the adorability and amusingness of animals and is thereby degrading the attractiveness of nature and agriculture. Furthermore, due to the division of agricultural labor, the formation of a well-rounded human character is hindered. Quite understandably, while also considering the economic aspects of life, it is observed that those who newly enter into agriculture from cities tend to emphasize non-economic aspects such as the beauty of nature, health, relaxation, excitement, appreciation and individuality that have been disappearing from modern agriculture and farm life.

In the case of Japan, this kind of agricultural process was carried out in a uniform fashion by farming families until the 1960s. In chicken raising, duties were split between family members. Water, grass, vegetables and seashell chips were provided to hens that were raised with hard work and care, and their eggs then collected. When they became unable to produce eggs, the father of the family would take them to the riverside and kill them while uttering words of apology. After processing, the chickens would be served on the dining table. Under this kind of a continuous and unified process that occurred right before the eyes of the family, things such as hard work to bring up animals in a caring way, excitement towards nature, adorability of animals and corresponding kindness, cruelty and guilt of killing, prayers, joy of eating, bodily health and appreciation were directly experienced by the family as a whole with the father at its centre. At present, especially in cities but also in farm villages, this kind of unity has been lost and the processes and totality of human action linked to production and consumption have been hidden from sight.

In this way, it can be said that we have lost many things that emerge from a close relationship with nature. These include various kinds of knowledge, the feelings of awe, wonder and sensation towards nature, consideration and prayer as well as appreciation for nature and hard work. This kind of totality of human existence and livelihood is reflected in many of the agricultural rites and festivals that remain in regions all around the world. It can furthermore be said that we are being reduced to simple-minded human beings who are, due to the abovementioned 'invisible' agricultural production process, neither sufficiently aware of nor able to embrace the contradictions inherent within the practice of 'raising, killing and eating'. Processes of simplification and division are separating into different places activities that should rightfully be continuous and integrated. By separating these contradictions to different locations, modern society is able to attain a certain limited level of consistency and rationality. By avoiding the placement of contradicting items in a single place, new and even larger contradictions emerge as a result. If allowed to continue, this will result in the flattening and narrowing of our views on nature and society. Furthermore, while living impassively in affluence, engaging in mass-consumption and mass-disposal, we are not only polluting the natural environment and causing a crisis of global environment, but also bringing about a 'crisis of humanity'. We are now hard-pressed to invent ways to restore our relationship with nature within our everyday lives. The reconstruction of the relationship between humans and nature is indeed a major issue for contemporary society.

Agricultural production, vermin and weeds

Vermin that attack pasturing livestock or that spoil crops, and weeds that compete with and invade crops, become targets of extermination and removal for the farmer. Vermin are shot and captured in cages or driven away with dogs, while harmful insects are eliminated with insecticide and weeds with herbicide. While there are plans to develop highly effective and non-toxic chemicals that easily dissolve in nature and have low levels of residuality in the future, various types of chemicals are already being used in great amounts. The larger the scale of an agricultural business, the greater is the amount of chemicals and the area where these chemicals are applied. At major American farms, not only agricultural chemicals but also seed and fertilizer are spread by light aircraft. Even in Japan and other regions of small-scale agriculture the use of light helicopters has become common.

In this system, it is indispensable for agriculture to remove vermin and weeds from sites of agricultural production by using weapons and chemicals in order to make production economically efficient, to sustain a supply capacity commensurate with rates of population expansion and mass consumption, to protect livestock and crops, to guard family life by securing a minimum amount of income, and finally to free workers from heavy labor in the fields.

Supplementary note. The facts of bird and animal damage

Here is an example of a farm family's battle with bird and animal damage in the town of Mizuho located in a mountainous area of the Chūgoku district. Arii Haruyuki, a farmer born in 1927, decided in his mid-thirties – at a time when Japan was going through a period of high economic growth – that agriculture should also be reformed and aimed at a considerable income of five million yen from his farm. The plan was to manage the five sectors of 1.1 hectares of rice field, 5–6 fattened cows which are raised by high degree of self-sufficiency in feed, one million shiitake mushrooms, 3 hectares of chestnuts and 25 hectares of forests using the labor power of four people, including eventually the elder son and his wife. The plan progressed smoothly and the elder son, upon realizing this, set his heart on pursuing agriculture, got married and started living in the same house with the family. However, just when the plan was about to be completed around the year 1975, damage wreaked by bears, monkeys and wild boars increased drastically. These animals infested the chestnut fields. Bears climbed the trees and caused damage, leaving behind ravaged gardens and at the same time making it dangerous for people to work in the area. Since the bear was designated a protected species, killing them was out of the question. Furthermore, a situation arose where wild boars and monkeys came out of the forests and wreaked havoc on rice paddies and fields right before harvest time, ate and scattered shiitake mushrooms, and even attacked humans and especially elderly ladies, robbing them of their belongings.

All this extended to three or four villages within the town of Mizuho. Monkeys used to only move within the valleys without invading agricultural spaces when their population stood at 30, but this rose to around 140 between 1979 and 1985, resulting in the formation of two groups. At this point, the eldest son of the family gave up and started seeking work outside the farm. Without being able to come to terms with his disappointment, Arii became a hunter and started shooting wild boars and monkeys despite being aware of the criticism he would face. Finally, the city authorities acknowledged the seriousness of the situation and began to offer bounties of 30,000 yen per animal shot dead or captured in a cage. Gradually, damage began to decline and at present the situation is balanced. The wild boar and monkey populations may have increased for the following reasons: 1) The heavy snowfall of 1963 as the turning point, the village rapidly de-populated, resulting in few energetic young people; 2) even those who stayed sought non-agricultural work outside, leaving the village empty of people during the daytime; 3) firewood was substituted by gas and petroleum, causing people to stop visiting the mountains; 4) as the population decreased and food products changed, persimmon and chestnuts and other fruits were left unpicked and were therefore available for animals to feed on; 5) after miscellaneous small trees were cut to allow for planting of new trees, the forest type changed with more nuts as well as insects and earthworms under piles of cut-down trees, thereby producing more nutrition for animals; 6) wild boars' food supply temporarily increased with more flowers and nuts growing on low bushes; 7) insects and small animals (food for larger animals) decreased in the mountain forests due to spraying of chemicals aimed at exterminating pine sawyers; and 8) the animals had acquired a taste for farm-grown fruits, vegetables and rice. A number of other social and natural factors can be suggested.
(Based on the survey of 1998–1999)

Similarly to Arii's case, farmers around the world are troubled by the damage wrought by weeds, birds and animals. In Japan, the advance of depopulation has deepened the problem and the agricultural production environment has grown worse, especially in the mountainous area.

Wherever human beings remain dependent on foodstuffs for their survival, struggle with nature, competition with specific animals and mutually exclusive relations are an unavoidable and solemn fact of life. Any particular state of equilibrium is arrived at through the rivalry, competition and struggle between agriculture – i.e. productive activities conducted for the sake of human survival – and animals' ability to survive. In the end, a state of spatial separation emerges. Yet, the borderlines of this equilibrium remain in a constant state of tension and are subject to readjustment at any moment. This problem has to do with how humans manage their economic lives, their views of nature, and of adjustment and reconciliation between humans and nature. Accordingly, the relationship between humans and nature in spaces of agricultural production depends on the so-called formative equilibrium that is consciously brought about through the application of human knowledge, wisdom and will.

This confrontational relationship between humans and plants and animals in agriculture often ironically results in an unfortunate relationship between humans and the ecosystem. Large-scale forest development and the conversion of forests into agricultural lands in order to respond to the population explosion are parts of this problem. More and more animals that lose the territory essential to their survival and are killed as vermin when found in settled areas, face a gradual crisis of extinction. Particularly, the use of insecticide and herbicide, and the construction of drainage canals out of concrete, as well as vinyl and other chemical materials have caused various plants and animals to perish.[2] Furthermore, this has also bred problems that affect human life and health such as the pollution of drinking water, the direct pollution of food, the accumulation via the food chain of toxic substances in plants and animals consumed by humans. The consumption of fossil fuel by machines such as trucks and tractors and facilities such as green houses are linked to global warming that is gradually leading to a crisis of the whole human race. It is clear that this separation of humans and nature causes a multitude of problems.

Deep ecology's view of nature

The philosophy of animal and plant rights

So far we have discussed the relationship and state of separation that exists between humans and livestock and crops, and between humans and vermin and weeds, and the problems that have emerged as a result. These relationships and problems are gradually becoming acknowledged more strongly worldwide and they have also given birth to views on nature and society that are critical not only of modern science but of modern agricultural science too. For instance, new concepts such as 'animal rights', the 'value of vermin' and the 'welfare of animals' are being advocated.

While the modern view of nature was established by Newton and Descartes, the latter strikingly compared animals to clocks and their springs, likening them to 'well-crafted machines' (Ferry 1992: 21). Nothing better represents the modern view of nature and animals than the idea that animals are nothing but 'automatic machines'. In fact, 'animals that were reduced to the level of mere machines were denied intelligence, feelings and sensitivity'. This view lies at the roots of modern society that has single-mindedly focused on accumulating ideologies based on economic rationalism and that has placed utmost priority on efficient production, *thereby becoming one of the factors that have bred the various problems examined above.*

The so-called deep ecology philosophy that rejects Cartesian thought thence emerged, overtaking the animal welfare theories of contemporary humanism. Deep ecology takes the view that animals are capable of thinking and of feeling pain and possess both emotions and intelligence. The philosophy holds therefore that animals and nature should be granted rights that at least match those held by humans. Moreover, deep ecology also asserts the value of vermin.

Numerous 'deep ecologists' such as Maupertuis, Réaumur, Condillac, and Clemenceau have made their appearance. Their strain of thought is associated with the reversal of the anthropocentric view of nature, instead placing nature in the centre – hence creating an 'ecotopia'. Conversely, from this perspective humans tend to fade from the picture. Of course, after examining the state of separation between humans and livestock and crops, and between humans and vermin and weeds, and ascertaining that this situation has bred problems, there seems to be some basis for arguing for the primacy of nature over humans. However, with the ever-continuing population explosion and industrial development, fundamental questions relating to the very survival of the human race, such as 'can human beings coexist with plants and animals without facing starvation?' and 'can human beings protect nature without starving?' arise. Deep ecology proposes a view that is extremely pure and clear; however, problems remain. I believe that this theory will result in the emergence of a division and deadlock of a reversed kind in the relationship between nature and humans.

As Ferry has argued, the issue of the rights of nature, plants and animals must be raised anew, separately both from Cartesian anthropocentric philosophy and the philosophy of deep ecology (Ferry/Katō 1994: 267).

Animal protection and the idea of animal welfare

Under the above circumstances, the pursuit of coexistence, the philosophies of animal welfare and biotope (that is concerned with the protection of 'habitats of life' such as marshes, coppices, ponds and rivers) have emerged and are solidifying.

By definition, animals and birds become vermin and plants become weeds basically only if they invade spaces of agricultural production. It is possible to capture or kill vermin and exterminate weeds through the application of agricultural chemicals, but we have seen that this causes problems from the point of view of sustaining and preserving the ecosystem and the rights of nature. Animals and birds that encroach upon agricultural space can be driven out of range for instance by producing explosive sounds, setting up scarecrows, throwing pieces of clay at them, or by having humans chase them away. However, this can often result in a 'futile cat-and-mouse game' sort of a scenario, consume plenty of labor and have limited effects. Recently, there have been efforts in Japan to take a 'coexistence-promoting mutually exclusive approach' check to the relations between humans and animals by surrounding fields and afforested areas by fences to prevent damages by deer and wolves, and by covering crops with nets to protect them from monkeys and birds, but this has caused production costs to increase.

While allowing that the slaughtering of livestock may be unavoidable, there is a school of thought that advocates the minimization of animal suffering. Concerned with the protection of animals from cruelty, this way of thought asks questions such as 'are animals given sufficient amounts of feed and water and are they provided with ventilation when transported from the farm to the slaughterhouse, and what methods of slaughter are being used?' In the EU, a new proposal concerning raising chickens in cages has been adopted. In short, this proposal stipulates that the current cage space of 450 square centimetres per chicken be expanded to 550–800 square centimetres by 2003, and that cage farming be completely abolished by 2012 resulting in free range chicken houses with 750 square centimetres secured per chicken (International Egg Commission 1999: 1–2). In the future, it is likely that this kind of trend will strengthen and come to apply to all livestock. Additionally, along with other issues, matters such as the treatment of test animals used for research, the nature of bullfighting, and the use of fur and ivory are now being debated.

Supplementary note.
At universities, guidelines regarding animal testing have been established with consideration to animal welfare, protection of animals from cruelty and safety management. For instance, when livestock, wild animals, fish, insects or others are used for testing, regulations demand the following be ensured: the minimization of use (of animals for testing); the drafting of an ethical, legal and appropriate testing plan; keeping of records; the minimization of stress to the animals resulting from transportation; the proper quarantining and handling of animals with irregularities; the provision of feed and water, and the proper adjustment of temperature, humidity and lighting in the cages (rooms) where animals are held; not causing unnecessary pain to the animals; the proper and comfortable treatment of animals after testing and appropriate disposal of corpses; the obligation to conduct suitable training prior to testing. (Animal Testing Guidelines of the Agricultural Science Postgraduate Program at the Graduate School of Kyoto University, 1999.)

'Formative equilibrium' as a new relationship

Humans, representing one of the many life forms on Earth, are natural beings that live within the natural ecological system and within the bounds of its laws. Humans are also cultural and social creatures that form families and societies within which they lead lives filled with a range of emotions. Furthermore, humans are also economic beings that make things necessary for their survival and livelihood. Making things are mainly industrial and agricultural production. Accordingly, the area that forms the sphere of everyday life for humans is at the same time an ecological place comprising 'an ecological unit', a living place that fulfils various human needs, and a place of production. The plants and animals that are utilized in agricultural production activities are organic life forms that in themselves are not mere 'objects' or machines but comprehensive living beings. The dynamic relationships between humans and plants and animals can be mutually dependent, competitive or confrontational by character. If we classify these relationships once more, the three previously examined aspects come to the fore:

1. The interdependent symbiotic relationship: livestock and crops beneficial to agricultural production – the symbiotic principle.

2. The mutually exclusive competitive relationship: pests and weeds harmful to agricultural production – the competitive principle.
3. The habitat segregative, coexistent relationship: plants and animals neither beneficial nor harmful to agricultural production – the coexistence principle.

Humans and livestock and crops stand in an interdependent symbiotic relationship based upon the 'symbiotic principle'. On the other hand, pests and weeds that invade spaces of agricultural production are in a mutually exclusive competitive relationship with humans, adhering to the Darwinian survival of the fittest-type 'competitive principle'. The plants and animals living outside agricultural spaces and that are neither beneficial nor harmful to agricultural production stand in what Imanishi Kinji has called a 'habitat segregation' relationship with humans.

Darwin argued that the natural world is based on the harsh competition of survival where only the species with the strongest variation can win. With a series of natural selections, mutation of species takes place. The struggle for existence is the selection mechanism. As a consequence, the principle of the survival of the fittest prevails, and living beings transform themselves into adaptive forms (Darwin 1859).

Imanishi squarely challenges the Darwinian position, maintaining that nature is not governed by the principle of the survival of the fittest but by a principle of habitat segregation and co-habitation where each species mutually complements, segregates and coexists in a holistic and coordinated society. Imanishi derived his theory from his observations of how mayflies segregate depending on different river velocities. Based on various differences in the natural world, species coexist spatially, temporally and seasonally (Imanishi 1994). Here we see a worldview totally different from the Darwinian ideas.

Imanishi, however, appears to have adopted a static view on the issue. The habitat segregation situation is merely a momentary 'frozen-time' equilibrium that may crumble at any time due to a variety of reasons and whose borders are subject to change. The competitive principle represents a dynamic view and the coexistence principle a static view. Each of them describes an aspect of nature. As was shown above in the case of Mizuho town and its struggles with bird and animal damage and spatial separation, the true reality is expressed at the borderlines and through their movements; i.e. where the competitive and static aspects meet.

For the time being, it may seem as if humans and plants and animals stood in a peaceful coexistence-based relationship neither harmful nor beneficial to one another under the habitat segregation principle based on the state of spatial separation. However, as the modern-day production and livelihood activities grow more enormous and gain vigor, and as environmental degradation proceeds in ever larger areas and on a global scale, the very survival of animals, plants and humans is called into question. In other words, even the fate of any possible peaceful coexistence is for the most part in the hands of humans.

When thinking in this way, I began understanding that something more thorough could be achieved by combining the above three human-nature relationships and subsuming them under one theory. I propose that the modern-day relationship between humans and nature should be founded upon the *principle of formative equilibrium* based on human 'imagination'. In the present age that has witnessed a population explosion and an enormous expansion in human activity and accompanying problems, and with help from general human creativity capable of self-reflection and self-awareness, there is no alternative but to seek for the best and most desirable point of equilibrium in the human-nature relationship at a particular point in time. A reflective, ethical world of 'formative equilibrium' must be built. The recently established academic discipline of environmental ethics promotes this kind of human awareness and reflection, and attempts to establish starting points for a new kind of human-nature relationship.[3]

Whether called symbiosis or coexistence, and whether considered good or bad, it is a stage of equilibrium at the end of the human 'struggle for life' for which survival and livelihoods have been risked. Whether symbiosis or coexistence, even if the phenomenon that emerges as a result is desirable and beautiful, it is a product of the process of human struggle. Only when we combine the processes and results of this kind of a human–nature relationship and grasp them as a totality, and achieve self-awareness with regard to them, can we first reach a total and original, human state of affairs where we feel excitement, joy, awe, appreciation, fear and prayer towards nature.

I propose that the human-nature relationship in agriculture and agricultural science must be reconstructed based on the above viewpoints. Furthermore, agricultural science must be transformed from a mere study of biological production to a holistic science of the life. There is a need to build a new mutual relationship between the three scientific and technological spheres of (1) human sciences that strive to understand humans and their societies, (2) the deepening natural sciences that strive to understand the natural world, and (3) the development of 'practical science' based on the progress of the former and used for the purpose of creatively constructing the human-nature relationship.[4]

3 The Development and Value Objectives of Modern Agricultural Science

This chapter aims to elucidate the development of modern agricultural science to highlight its process after World War II and to clarify the issues, concepts and value objectives that it has to deal with. For these purposes, we: (1) examine relationships between science and values; (2) analyze value questions in agricultural science; (3) investigate a group of problems, a cluster of roles and new value objectives that face modern agricultural science – while we deliberate primarily on the value objectives of Japanese agriculture after World War II, contemporary agricultural science represents both the culmination and the weaknesses of modern agricultural science since the Industrial and Agricultural Revolutions, a situation which has prompted us to present a new paradigm in this book; and (4) focus on 'local communities' as places to achieve these objectives.

Development of science and values

Achievements and limits of modern science

Using 'facts' verified by 'empirical methods,' as its major weapon, modern science liberated humankind from the limitations of the religious world which had prevailed in the European Middle Ages. Modern science has also made great contributions to the development of civilized society since the Industrial Revolution. Benefits of science and technology have enabled mankind to acquire more material wealth than ever before.

Modern science is based on two views of nature. One is the mechanical view of nature, based mainly on physics and typically represented by the work of Newton. The other is the anthropocentric view, which separates humanity from nature, and is typically credited to Descartes. Combined, these two positions form the 'mechanistic and anthropocentric view' which is often called the 'Cartesian–Newtonian paradigm' (Fujisawa in Uzawa 1989: 27). Based on this paradigm, modern science has developed through specialization and differentiation, promoting elementary-reductivist thinking, a mode of reasoning that analyzes phenomena by reducing it to its constituent elements.

With its cumulative success, modern science became increasingly compartmentalized, isolated and self-righteous, losing sight of its social position. Science and technology was gradually co-opted by 'econo-centric' thinking, with its 'values' determined by the economic market. Consequently, modern society has developed with the integration of the nation, economy, and science and technology. In pursuit of economic efficiency and the superficial comforts of everyday life, we have destroyed our ecological environment and repressed our human nature. Nature's adaptive and self-correcting capacities are at a critical point that threatens human well-being and life itself.

As Tamanoi Yoshirō points out, the situation described above prevails in social sciences in general and economics in particular. As symbolized by the appearance of environmental problems, economics has become increasingly ineffectual with the discipline tending to be excessively technical and dehumanized (Mizuta and Tamanoi 1978: Preface 1; Tamanoi et al. 1978).

That is why many philosophers and scientists regard the contemporary period as an age of unpredictable crises – with conflicts, severances and rifts appearing everywhere – and believe that it is time for comprehensive changes in the current state of society and science. Because of the extreme specialization of science, scientists often get lost in the narrow and reductive scientific world, forgetting about the human basis of science. It appears to be all too easy to fall into the illusion of equating the partial knowledge of a scientific discipline with the comprehensive knowledge of the universe. The limitations of scientific knowledge are clearly evident in the current social context in which the development of compartmentalized science, guided by the principles of the market economy, has destroyed the ecological environment and inhibited the achievement of human well-being.

Fujisawa Norio (in Uzawa 1989) calls for the restitution of ecological and holistic views of nature to replace the mechanical /anthropocentric perspectives of modern science. He states: 'It is true that, in the development of modern science, the Cartesian–Newtonian mechanical view of nature, which is based on the analysis of motion and disposition of 'things', has now revealed its limitations, although it should not be totally rejected; the situation simply suggests that it does not squarely correspond to the ultimate reality of nature. In all senses, this doesn't mean that it should be totally rejected.' Kira Tatsuo also points to a future direction where the problems of 'compromise of man and nature' as well as 'agreement between the natural system and the civilizational system' ought to be resolved (in Uzawa 1989).

These considerations compel us to face a fundamental problem of what science has been for society, and what we should do in the future.

Specialization in science and social responsibility

Modern science, especially the natural sciences, has claimed value-neutrality to be a guiding principle and distinguishing characteristic. From this perspective, value is 'the standard of choice necessary for the existence of the subject', the benchmark which also satisfies his or her material and spiritual needs. Therefore, value involves some satisfaction or dissatisfaction, as well as advantage or disadvantage. These subjective values differ from, and are incompatible with, the objective observation of lawful regularities that are sought by scientific research. From this perspective, scientists should therefore take a value-neutral stance in order to distinguish facts from values. On the basis of their element-reductive method, scientists have increasingly specialized in highly limited areas, actively asserting their own value neutrality and ignoring a variety of problems produced by modern science, and thereby denying any troublesome responsibility. It is doubtful how neutral they can be with respect to social impacts of scientific findings.

Many scientists appear to live in small self-satisfying worlds of extremely specialized research fields which they presume to have no connection with power or economic interests and take pride in their contributions to the investigation of scientific truth in this vacuum. However, it is impossible for human acts to produce completely value-free facts. Science has grown enormous and, in the process, has significantly influenced society and nature more and more. The scientist's social responsibilities have correspondingly increased for the results of their research findings. In choosing the subject of research and implementing it, even though the research methods themselves may have little to do with values, the starting point of research and its results, more or less, cannot be free of values. When we contemplate both the positive and negative contributions of modern science, we should start with the realization that scientific endeavors are always either directly or indirectly, more-or-less value-laden.

On value-freedom in science

Max Weber makes an important point about facts and values. In his treatise on objectivity in social scientific and social political research, he analyzed the concept of 'value-freedom' (*Wertfreiheit*) in science (1904/2005). His argument has been frequently misunderstood to be a claim that scientists should be neutral from and should distance themselves from any values. The question of values is all the more significant in agricultural science which is both a branch of the integrated sciences of living beings (which includes social sciences) and an action-oriented practical science. Weber sharply criticized Shmoller – the doyen of the historical school and a leader of the German economic policy association – for too readily associating himself with the political power of the times, thereby making science a servant of politics. Weber argued that scientists should not bring the world of 'oughtness' (*sollen*) into that of 'being' (*sein*). This does not mean that scientists must be totally distant from values or totally immersed in them. His point was that we ought to take a flexible stance by choosing between competing values while not being obsessed with our choices. Value freedom, then, points to (1) the freedom to choose particular values; (2) a free and objective epistemological stance uncontrolled by the chosen values; and (3) a guarantee of free debate over the validity of various values (Sumiya 1970).[1] In other words, scientists should always carefully refrain from the temptation to form conclusions favorable to their own value perspectives, by consciously or otherwise choosing only convenient facts that support their preconceived position.

Weber constantly pursued research morals based on intellectual integrity, making a clear distinction between the researchers' value orientations and scientific epistemology. This must be the basic stance of our agricultural science research. Agricultural science attempts to contribute to the welfare of farmers as well as that of the general public through the development of agriculture, forestry and fisheries as well as their related industries and rural communities. General natural science as basic science has tended to be thought of as separated from human subjective values, with the aim of finding and explaining the dispensations of nature and discovering the laws governing it. In contrast, human science has tended to be seen as the science that understands and interprets human and social facts and examines their particular meanings in value contexts. As will be detailed in Chapter 10, similar to the disciplines of engineering, pharmacy and education, agricultural science should actively set, create and promote value objectives as an imaginative science. In other words, it is applied, practical and action-oriented science. General natural science cannot operate in isolation from value issues.

Though we suggest that agriculture is a practical science, it cannot ignore basic research. While agricultural science is oriented to certain value objectives, it conducts its own fundamental research and achieves those objectives by applying and developing technology. The influence of research outcomes is concrete and direct, which makes it all the more important for us to be reflective and self-conscious about our research results. Agricultural science deals with the cultivation of living things, and is thereby closely connected with nature. It is the science that is required to fulfill the requirements of human existence, and is thus firmly and directly concerned with value issues. Accordingly, the *raison d'être* of agricultural science rests in the kinds of values it pursues and their validity in real life.

The main purpose of the philosophy of agricultural science is to grasp the realities and the problems of agriculture, to clarify the history, the present conditions and all the issues of agricultural science, and to show its directions by promoting self-reflection and self-awareness based on agricultural realities. In the philosophy of agricultural science, self-examination and conscious self-searching is far more important than specialist debates.

To reiterate the point, as agricultural science is a value-seeking practical science, we should conduct research with the clear specification of its value objectives. When research objectives and goals are formulated, their ethical standard must be put under the microscope. As Miki Kiyoshi stated: 'Technology must be moral in its invention of goals' (1967: 285). In the next section, I focus on the specific value objectives that agricultural science and modern agriculture should pursue.

Values in agricultural science

There are different definitions of the term 'value.' According to Stern, Platonists define value as trans-temporal ideals that exist independently of human desire. In contrast, Kantian's take it to be the ideal expression of oughtness, which has validity but no real existence. Furthermore, Protagoras, Nietzsche, Sartre, Russell and Perry consider it to be no more than the 'creation of subjectivity.' Stern defines it as the relationship between the objects and appreciating subjects (1962: 101). Reflecting on these different perspectives, Ueyama Shunpei defines it as 'the criterion of the selection act necessary for the existence of the subject' (Kurita and Ueyama 1968: 295).

The differences in these definitions concern whether values are subjective, objective or relational. There is some truth in Stern's definition to the extent that values reflect the nature of the object, be it a real thing or an abstract idea. Ueyama argues: 'I object to the theory which defines values as reflective of the quality of the object. The relational theory which sees values in terms of connections between the subject and the object appears safe, it is no more than a futile theory which avoids touching the heart of the problem' (Kurita and Ueyama 1968: 353). I ultimately support Ueyama's argument that values are 'the criteria of the selection act that is necessary for the existence of the subject; values start to have concrete meanings only after the subject becomes consciously aware of them. However, we do not necessarily have to deny that values are connected with the quality of the object.

In the sphere of agriculture, for instance, we often state: 'Agricultural production not only supplies people with food and materials for industrial production, but plays an important role in the preservation of the environment, and has social and cultural meanings as well.' This presumes that agriculture has a concrete quality. Only after this idea is realized and expressed by an individual or a group does the quality of agriculture come out as a substance with certain values. However, if it is not realized by individuals and does not become the common understanding for many people on a regional or national scale, policies for preserving agriculture will not be adopted and, as in Japan, agriculture will be in constant decline. The question is how far various qualities of agriculture are evaluated and how widely and consciously they are accepted by subjects as meaningful for our lives. A comprehensive value system concerning agriculture and farming villages begins to emerge when subjects sublimate separate interpretations and significations about them into coherent evaluations. If an extant quality of agriculture is neither recognized nor embraced by the subject, it has no value and will be neglected and decline. These are the circumstances in which agriculture finds itself in contemporary Japan, and is a source of imminent crises. In this sense, values are eminently historical and subjective.

Moreover, we must make two points. The first concerns the levels of the subject. Stern classifies values at (1) individual, (2) collective and (3) universal levels. Individual values are dependent upon the individual peculiarities of the evaluating subject. Group values depend on the collective peculiarities of the groups, such as nation, class, caste and political party. Universal values are shared by humanity in general, surpassing individuals and groups (1962: 132–3).

The second point concerns the process of value selection, of which there are two types. On one hand, there are subjective and direct expressions of likes and dislikes, pleasure and discomfort. For instance, one may like or dislike farm life and village living. On the other hand, there are objective and indirect styles of expressions of truth and falsehood, fairness and unfairness, right and wrong, loss and gain. For example, one may claim that agriculture serves a function of preserving national land and the environment and should therefore be protected. Or, one might argue that, because Japan's agricultural products are expensive, cheap foreign products should be imported.

Notions of values are wide-ranging and diverse, depending upon (1) the level of the subject (whether it is an individual, group or humanity) and (2) whether the values in question are subjective and direct or objective and indirect (Kurita and Ueyama 1968: 353). Especially on the individual level, the definition of values is complicated and involves questions of consciousness (whether they are consciously realized or not) and systematization (whether they are systematic views or not).

Thus, when we examine how people value agriculture we must investigate both their direct and indirect expressions. At the direct level, we can examine the views that are consciously expressed, both orally and in writing – such as various quantitative surveys and in-depth interviews as well as books, magazines, newspapers and other publications that re-present individual or organizational views on agriculture. Although not-consciously expressed, the indirect level involves what can be understood or interpreted from actions, attitudes or situations, or read between the lines. In the second area, as a result of the accumulation of human knowledge, we find the accepted customs of individuals and society, latent beneath the conscious layers. We also find general expressions of value orientations in various other pieces of information in society and attempt to comprehensively interpret and understand them (Soda and Ōhara 1994).

Many and diverse values are expressed and held about agriculture and rural communities simultaneously. These diverse values are communicated widely and then converge into central consensus-forming values that are consciously emphasized at a variety of collective levels, such as organizations, districts and nations. Ultimately, the diversity of values and their integration are not incompatible. Value diversity enables us to engage in serious dialogue, debate and other interactions to solve pressing problems about the economy, life, the ecological environment, and to sustain human existence. This process makes it possible to develop more valid, higher-level, comprehensive values. But first, individuals must strongly and consciously articulate values and views on agriculture before we can develop district-based or national consensus. In this process, we find the beginnings of solutions to the problems facing agriculture today.

In the following section, we will examine the national values of agriculture and rural communities in postwar Japan and corresponding values toward which agricultural science has aimed.

Value objectives of agricultural science in Japan

What are the value objectives of agriculture and agricultural science in Japan today? Without reflecting the extent to which they have been realized by agricultural practice, the objectives that have been proposed and expected from agriculture and agricultural science in postwar Japan are summarized in Table 3.1. This table also serves as a condensation of the concept of agriculture and agricultural science in this book (Soda and Ōhara 1994: 323).

'Agricultural science of production' and economic values

The central task facing Japan in the late 1940s and early 1950s was to restore the economy, whose productivity had fallen to one-third of the prewar level, and to meet the minimum needs for its people's survival. In this context, agriculture, rural villages and agricultural science were expected to play important roles.

During the late 1950s and early 1960s, Japan entered and consolidated its now famous rapid economic growth period, experiencing rapid industrial development and the increasing concentration of population and industries in large cities. Under these circumstances, agriculture's expected roles included: the 'selective expansion' of production to promising new fields, such as fruit cultivation and stock farming; delivering stable food supplies at low prices by improving production efficiency; and improving the income of farmers to levels equivalent to town workers. In other words, agriculture was expected to contribute to bringing the economy from the level of simple survival just after the war's end to a level which ensured a healthy standard of living.

Thus from about 1945 to 1965, agricultural science was in a stage of increasing production, whose primary goal was the 'realization of economic values' and which was intently focused on increasing and

Table 3.1: Development of Japanese society and the main functions of agriculture and rural communities

Period	1945-54	1955-64	1965-74	1975-84	1985 forward
Major trends	Restoration period	First half of rapid economic growth period Expansion of manufacturing industry Expansion of cities	Latter half of rapid economic growth period Frequent occurrences of environmental and pollution problems	Period of low economic growth Frequent occurrences of urban and local problems Emphasis on quality of life	Maturation and computerisation Frequent occurrences of trade and international problems International exchange
Changes in, and multipolarisation and stratification of, roles of agriculture and rural villages	Survival-level economic roles	Living-level economic roles Survival-level economic roles	Ecological roles Living-level economic roles Survival-level economic roles	Social and cultural roles Ecological roles Living-level economic roles Survival-level economic roles	Comprehensive roles Social and cultural roles Ecological roles Living-level economic roles Survival-level economic roles
Trends in agricultural science	Agricultural science of production		Agricultural science of life (*sei no nōgaku*) Agricultural science of life (biological) Environmental agricultural science	Agricultural science of living (social) Social agricultural science	Agricultural science of *ba*
(Main values pursued)	(Economic values)		(Ecological values)	(Life values)	(Comprehensive values)

expanding production. By clearly pursuing these goals, agricultural science during this early postwar period produced excellent results.

For example, as per Figure 3.1, working hours in rice farming in the four decades between 1951 and 1990 decreased by approximately 80%, from 200 hours to 43.5 hours per 10 ares (0.24 acre), due to the increased use of machines and materials. Successful development of improved varieties and the use of fertilizers resulted in an approximately 50% increase in yield, from 330 kg to 509 kg, as well as a northward shift of cultivation limits, greater stability of production, improved qualities, and many other successes. These results show an overall improvement in land productivity and labor productivity through the development of relevant technologies. However, the small scale of Japanese farming often required over-investment, which appears to have resulted in poor capital productivity.

The annual distribution of tomato shipments detailed in Figure 3.2, indicate that new production facilities enabled supply of certain products throughout the year at steady prices, leading many people to forget the seasonality of such products. Many other technological successes were enjoyed in agriculture-related industries, such as

Figure 3.1: Changes in paddy yield per 10 ares (0.24 acre) and working hours

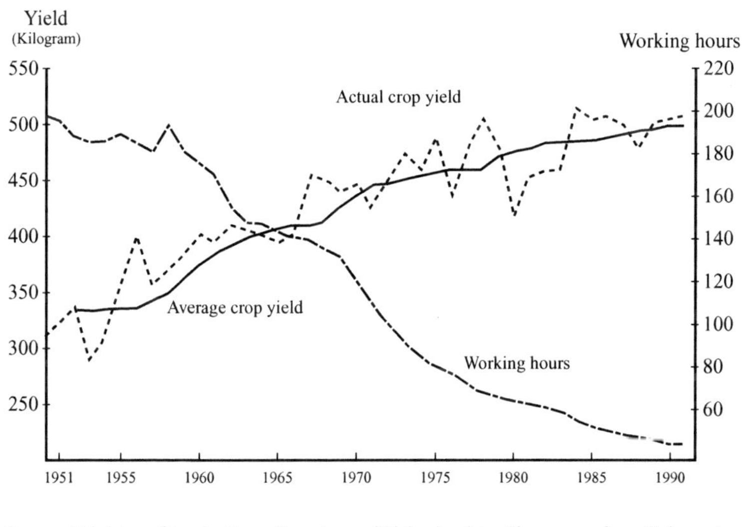

Source: Ministry of Agriculture, Forestry and Fisheries data. Figures are from *Sakumotsu tōkei* (Crop statistics) and *Kome seisan-hi chōsa* (Survey of rice production costs).

Figure 3.2: Increased stability of agricultural product supply in a year – cases of tomatoes (percentages shipped each month)

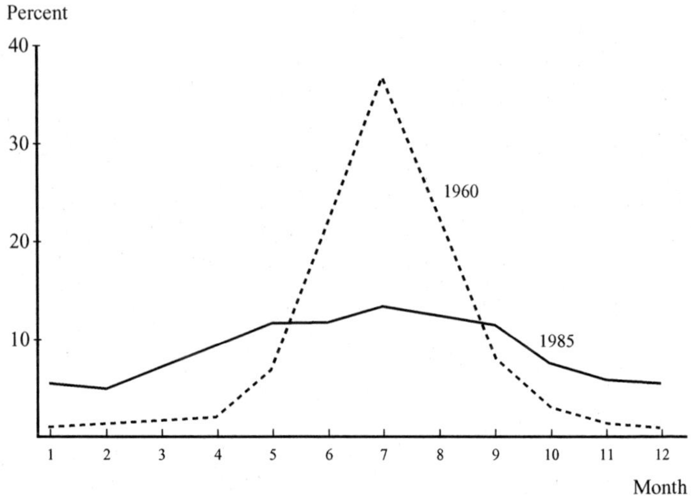

Source: Shokuryō to nōgyō Q&A, 1987: 59.

new agricultural machinery, fertilizers and agrochemicals, food processing and distribution, and various materials. Agricultural science contributed countless epoch-making technologies in various fields, including stock farming, fruit cultivation, forestry and fisheries, and achieved 'increased efficiency, stability, diversity and quality' of agricultural production, thereby meeting market demands.

Kashiwa Sukekata defined agriculture as 'human activities intended to realize higher economic values by rearing and raising crops and livestock, which are thus intentional activities' (1962: 162). Kashiwa considers farmers to be parties seeking not only acquisition (private interest) but also commonweal (public welfare) and ordination (creation of order within oneself) recognizing their public functions (1968: 86ff.). He locates the realization of economic values at the core of his theory, however. Nevertheless, for his time, this econo-centric perspective rather accurately reflected the primary objectives of agricultural science.

Agricultural science of bio-life and ecological value

In the late 1960s, Japan's economic growth began to decelerate and myriad problems generated by rapid growth began to emerge.

Among these, environmental and pollution problems were very serious, particularly in the larger cities. These problems included noise, air pollution, pollution of rivers and drinking water, small houses and lack of sunlight, and health problems attributable to these factors. Prominent examples of these health problems include Minamata mercury poisoning, *itai-itai* disease, and Kawasaki and Yokkaichi asthma. Agriculture became highly industrialized during this period, using ever-more machinery and chemicals. Population increase, the movement of labor from agriculture to manufacturing industries, and from rural to urban environments, and the previously mentioned development of various land- and labor-saving industrial technologies all contributed to expanding mass production and mass consumption and, the inevitable yet too often neglected, mass disposal. Increased farm specialization, monoculture and continuous cropping reduced soil fertility, leading to soil sickness and reducing the vital force of crops. This resulted in more crop damage from insects and disease, which, in turn, fueled the use of ever-greater quantities of agrochemicals. The results have been widespread environmental pollution and increasing doubts about food safety.

During this period, however, an increasingly strong voice began to point out that agriculture and forestry as green industries have national land conservation functions, such as water conservation, air purification, the prevention of soil erosion, environmental preservation and flood control. Agriculture and forestry thus were increasingly expected to play an ecological role, by recognizing the externalities of agricultural production and minimizing the negatives while increasing the positives. This was expressed by two new schools of thought: the 'agricultural science of life' (Sakamoto Keiichi 1981; Tsuno Yukindo 1975 and others) and the 'agricultural science of environment' (Matsuo Takane 1974 and others), both of which advocate pursuit of 'ecological value' in order to maintain and protect human lives. This turn has also led to the concepts of 'low input sustainable agriculture' and environmentally sound agriculture (see Kyūma and Soda 1995; National Research Council 1992). Of course, some have taken this direction to an extreme, advocating concepts such as the 'value of noxious animals' or 'animal rights' (Worster 1985: 258), which assert that wild animals should not be killed even if they are noxious to agriculture, thus complicating efforts to plan agriculture in intermediate and mountainous areas where farmers are continually affected by wildlife damage caused by birds, monkeys, wild boars, bears and so on.

Agricultural science of socio-life and life value

Following the oil crises of the 1970s, Japan entered a period of low economic growth which continued for more than a decade. In the late 1970s and early 1980s, numerous urban and local problems began to appear, and 'quality of life' became an issue of growing concern. Increasingly overpopulated cities experienced, in addition to the previously mentioned environmental problems, a wide range of pathological phenomena – both physical and psychological. These included: soaring land prices and housing shortages; insufficient amounts of parks and green spaces; garbage disposal problems; traffic problems; community breakdown and solitude; increasing incidence of mental disorders; a diversification of criminal activities and increasing incidence of crimes committed by younger people. At the same time, increasingly under-populated rural villages experienced: declining levels of community activity; deterioration of the traditional system for locally controlling nature, accompanied by more frequent disasters and increasing difficulty in restoration; shortage of agricultural labor; aging of farmers; split families and increasing numbers of households composed only of elderly people; and shortages of marriageable women. The situations in both cities and rural villages directed people's attention to the social and cultural functions that should be performed by agriculture and rural villages. These functions include: recovery of humanity, educational functions and welfare functions. This criticism of urban concentration led to a regionalism movement that emphasized local regions, from which emerged the 'agricultural science of living' or 'social agricultural science' (Yoshida and Okabe 1985). Advocates of these 'sciences' intend to reconstruct rural villages from the perspective of community life. The primary objective of these concepts of agricultural science is the 'realization of life (social and cultural) values'.

I refer to these concepts collectively as the 'agricultural science of life'. These concepts are clearly articulated by Sakamoto Keiichi (1989). Sakamoto defined agriculture as 'active, systematic human activities to obtain materials and information essential for the realization of human "life", through utilizing, rearing and raising certain organisms constituting a life system which has been established based on the global ecosystem' (1989:1). In this context, human life refers to 'bio-life, living lifetime, and ways of life in their entirety'. Sakamoto reached the conclusion that 'agriculture is equal to life' and the 'principles of agriculture are equal to the principles of life'. This represents the very substantial conversion from 'agricultural science of production' to 'agricultural science of life'. It can be said that in the latter concept the relative importance of the pursuit of economic values is greatly reduced and the pursuit of ecological and life values is pushed to the front.

'Agricultural science of *ba*' and comprehensive values

The Japanese economy, agriculture and villages

In the late 1980s, the Japanese economy entered a 'mature' period. The trend toward globalization strengthened and the pressures on Japanese agriculture intensified. After the long period of rapid economic growth, Japan had become 'the factory of the world'. Globalization was of course crucial to Japan's success, which was achieved under the economic doctrine of the international division of labor, based on the principle of comparative advantage. Japan's great technological power, combined with its people's attitude toward work, overwhelmed the world market. Its exports first hit the US, then the European and other Asian countries like 'torrential rain', causing intense economic friction. This was epitomized by Japan's huge trade surplus and the US's huge trade deficit. In their efforts to redress this imbalance, the US urged Japan to at least import more military supplies and agricultural products. In a sense, then, Japanese agriculture was caught in a pincer between its own world-dominating manufacturing industry and the US's world-dominating agricultural industry. The US's agricultural production costs were low and their export volume very large. In the 1980s, Japan was already the largest net importer (its import value exceeded its export value) of agricultural products in the world, and its imports of agricultural products continue to increase, pushing its domestic agriculture further into decline. The relative importance of agriculture in Japan's economy is minute, with agricultural products accounting for a mere 1.95% of GNP and the agricultural population accounting for 6.4% of its workforce in 1989. The self-sufficiency rates for Japanese agricultural products were 40% in terms of calories and 30% in terms of grain, an unprecedented situation in world history.

Furthermore, globalization has produced new trends, such as: multinational corporations; trade expansion and economic friction; aggravation of the North-South problem and ethnic conflicts; increased need for international cooperation such as technical assistance and food aid; and international communication between cities and rural villages. Japanese agriculture is under pressure from global competition to move toward free trade, and to reduce the price difference between domestic and overseas markets. At the same time, it is being urged by various domestic forces to play social and other roles such as the conservation of life and environment by environmentally sound agriculture, exchange between towns and rural villages, and decentralization of population, industries and administration. In other words, Japanese agriculture is urged to achieve economic, ecological and life values all at the same time.

The tasks of agricultural science are to solve the problems listed in Table 3.2, to promote the functions of agriculture as an industry of multiple values, and to promote rural villages as multifunctional spaces.

Realization of comprehensive values

The values to be pursued by agricultural science have become diverse and multilayered in the course of history. These often conflict and demand trade-offs with each other in the present socio-economic environment.

First, where economic value is prioritized ecological and everyday life issues tend to be over-looked, ignored, or degraded. One example of this is the frequent and widespread application of agricultural chemicals in large quantities to achieve higher yields in production and, at the same time, to increase the market value of the product by controlling its appearance. Another example: In highly productive farming regions in Japan, family income is as high as 10–20 million yen (US$83,000–166,000) per annum; but if the farmer suffers health problems due to excessive working hours or exposure to agricultural chemicals etc, or remains single and alone due to the shortage of marriageable women in the region, this income does not translate into a satisfactory standard of living. A third example would be a region that has developed its economy at least in part through monoculture farming practices, and as a result is experiencing reduced soil fertility and soil sickness, and thus falling productivity.

Second, where ecological value is prioritized, production efficiency and/or people's everyday life activities are curtailed. For example, if farmers' prioritization of ecosystem conservation means a refusal to use any and all agricultural chemicals regardless of the circumstances, this may require long hours of hard toil for weeding, etc, and still not produce enough food to feed the nation's population.

Third, when the values of human life, society and culture are emphasized, economic issues are often forgotten. Neglecting economic issues to nostalgically preserve one's hometown, traditions,

Table 3.2: Major issues surrounding agriculture, forestry, fisheries and rural districts

Economic issues	Ecological issues	Living issues
Population explosion and starvation	Energy saving and energy development	Shift in lifestyles
Food production increase	Depletion of resources	State of dietary habits
Increased meat eating and feed production increase	Deforestation	Concentration in large cities and town-rural village relations
Low prices and stable supply	Excessive cultivation and excessive grazing	Secure shopping and leisure spaces
Efficient production	Desertification	Hollowing out of local community
Decreasing self-sufficiency rate	Accumulation of salts	Changes in family system
Rationalisation of distribution process	Soil erosion	Establishment of recycling system
Promotion of LISA (Low-Input Sustainable Agriculture)	Continuous cropping hazard	Nature and agriculture education
Coordination between full-time and part-time farming households	Frequent use of chemical fertilisers and agrochemicals	Lack of marriageable women
International trade disputes	Pollution and deterioration of soil	Aging society and falling birthrate
Food security	Endocrine-disrupting chemicals	Sewage disposal problem
North-South problem	Dioxin	Improvement of women's position
Economic activities and environment	Water pollution and shortage	Remaining convention
Wildlife damage control	Air pollution	Decline of traditional events
Lack of successors	Global warming	Sprawling suburbs
Farmland shortage and conversion	Abnormal weather	Abandonment of public transportation
Limits of production per unit area	Genetically-engineered food	Shortage of medical facilities
Use of food as weapon	Environment rehabilitation	Long-distance commuting to school
Fishing ground disputes	Increasing number of extinct species	Employment of foreign workers in agriculture
Abandonment of cultivation, etc.	Recycling of resources	Difficulty maintaining fire-fighting team activities, etc.
	Production and processing of safe food	
	Environmentally sound development, etc.	

and the social and cultural significance of villages would fail agriculture and would have no currency in today's global society. Human culture, society and life require a certain material basis that meets the demands of the time.

These examples only scratch the surface of the trade-off relationships between these three sets of values; but are sufficient to evoke the difficulties in solving problems in this regard. Nevertheless, we must realize all three sets of values simultaneously and harmoniously. In other words, the goal of modern agricultural science is to maximize total welfare by overcoming the intrinsic difficulties of achieving comprehensive values. This idea is represented in Figure 3.3 and is the central concept driving this work. The development of technology, agricultural management and economics, and research into life in rural villages should be conducted from this perspective.

Similarly drastic changes in the priorities of values were common to many countries when their rapid postwar economic growth came to an end. Let us take a look at the changes that occurred in Germany from the latter half of the 1970s.

Table 3.4 lists the changes in value concepts from industrialized society to post-industrialized society as observed by Harmann in 1975. Although these changes were expressed differently in Germany than in Japan, the keywords in the table provide a meaningful outline of such changes (cited Hübler 1987: 35). Specifically, Harmann points to changes: from a drive for material growth and material contentment to a drive for spiritual or personal growth; from a confrontation with nature to coexistence with nature; from selfish individualism to harmonious individualism; from a large scale-orientation to a small scale-orientation, and so on. These largely correspond to the changes in Japan, which have been expressed in terms such as changes: from material to spiritual; from heavy, thick, long and large to 'smaller is better'; and from economic growth-priority to true richness.

Affeldt understands these changes as the conversion from the world of having (*Haben-Welt*) to the world of being (*Sein-Welt*) (Hübler 1987: 21). Hübler observes that in the 1960s the economy-oriented priorities gradually began to change as environmental issues became increasingly apparent, and in the 1970s environmental issues were increasingly prioritized along with socio-cultural perspectives (1987: 84). He also says that a silent revolution (*stille Revolution*) toward post-industrialized society was under way, in which the value of maximizing income and, in turn, consumption, shifted towards an emphasis on human and cultural values, such as freedom, well-being and love (1987: 26). Interestingly, Hübler's conclusion roughly parallels my own observations that the harmonious realization of comprehensive values (economic, ecological and life values) has become the overriding objective of modern society and, in turn, modern agriculture and

Figure 3.3: Process for increasing total welfare

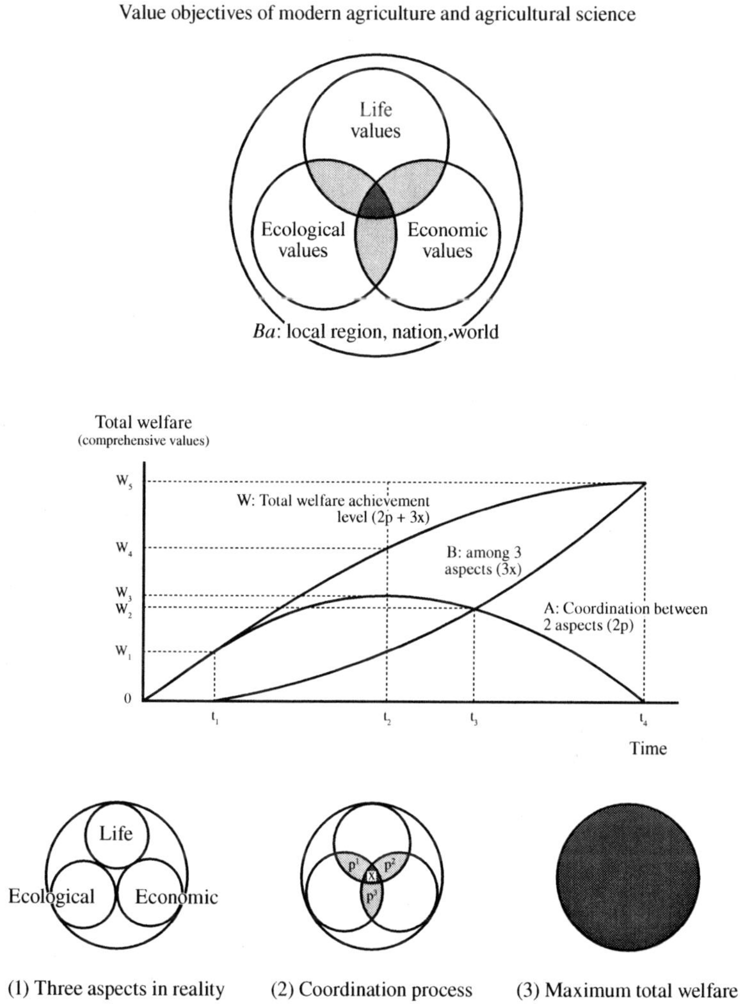

Explanation. (i) The three aspects have their respective problems and require coordination among them. Increase of total welfare is regarded as the process for expanding and developing each aspect and settling problems in respective aspects by mutual coordination. (ii) The process for coordinating the three aspects takes place first as the process for coordinating two aspects (increase of domain p) and is represented by line A. (iii) Comprehensive coordination among the three aspects (increase of domain x) then follows and progresses, whose process is represented by line B. (iv) The level of total welfare achieved is expressed as (A + B) and describes a curve like curve W. (v) Bottom figures (1), (2) and (3) show the reality of the three aspects before coordination, the process of coordination, and the maximum total welfare after perfect coordination, respectively.

Table 3.3: Functions of agriculture, forestry and rural communities found in different literature

Economic roles	Ecological roles	Social and cultural (living) role
1. Efficient food production	*1. Land conservation*	*1. General roles*
Supply of low priced food	Ecosystem maintenance	Diversity, stability and perpetuity of society
Stable food supply	Water resource maintenance	Maintenance of local community
Supply of living and housing materials	Soil conservation	Overcoming increased division and simplification of labor
2. Supply of good quality food	Function as natural dams	Overcoming increased standardization and totalism
Fresh food	Surface water storage	Socially stable class
Tasty food	Ground water storage	Social solidarity
Various food	Flood prevention	*2. Social interaction*
Year-round supply	Erosion prevention	Town-rural village interchange
3. Roles in national economy	Wildlife conservation	Farm-fresh product campaigns
Contribution to growth of labor, land and capital	*2. Conservation of living environment*	Organic farming movement
Food security	Water conservation and purification	Cooperation between cooperative associations
Stability by storing	Air conservation and purification	Sister municipalities
Stable economic growth	Noise prevention	U-turns and new entry
Buffer against risk	Odor prevention	*3. Welfare roles*
4. Local economy stimulation	Natural scenery	Meaning to life of the elderly in aging society
Diversity and stability of local economy	Green space	*Ba* of employment and work
Effects of elderly employment	Pastoral landscape	Age-appropriate work
Possibility of improved energy productivity	Evacuation sites	Lives of the disabled
	3. Possibility of sustainable agriculture	*4. Educational roles*
	Safe food	Understanding nature
	Avoidance of food additives	Harmony and cooperation
	Biological control (selective breeding, new breed development, natural enemy)	Patience and sensibilities
	Reduction of agrochemicals and chemical fertilisers	Creativity
		School farms

Economic roles	Ecological roles	Social and cultural (living) role
	3. Possibility of sustainable agriculture (continued)	*4. Educational roles (continued)*
	Recycling of resources	Rural village stay and study programs
	Utilization of local energy	
	Organic farming	*5. Humanity recovery function*
	Conservation of biodiversity	(1) Provision of *ba*
		• Recreational natural forests
		• Hobby farms
		• Tourist farms
		• Country forests
		• Second houses
		• Allotment gardens (*Kleingarten*)
		• Farms for farm work experience
		(2) Humanity recovery
		• Serenity and relaxation
		• Improvement of personal relations
		• Improvement of family relationships
		• New richness in society turning away from material possessions
		• Freedom and independence in agriculture
		• Changes and diversity of life (One person has multiple jobs at one time or in his/her lifetime)
		• Art and agriculture
		(3) Medical effects
		• Nature and health
		• Relaxation
		• Forest bathing
		• Alleviation of modern disease
		6. Agriculture as a lifestyle

Note: Discussions of functions appearing in literature after the Second World War were collected from various sources and, taking my personal view into account, categorized and organized from three aspects: economic roles, ecological roles, and social and cultural (living) roles. All roles are listed even though some may overlap or conflict with others. Some are roles currently being played and others are roles expected to be played. The term agriculture in this table is used in a broad sense to include full-time and part-time farming as well as private vegetable gardens.

Table 3.4: Contrastive changes in concept of values (1975)

Concepts of values in industrial society	Concepts of values in post-industrial society
Material growth	Growth of contented spirit and life
Confrontation with nature	Coexistence with nature
Ideas of competition	Concern for conscious individuality
Self-centered individualism	Cooperative individualism
Rationalism	Rationalism and intuition
Conventional society	Unconventional society
Large-scale housing and labor corporate body	Small-scale simple living and labor environment
Increasing material centralization	Decreasing material centralization
Space technologies	Appropriate technologies
Identity formed by consumption behavior	Identity formed in the process of mutual valuation
Concentration of authority and control at the level of nation	Dispersive local governing able to communicate at global level
Specialists and division of labor	Possibility of comprehensive multi-faced labor
Undifferentiated short-term mass production	Hand made lasting original production
Cultural homogeneity	Cultural heterogeneity
Ideas of efficiency, stress	Ideas of harmony

Source: Karl-Heimann Hübler, 1987: S.35

agricultural science (Hübler 1987: 27; Soda and Ōhara 1994: 326). In fact, Hübler concludes his study of the changes in values in modern society from three aspects – the economic, ecological and social and psychological values – and states that the well-being (*Wholstandniveau*) can only be achieved by realizing these values comprehensively (Hübler 1987: 27). Other authors adopting a similar position include Stratmann-Mertens (1991) and Thiede (1988).

Quoting the philosopher Oldemyer, Hübler also points out that the period after 1979 was a time of 'conversion to pluralistic values (*die vielfältigen Werten Veränderungen*)'. This coincides with my identification of agriculture as an 'industry with pluralistic values' (Hübler 1987: 27; Soda and Ōhara 1994: 332). Another important point of Hübler's is that the harmonious realization of comprehensive or pluralistic values is a theoretical task, as it is impossible to demonstrate empirically when we should actually balance and harmonize these values with each other.

These pluralistic values were incorporated into the principles of social market economy (*soziale Marktwirtschaft*) which have consistently underpinned government policies in the former West Germany and present-day Germany since World War II.

As described above, the German economy has gradually transformed the values in its policies, eventually establishing policies that aim at the comprehensive realization of pluralistic values. Following this approach, Germany has promoted policies on environmental improvement and local regions.

'Local communities' as *ba*

I assume that 'local communities' are *ba* where the possibility of realizing comprehensive values is maximized. It is the place of production (economy), place of ecosystem, place of daily socio-life, and *ba* of comprehensive 'life'. In addition, it is ultimately a local community which is a 'small/medium town-rural village complex' rather than a rural village community. I have explained the basic perspective here in the introduction of *Jizokuteki nōson no keisei* (The formation of the sustainable rural village) and, through studies of German ideas of combining city and country, in *Toshi to nōson no ketsugō* (The combination of city and country), and other works.[2]

In the modern market society, a local community is a relatively intimate lifeworld, where anonymous relationships constitute a minority. In such a world, efforts to overcome the harmful effects of excessive market-orientation, and market failures, are recognized more concretely and sincerely and are therefore more likely to arouse wide sympathy. It is a world where excessive industrialization, urbanization, and economic rationalism, are checked, the surrounding ecosystem is preserved, and the logic of everyday life is emphasized. In other words, this world aims at creating a living environment that is reasonably and realistically rich in diversity, stability and sustainability. An ideal situation to follow would be where such local communities directly appear in the world, unite with each other on the lifeworld level across national borders, and solve various problems of mankind (see Tamanoi 1977).

I thus designate the standpoint that aims at realizing the satisfaction of a happy, human and comprehensive lifeworld, and intends to reorganize agriculture and rural village communities accordingly – or create *ba* of human life – the 'agricultural science of *ba*'. *Ba* means, as explained in Chapter 1, not merely space in terms of geography or location but also a space of various conditions and situations. Thus it is also concrete '*ba* of conflicts' and '*ba* of problems', where our three main value objectives are pursued in a balanced relationship with each other.

As discussed earlier in this chapter, however, it is only when conflicts or problems are located in the same place that they are clearly recognized and may be solved, integrated or overcome. Such a place will become new '*ba* of creation', where experiential, everyday efforts of farmers and rural village inhabitants, scientific research efforts and policy efforts to solve these problems are gathered together, to make it '*ba* of problem solving', '*ba* of liberation' and '*ba* of better living'. Agricultural science is thus a science of conceptualization and creation for understanding, and resolving, problems. The mission of studies of agricultural science is to attempt to identify and solve problems surrounding agriculture and rural villages. This is my basic conception of agricultural science (Soda 1990; cf. note 2 this chapter).

I therefore define agriculture as 'intentional, social human activities intended to realize the economic, ecological and life values in a harmonious manner by conserving and utilizing local resources and managing, raising and rearing organisms useful for human beings'. I argue that such activities are most feasible in the aforementioned *ba*.

In this context, realizing ecological values may seem to be only indirectly related to, and thus remote from, individuals' production activities, but this is not the case. Historically, local agriculture and forestry developed through the contrivance of effective ways to prevent flooding and other disasters in local communities. In metropolitan suburbs, the mere presence of agricultural spaces is socially appreciated in maintaining a 'productive green space'. In other areas also, agriculture plays a major role in preserving the environment – a contribution which must be accorded financial value and reflected in policy. This is the assessment of so-called external economic effects. Furthermore, there is growing demand for agriculture to consciously change those of its practices that are stressful for human life and the environment by promoting sustainable or environmentally sound agriculture, and this is significantly changing farmers' perspectives and attitudes.

When it comes to production, however, apart from cases where a local community has voluntary restrictive rules for the conservation of its environment, there are real limits to the extent to which producers can self-limit their activities to protect the environment while delivering low-cost goods to consumers – for such efforts typically mean reduced incomes and increased hard labor. There is no doubt that the pursuit of economic values to earn a living has immediate priority for producers. Realizing ecological values therefore depends on whether or not people will accept higher prices in recognition of agriculture's external effects (Soda 1991; Mitsubishi Research Institute 1997, 2003), or a financial (tax) burden to directly compensate farmers for reduced income, as in the decoupling policies in Europe.[3] In other words, it depends on whether or not a national consensus is formed that recognizes the multiple values of agriculture. Fortunately, the initial steps in this direction have been taken in Japan by the introduction of a direct payment policy for disadvantageous conditions in intermediate and mountainous areas under the Basic Law on Agriculture, Food and Rural Areas, 1999.

As we have seen, historically agriculture has developed in response to the issues and demands of the times, aiming to realize the values that characterize each historical period and local community. In other words, agriculture has developed with the historical development of related issues. Through resolving such issues, the welfare of each individual and society increases. Agricultural science is a practical science concerned with biological production as practice, but also an intentional science to conceptualize and realize the values of the times. With local communities as the stage, modern agricultural science must strive to harmoniously realize the three sets of values – economic, ecological and life values – to maximize total welfare, reforming agricultural technology and management towards this goal.

4 Agriculture and the Economy

In this chapter, I will consider the relationship between agriculture (including the forestry and fishery industries) and the economy, focusing on the post Second World War era, and attempt to clarify how the position of agriculture should be considered in a market economy society. The capitalist, market-economy based society regenerated and expanded after the Depression and World War II. First, I examine the two economic approaches that have supported Japan's postwar high economic growth: Keynesian economics and the theory of an international division of labor. Second, I outline the outcomes and problems associated with these two economic approaches. Third, I demonstrate how approaches that emphasize economic growth above all else have generated problems, particularly for Japan's agriculture. I argue that the severity of these problems can be attributed primarily to ignorance of three distinctive features of agriculture and examine the relationship between these features and economic principles. Fourth, considering the self-corrections of the capitalist economy and the market liberalization of the socialist economy of the 21st century, proceeding from the supposition that the mainstream approach in the contemporary world economy is the social market approach (*die soziale Marktwirtscharft*, or 'the third way'), I consider the current state of agriculture and agricultural policy from this perspective.

Market principles and the postwar world economy

It is probably uncontroversial to say that economic policies of developed countries, especially Japan and the United States in the postwar period, were primarily grounded on Keynesian economics, on the one hand, and the principle of international division of labor based on the idea of comparative advantage, on the other. Both of these approaches are based on the principles of the free market.

Keynesian economics and the idea of high economic growth

It is impossible to talk about Keynesian economics without discussing the Great Depression that started to affect capitalist countries in 1929. Faced with massive unemployment, stagnant consumption and production, an increasing gap between rich and poor, and other characteristics of the Great Depression, John Maynard Keynes set-out to formulate a new economic theory, publishing *The General Theory of Employment, Interest, and Money* in 1936. Keynes had previously outlined the central idea behind this theory in *The End of Laissez-faire* (1926). Here, Keynes declared that the *laissez-faire* approach featuring a small government, which had been dominant since Adam Smith, was not capable of addressing the new economic circumstances. Specifically, Keynes outlined the distinguishing features of the new economic world: (1) the transfer of large firms from private (individual/family) to public (shareholder) ownership; (2) the declining influence of capitalists and the rise of professional managers (that is, the separation of capital and management), (3) as a result of those two features, increasing emphasis on the stability and reputation of the firm rather than the pursuit of profit for shareholders (i.e., long-term sustainability versus short-term profit-maximization); (4) the limitations of the Gold Standard; (5) a lack of investment opportunities comparable to those of the nineteenth century; and (6) the desire of rentiers to accumulate profits derived from economic recessions and social ills (Keynes 1926).

In other words, Keynes proclaims the end of *laissez-faire* and hence the world of classical economics. He argues that the cure for these problems is to be sought through the deliberate control and reform of the capitalist system by government. Thus, while basing his theory on market mechanisms, Keynes maintains that corrective government policies are nevertheless necessary. He therefore argues for a kind of social reformism that would provide for a fusion of the principles of planning and control with the principles of the free market, which would require moving from a small government to a large one.

During the time of his writing, Keynes was accused of socialist beliefs, but he was in fact a proponent of a capitalist free market economy. Although he explored possibilities for reviving and revitalizing capitalism, he did not approach the problem from the socialist perspective of moving away from the early stages of capitalism towards government planning and state control. Nevertheless, he advocated planning and control as a means for overcoming the 'market failures' that had led to the Great Depression – the dark side of the very core of the free market mechanism.

Keynes' economic theory provides for policies such as directly addressing unemployment through the creation and expansion of demand. In my understanding, Keynes' reasoning regarding policymaking can be summarized in four points (Keynes 1936; Itō 1962; Dillard 1950; Kawaguchi 1977; Miyanaga and Soda 1981)

1. Expanding demand through changing the consumption propensity of the whole population to increase their levels of consumption (with the consequence of simultaneously decreasing their saving levels/habits). This is to be achieved through progressive taxation, which will result in a rather egalitarian income distribution.
2. Decreasing the official discount rate and engaging in open market operations to reduce interest rates, thereby stimulating private investment and creating effective demand.
3. The government should not fear deficits, and fiscal policy should include public investment in public works projects (which inject money into local economies by creating employment).
4. A prerequisite for the above is the abolition of the inflexible Gold Standard and the establishment of a managed currency system.

These four principles, although perhaps not all with equal fervor all the time, served as the guiding principles and theoretical basis for President Roosevelt's 'New Deal' program. The Second World War resulted from the clash of two clusters of states, distinguished by their different approaches to overcoming the Great Depression. One was the 'New Deal' group of countries – known in the West as the 'Allied Powers' – who embraced the idea that domestic policies could best achieve this objective. This group included the UK, the US, France, and others. The second, Fascist group of countries – known in the West as the 'Axis Powers' – included Germany, Japan, and Italy, and chose international aggression as their strategy (Kawano 1967). After World War II, under the US umbrella, Japan became an 'Ideal "New Deal" country' (Chō and Sumiya 1969). After the democratic reforms of the postwar period, Japan entered an era of rapid economic growth.

International division of labor: comparative advantage principle

Another economic theory that was embraced during – and perhaps contributed to – Japan's rapid economic growth was originally proposed by Ricardo. After further development it is commonly known as the Heckscher-Ohlin theory of international division of labor, based on the idea of comparative advantage (Itō and Oyama 1985; Kanemitsu 1991). While Keynes' theory was highly influential in domestic economic policies, the theory of the international division of labor has become predominant where international trade relations are concerned. The Great Depression of 1929 resulted in the closed regionalism of the Monroe Doctrine, which triggered the economic and political antagonism that climaxed during World War II. The heightened interest in the theory of free trade based on the idea of the international division of labor and the desire to implement it was stimulated by these regrettable historical developments. After World War II, as mentioned, domestically, many countries embraced planning and regulation to reduce and redress the negative aspects of capitalism, while in international economic relations, they tended to reaffirm the principles of freedom and cooperation. In short, the theories of Keynesian economics and international specialization met in the policies that were being implemented at the time.

The Heckscher-Ohlin theory is based on the concept of relative prices. As mentioned, this was developed by Ricardo and then later augmented by Eli F. Heckscher and Bertil Ohlin. In the original model, Ricardo viewed labor as the only producer cost factor. Heckscher and Ohlin's analysis considered three factors that contributed to the comparative cost of production: labor, capital, and land. In simple terms, the Heckscher-Ohlin model of relative costs can be explained as follows.

Let us consider, for example, rice and car production in the US and Japan (refer to Table 4.1) (Soda 1987, 1989). We assume that both countries can produce only two goods and that production costs remain fixed, regardless of production scale. We assume further that in Japan, on the one hand, the cost of producing either one car or three tons of rice is ¥1,000,000. In America, on the other hand, the cost of producing one car or 20 tons of rice is $6,000. In Japan, therefore, the

Table 4.1: The relationship between cars and rice in the US and Japan (an example)

		cars	rice
Cost of production	Japan	¥1,000,000 per car	¥330,000 per ton
	America	$6,000 per car	$300 per ton
Exchange rate of goods	Japan	1 car = 3 tons of rice	
	America	1 car = 20 tons of rice	

relative price of one car is equal to three tons of rice, while in America you can exchange one car for 20 tons of rice. However, if America exports those 20 tons to Japan, it can buy (almost) seven Japanese cars (omitting the costs associated with exporting). Thus, if Japan specializes in the production of cars and America specializes in the production of rice, both countries will profit from trading with each other and both will prosper.

Postwar Japan has maintained extremely close relations with the US economy since it embarked on the path to recovery and development; so close, in fact, that it has often been said 'when America sneezes, Japan catches a cold'. However, in the 1980s a great number of Japanese industries, especially automotive manufacturing, consumer electronics, textiles, and machinery, developed to such an extent that the US came to see them as a threat. Even in high-tech industries that were for a long time considered to be the prerogative of the US, Japan has become a powerful competitor.

Under the motto of free trade, the US, despite being overwhelmed by Japan's rapid development, allowed this process to continue. In the second half of the 1980s, before the expansion of the EU, the US began to face a significant trade deficit, the major part of which resulted from trade with Japan. In an effort to decrease its trade deficit, during the trade negotiations of the GATT Uruguay Round, America – whose agricultural industry is the strongest in the world – put the topic of agriculture on the table and argued for adherence to market principles, for a liberalization of trade, and for deregulation. Although not targeted exclusively at Japan, this move demonstrated America's willingness to take measures necessary to protect its interests (Soda 1993). The Cairn's Group of the world's largest producers of agricultural products, including the USA, Canada, and Australia, were in favor of indiscriminate deregulation of agriculture and manufacturing industry, while Japan and other large-scale importers of agricultural products opposed this position, striving to protect their domestic producers by retaining protectionist policies. The forum for resolving international trade disputes has moved from GATT to the WTO. However, although economic and market principles regarding agricultural products continue to be in the spotlight, the multiple non-trade functions of agriculture have also come to have a place in negotiations.

Supplementary Explanation

I would like to briefly summarize the main points regarding the history and items of dispute between the theories of free trade and protective trade

(1) Free Trade or Protection? – Smith and List

The views of Adam Smith and Friedrich List regarding free and protective trade are considered to be the most representative of the debate. Adam Smith published *The Wealth of Nations* in 1776, a time of industrial revolution that triggered various inventions and discoveries. He stresses the importance of free economic activity, based on the idea of natural law. Smith sees both the feudalistic system and moral sentiments as a constraint. He therefore ignores them and constructs a model where economic activities are motivated solely by the pursuit of individual self-interest. The harmonization of these interests ultimately result in the creation of a society where economic affluence is increasing and reaches all members of the society – and thereby the achievement of social justice. Extending this theory, Smith argues for giving freedom to colonies and for free trade.

List, in contrast, claims that free trade is only for developed countries. Developing countries, where the process of industrialization is not yet complete, should adopt protectionist policies; otherwise, they will end up being dominated by the developed countries, in which case the world economy will suffer. Whereas Smith ignored considerations such as borders between countries and a country's stage of development, List argues in *Das nationale System der politischen Oekonomie* (1841), that as soon as Germany, whose industrialization was rather late, achieved a certain level of development through mobilizing the physical and spiritual productive power of its people effectively, it should switch to free trade. But at the early stages of capitalist development, List maintains, protectionist policies allow states to overcome the gaps in industrial development with their trade partners and thereby to achieve independence. This view still accounts for the issues in dispute between the developed and developing countries today.

(2) Free Trade or Protection of Agriculture? – Ricardo and Malthus

In addition to issues regarding inter-state relations, debates have included discussions on the relationship between agriculture and industry. In 1815, Ricardo wrote *On Protection to Agriculture*, in which he argued that in order to develop economically, Britain needed to abolish the corn laws and liberalize agricultural trade. Malthus, on the other hand, in *An Essay on the Principle of Population*, argued for the need to protect agriculture. These arguments reflect the opposite interests of entrepreneurs in commerce and industry on the one hand, and agricultural landowners on the other. Ricardo, based on his observations of, first, the decline of agriculture in Britain, brought about by progressing industrialization and population growth, and second, falling wages coupled with rising grain prices, emphasized the need to provide stability for workers and a chance for those in the fields of commerce and industry to make a profit by allowing cheap grain imports into the country. Ricardo is thus credited with the idea of comparative costs examined earlier, and was a proponent of the international division of labor.

Malthus predicted that at a certain point in time, food production would not be able to sustain the level required by the growing population, and put forward arguments for protection: (1) in order to secure the availability of food. Specifically, he noted that large powerful countries will become the object of envy for smaller countries, and argued that relying on foreign countries for food is too risky; (2) to ensure the harmonious development of the industrial sector and of society in general. Specifically, Malthus stated that industrial development has expanded intellectual horizons and caused a revitalization and affluence of things, but that the increase in the proportion of population working in the industrial sector is not good for the stability, happiness, health, and morality of the people; (3) as a means to achieve the stabilization of the economy. Malthus envisioned that the liberalization of grain imports would result in violent fluctuations in prices, which would have a negative effect on farmers.

Some critics have observed that Malthus's analysis granted first priority to the profits of the landowner class in Britain and that his writing was not necessarily based on his concern for farmers' welfare. Others have argued for the need to revise his population growth theory, and still today there is no consensus on this issue. Nevertheless, Malthus's three arguments have never ceased to be considered to be important.

Market failures and government failures

Rapid economic growth in developed capitalist countries

These two theories – Keynesian economics and the international division of labor have become mainstream approaches and served as the driving force around the world for postwar economic development. It was against this background that Japan enjoyed its rapid economic growth from the 1950s to the 1960s (Miyanaga and Soda 1981: 162–81; Soda and Ōhara 1994: final chapter).

A Japanese government economic white paper for 1956 declares: 'Japan is no longer in the immediate postwar period. The country is about to face a completely changed reality. The period of economic growth derived only from economic recovery is over. Growth from now on will be supported by modernization'. Beyond suggesting that the best prospects for economic growth lay in technological innovation, the paper also reflects perennial fears of a stagnating economy as the period of recovery and reconstruction draws to a close. After 1955, progressive economists promptly declared the Japanese economy to be in a deadlock at every sign of economic contraction. In reality, however, although there were of course occasional drawbacks, the economy continued to experience one economic boom after another – the Jinmu Boom (1955–1958), the Iwato Boom (1958–1961), the Olympics Boom (1963–1964), and the Izanagi Boom (1965–1970). From 1960 until 1972, as Figure 4.1 indicates, Japan posted real GDP growth of more than 10 percent per annum. This was unprecedented for Japan, and quite remarkable by international standards.

Japan's Gross National Product (GNP) grew rapidly, rising 20-fold from 6.7 trillion yen in 1955 to 136 trillion yen ($465 billion) in 1974, at which time it assumed the second position in the global GNP ranking behind the US's $1.397 trillion. That year, West Germany ranked third with a GNP of $412.4 billion, followed by France ($298 billion) and Britain ($191.2 billion). Japan's impressive economic growth was primarily due to increasing production in the mining and manufacturing industries: specifically, the steel, petrochemical, shipbuilding, automotive, precision machinery, electronic appliances and other heavy industries, all of which recorded substantial growth.

Figure 4.1: Net economic growth rates during Japan's postwar rapid growth period (1961–1974)

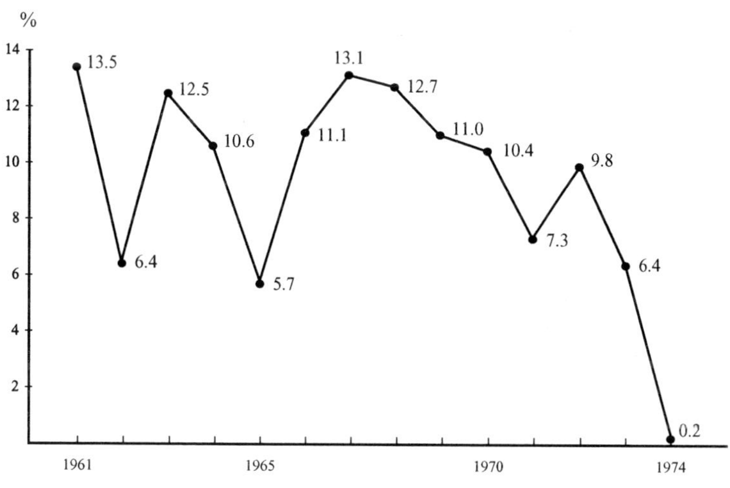

Source: *Nihon kokusei zue* (Collection of pictures of Japan's national resources), 1977 edition: 28. Growth rates over previous year. The years are fiscal years.

Furthermore, the buying power of Japanese consumers also rose rapidly to unprecedentedly high levels. For instance, durable consumer goods such as televisions and washing machines found their way into most homes, while the diffusion rate of automobiles reached 40 percent and air conditioners hit 12.4 percent. Electric vacuum cleaners, sewing machines, pianos, stereos, electronic calculators, and cassette recorders also spread quickly. This rapid proliferation of electronic appliances and automobiles among households could not possibly have been predicted in the 1950s. Japanese citizens were ecstatic about their continually increasing affluence and the increasing availability of goods. They were quite proud of the fact that Japan had assumed a prominent place among the world's most developed countries and praised the high rate of economic growth.

The Japanese economy continued to enjoy substantial growth until the country found itself facing environmental problems, the Nixon shock (i.e, the float of the US Dollar), oil shocks, and stagflation of the 1970s. Government policies during the period of rapid economic growth were predominantly short-term measures aimed at stimulating the economy. These policies were generally approved despite the predictable inflation and inefficiency that accompanied them, for these were considered to be 'natural' to some extent. Postwar Japan thus employed Keynesian economic policies over a long period of time, setting a course in which 'one investment brings another' (Economic White Paper, fiscal year 1960). In the process, Japan has become one of the world's leading industrial countries.

Rapid economic growth and its problems

However, Japan also experienced the negative affects of Keynesian economics. These included stagflation – that is, the concurrence of inflation and economic stagnation – waste of resources, environmental degradation, urban problems, and growing financial deficits. In response, the government policies of the rapid economic growth period have been criticized for putting GNP expansion above all other objectives. This 'growth at all costs' mentality is perhaps exemplified in a slogan adopted by an advertising company at the time for training their new employees: 'Make people use more, throw away more, generate more waste, forget about seasons, give more presents, buy more and so on' (Morita 1971: 61). In other words, the Keynesian idea that 'consumption is virtue' soon gave birth to the idea that 'waste is virtue'. This is the defining characteristic of the current global economic system and is directly responsible for the problems listed above. In the US, this 'institutionalization of waste' (Itō 1962) can be seen in the massive investment in public works, especially in military goods and facilities.

Thus, economic development, which is typically represented as a path towards healthier and happier lifestyles and for advancing cultural development, has in practice become a system for controlling people and imposing restrictions. Rather than a means to an end, the economy has come to be seen as an end in itself, and the great diversity of human beings have come to be judged solely on the basis of their economic productivity. Labor and land are seen solely as commodities to be bought and sold. All other human values have been deemed to be irrational. Rationality has come to mean 'economic rationality', implying that all human action and decisions must be rational in the economic sense. Thus, society has been reduced to a stage for economic activity, and humans themselves to economic animals. This emphasis on economics in human society to the exclusion of every other value is deeply and profoundly malignant.

We are intoxicated by the achievements of modern scientific civilization and the affluence that it has produced, and we rely on the market mechanism in our continuing economic struggle. I do not deny the achievements of modern civilization – economic development has greatly enriched our lives; this is beyond dispute. However, it is also undeniable that we have subordinated many of our diverse wants and potentials for the sake of pursuing material well-being and thus have limited the scope for happiness to the world of material possessions (*Habenwelt*). We have forgotten that economic happiness is only partial happiness (I return to this later in this chapter).

This economic reductionist approach poses two serious threats to contemporary society. First is the direct threat to life from the destruction of the ecological environment. Second is the threat to human well-being, due to this truncation of our recognized human needs.

By the end of the twentieth century the Japanese economy was characterized by massive exports, trade friction, revitalized (Tokyo-)centralization, dramatic fluctuations in land and stock prices, fluctuating exchange rates, and speculative investment activities. The US responded to the pressure from Japan's rapid economic expansion by implementing protectionist policies – which in turn affected Japanese agriculture – while attempting to revitalize its own manufacturing and industrial sectors.

If we regard only market mechanisms and the principles of competition to be important, our understanding of capitalism will be of 'a system of wants and perpetual actions aimed at maximizing capital returns' (Yanai 1992: 132). Any solutions derived from such a limited view point would not be sufficient to save the human race. Those states that implemented Keynes' prescriptions for redressing market failures such as economic depression, were soon faced with numerous other problems that resulted from these policies. For the most part, they not only accepted the generation of these problems, but did nothing to stem their continuing generation. This represents not only a new type of market failure but also a clear government failure.

Government failures in socialist countries

The second type of government failure is the excessive economic rigidity of the socialist countries – most notably the former Soviet Union and China. In the beginning, socialism aimed to contribute to the freedom and equality of workers and to increase productivity. However, although we cannot directly compare government failures in a capitalist economy to those in a socialist economy, it is probably fair to say that the comprehensive policy control of the centralized socialist governments sapped the economy of its vitality, often through simple carelessness. In effect, under socialist rule, a new class of 'worker-aristocrats' came into being, bureaucracy became more rigid, and political opposition was suppressed along with peoples' freedoms.

The socialist countries' approach to environmental problems was no better, and perhaps even worse than the consumerist societies discussed above. Socialism also strives for increased production and productive efficiency, seeing itself in competition with capitalist countries to provide their citizenry with the greatest material affluence. Meanwhile pollution is viewed as an inevitable by-product of scientific and technological development. Because the central government attempts to control all aspects of the economy and society, the solution or absence of solutions for environmental problems depends solely on the government (Nihon Keizai Gakkai Rengō 1975). The acute environmental problems that are the legacy of this approach have become increasingly apparent since the collapse of the Soviet Bloc, as many Eastern European nations have turned down the road towards freedom and openness. I suggest we refer to these problems as indications of government failure. In December 1991, the Soviet

Union ceased to exist, China is attempting to create a socialist market economy, and Marxist-Leninist socialism has disappeared.

The tasks of economics

Some argue that economics (including the economics of agriculture) is now facing a crisis that demands a paradigm shift. For instance, in 1971, at the American Convention of Economic Associations, John Robinson presented a paper entitled 'The Second Crisis of Economic Theory' in which he argued that the original goal of economics has been forgotten and that the economy is increasingly dependent on the military-industrial complex, which is dangerous, and thus presents a fundamental crisis to economics as a field of study. *The Limits to Growth: A Report for the Club of Rome's Project on the Predicament of Mankind* (1972) also offered a strong critique of approaches that regard economic growth as the ultimate priority. The problems that arose after the period of high-economic growth, such as those associated with the inevitable slowing of the economy and stagflation, depletion of resources, environmental pollution and urban problems posed new challenges to the science of economics, leading many economists to discuss a crisis in the field.

As Mizuta and Tamanoi (1978) observe, 'It is certain that economics is facing a crisis. We are neither turning to Marx nor Keynes. In fact, it seems that both theories are no longer useful for economic policies... The crisis that economics is facing is not about which economist's theory to apply, but rather about economics as a discipline itself that has become more technologically advanced and less humanistic'. Although the Soviet Union – and thus the socialist paradigm – has collapsed, it was in response to socialist criticisms, most notably concerning exploitation and poverty, that the capitalist system strove to reform and improve itself. The mutual tension between the two competing socio-economic systems has undoubtedly fueled changes in the capitalist system. Nevertheless, the present capitalist system still needs to deal with problems such as pollution, quality of life, income disparities between North and South and between urban and rural dwellers.

In Japan, a decade from 1965, energy consumption and environmental pollution have been acute social issues, which have prompted people to question the exclusive prioritizing of GNP growth. This has further induced criticism of modern science and technological development. Tamanoi Yoshirō has argued that regardless of whether we follow modern economic approaches or Marxist theories, there is an inclination to prioritize production capacity above everything else. He goes on to argue that following this approach, the life-denying economics of the present era has caused innumerable problems for modern industrial civilization. It is therefore imperative, he concludes, to focus on those principles of agricultural production that sustain organic life and to move towards creating living economics (Tamanoi 1982). This is a move towards ecological and entropy economics. Recent developments in economics have included an increasing focus on issues such as natural resources, environmental aspects of production, externalities, social costs, and economic systems – which are analyzed both separately and in combination.

In Japan, especially in the decade from 1975, amid urban problems and disparities in regional development, questions arose about quality of life, including the impacts of pollution and environmental degradation. In 1983, the Economic Planning Agency of Japan began to monitor a quality of life index in addition to the GNP. Also, the Prime Minister's office has conducted a survey into citizen's views about the concept of 'affluence'. Respondents were asked if they placed more importance on material or spiritual affluence. From 1977, the number of people giving preference to spiritual affluence began to grow, and by 1990 it totaled 53 percent of all respondents, while those reporting that material affluence is more important constituted a mere 30 percent. This was a complete reversal of preferences compared to the pre-1977 data. William Leiss, in *Limits to Satisfaction: An Essay on the Problem of Needs and Commodities* (1976), argues that during the era of rapid economic growth the citizens of the developed world predominantly sought economic happiness, which is only partial happiness. He argued that we must expand the scope of our quest for satisfaction in order to seek complete human happiness. This would entail a change from perceiving development in purely economic growth terms to a more comprehensive vision of development as a process that affects the whole society. Karl Polanyi also observed a need to formulate an economics that is conceived to be a part of human activity in the broad context of society. In 1900, Robert Schmidt, who was engaged in regional development in Germany, observed that human beings in industrial society are often considered only as a labor force, and argued that we must embrace broader conceptualizations that see human beings as having basic needs to work, perform their daily life activities, and for entertainment, and must develop an approach that provides opportunities for satisfying all of these needs (see Chapter 7 for more on Leiss, Polanyi, and Shmidt).

As we have seen, in the face of new market and government failures, contemporary economies and the study of economics are faced with a need to move away from the narrow definitions of economic rationality and economic reductionist values and expand the debate to include issues such as the environment, quality of life, welfare, and the economic disparity between North and South and between urban and rural areas.

Agriculture, the primary focus of this book, is one field where the aforementioned market and government failures have been highlighted. Further analysis requires a clear understanding of the differences between agriculture and industry.

Market principles and the features of agriculture

The distinguishing features of agriculture and forestry

The distinction between agricultural production and manufacturing is exceptionally important. Although many argue that market principles are not suitable for agricultural production, market societies must nevertheless manage agriculture. Hence market failures quickly become apparent in agricultural management. As Adam Smith noted, the birth of a full-fledged market society and the growing recognition of the importance of market principles accompanied the formation and development of capitalism and industrialization. Some have therefore argued that agricultural problems can only be explained in terms of the distinguishing features of agricultural production.[1]

Eduard David's discussion of the characteristics of agriculture centered on his debate with Kautsky regarding Marxist Social Darwinism and the question of whether there is a fundamental difference between agricultural and industrial production. David contrasts the two, referring to the latter as mechanical production and the former as organic production. His views on organic production can be summarized in nine points, which provide the basis for the development of subsequent theories on the topic.

Specifically, he argues, organic production has the following characteristics: (1) the working process is seasonal, not continuous; (2) diverse labor processes that are constantly changing are required to complete production; (3) both laborers in the fields and the tools they use need to be mobile; (4) the production cycle needs to follow the natural cycle; (5) due to the vastness of the production space, close supervision of the employees' work is not feasible; attempts to closely supervise them will be costly in terms of both time and money; (6) in the process of managing agriculture, human beings are totally committed to nature; (7) increasing the scale of agricultural production is limited by natural constraints; (8) the law of diminishing returns applies to productive land; and (9) expanding one's productive lands (external expansion) implies rivalry with the other farmers.

In what follows, I examine what those writing after David had to say about these issues, and add the point that agriculture is a multi-value industry. I also add a regional perspective to the picture, and consider how the principles of market economy and agriculture are related, focusing on the following points: (1) The technological and managerial (production and distribution) issues; (2) agriculture as a multi-value industry and its non-economic value (the ecological and everyday life values); and (3) the regional characteristics of agriculture. The everyday life value of agriculture is the subject of Chapter 6, where I discuss related issues in greater detail.

Technological and managerial properties

Features of agricultural, forestry and fishery production

Agricultural production as organic development

Agricultural production employs natural resources such as air, water, and sunlight, and is therefore an organic development faced with time and spatial constraints – any expansion of production is dependent on finite amounts of land and water. In contrast, industrial production is, from a technological perspective, inorganic, based on 'dead nature'. The processes of disposing of and processing materials in the course of industrial production are inorganic, and thus separated from natural time and space dimensions.

Long-term production and sequential production dimensions

Industrial production is easily divisible into many steps or stages that need to be completed, and the production of parts occurs in a sequential manner. If we take automobile production as an example, assembly follows this sequential manner. However, unlike agricultural products the final product can be produced year-round, while all the components can be produced simultaneously. In other words, in industrial production, it is possible to replace sequential production with simultaneous and continuous production – i.e., parts can be produced without disruption by working the machinery non-stop.

Agricultural production, in contrast, is completely sequential; the cultivation of crops and livestock proceeds in a predetermined sequence (like seeding–transplanting–weeding–harvesting) and takes a fixed amount of time. Typically, agricultural production uses a calendar year as a unit of time. However, some crops, such as persimmons, tangerines, other fruit, and tea-leaves, require as much as a decade from planting until the first harvest. Cattle, especially dairy cattle, also require a number of years to become productive. Furthermore, in many cases, as in fruit production, it is not possible to adjust the number of trees in response to market price once the initial investment has been made, since doing so would result in a total loss for the farmer (see Figure 4.2). In the forestry industry, it can take 30 to 50 years for a usable tree to grow (e.g., Japanese cedar); changes in supply and demand and the resulting fluctuations in market price cannot be foreseen. In the fishery industry, the depletion of resources and long-

Figure 4.2: Growth, yields and net profit of fruit trees

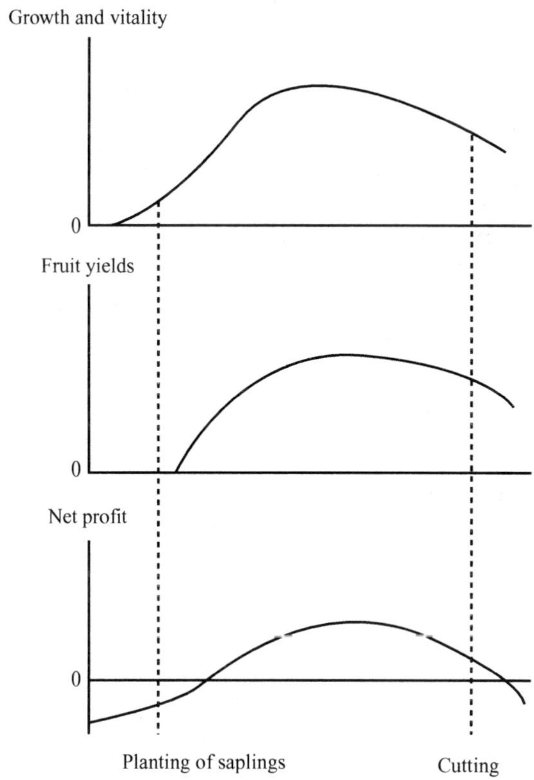

term problems related to planning, management, and development can occur. In short, agriculture (including aquaculture) entails numerous difficulties and risks that are unique to this industry.

The seasonal nature of production and fixed production cycles
Agriculture consists of many different labor activities. The production cycle of plants is fixed, including sowing, growing seedlings, transplantation, fertilization, weeding, harvesting, sorting/processing, and sale. Escaping the constraints imposed by limited amounts of land is difficult. Furthermore, these processes are frequently limited to specific seasons. Of course, it has become increasingly possible to use greenhouses that employ fossil-fuel energy to produce some vegetables, flowers, and fruit 'out-of-season'. But this type of production entails high costs and therefore cannot satisfy the needs of a large population.

Limitations and inefficiencies of implementing technology
In the industrial production of goods, a driving objective is the continuous reduction of the production cycle and working hours. The amount of time required for production processes is determined by considerations such as the stage of technological development, and the scale of capital investment. It does not have any seasonal constraints. In agricultural production, however, while it is possible to reduce the number of labor-hours through mechanization, it is not feasible to reduce the overall time required to produce rice, for example, to less than six or seven months. That is, production cycles are fixed and typically long. Furthermore, the amount of time that agricultural machinery is actually in use is limited to one week per year for a tiller/plow, five days for a rice-planting machine, and ten days for a threshing machine. That is, the amount of time that the machinery is actually in use is quite limited and such machines are therefore not very capital efficient. Hence, reducing total labor-hours requires excessive capital investment.

Importance of land expansion
Industrial production can be conducted in limited spaces and therefore is not as dependent on land size as is agriculture. Agricultural production, in contrast, especially grain production, the cultivation of citrus fruits and rearing large domestic animals, requires vast amounts of land. Agricultural production is thus far more dependent on the availability of land than is industrial production.

Land

There are two ways to expand the scale of agricultural production – either by an external or an internal expansion of land use. That is, either expanding the area used for agriculture – often through cultivating more marginal land, which has less fertile soil or is less favorably situated – or more intensively using the existing areas. Clearly more marginal land is less productive, while more intensive use raises the problem of diminishing returns. If a country expands its agricultural frontiers by relying on other countries for agricultural products, it could increase environmental pollution by external expansion or intensifying the use of the territory that it possesses.

Supplementary note: The law of diminishing returns

The law of diminishing returns applies to agriculture in two ways – that is, with regard to (1) land and crops, and (2) in the management of a firm.

(1) Land is the most important resource for agricultural production. The quality of land and its biological functions are therefore constraining factors for production, as they significantly affect the input-output ratio. In other words, there are limits to how much soil quality can be improved, to how close together seeds can be sown, and to the amount of sunlight that falls on any given parcel of land (consequently, to the photosynthetic process as well). Figure 4.3 illustrates that if we assume all other conditions to be constant and increase capital investment (for example, the amount of fertilizers used) and the amount of labor applied, production becomes more intensive. As illustrated, at stage one (from point 0 to e), an increase in investment results in increased output. At stage two (e to h), an increase in investment generates an increase in output that gradually diminishes. In other words, the use of additional fertilizers and labor at this stage does not result in the same rise in output as was achieved in stage one – the input-output relationship is no longer proportionate, and efficiency decreases. In the third stage, any additional investment, for instance, further fertilization and/or labor, may be harmful in the physical and biological sense, resulting in decreased output. The TPP curve shows the Total Physical Product. The APP curve shows the Average Physical Product, which measures output per unit of input. Finally, the MPP curve shows the Marginal Physical Product, which measures the additional output expected from an additional unit of input (Fukuda et al., 1971: 54–6; Kashiwa 1962: 188–94). Only from points 0 to h, is an increased investment of resources efficient.

(2) In both the industrial and agricultural worlds, at the managerial level, profit is more important than total output. Managers evaluate the

input-output ratio in terms of market prices, comparing the unit price of inputs to the value of the additional output that will be produced; if the latter exceeds the former, there will be increased profit and the firm would invest in additional inputs. In agriculture, the restrictions of the law of diminishing marginal returns may be overcome by expanding the amount of land used and by introducing new technologies. Thus, as Penrose and others have noted, the ultimate constraint for agricultural production lies in managerial competence (Penrose cited in Asami 1989). However, constraints differ between large-scale agricultural countries, with vast amounts of flat and fertile land, and small-scale agricultural countries, such as the EU and Japan, which have a small amount of more marginal land (note also that where there are vast amounts of space, land is considerably cheaper than in smaller countries with higher population densities, in accordance with the basic law of supply and demand). Furthermore, the level of technological sophistication varies according to the stage of development. Finally, ecological problems have come to be recognized as a constraining factor for agricultural production, a realization that other long-term changes may occur.

Figure 4.3: Increasing and diminishing returns

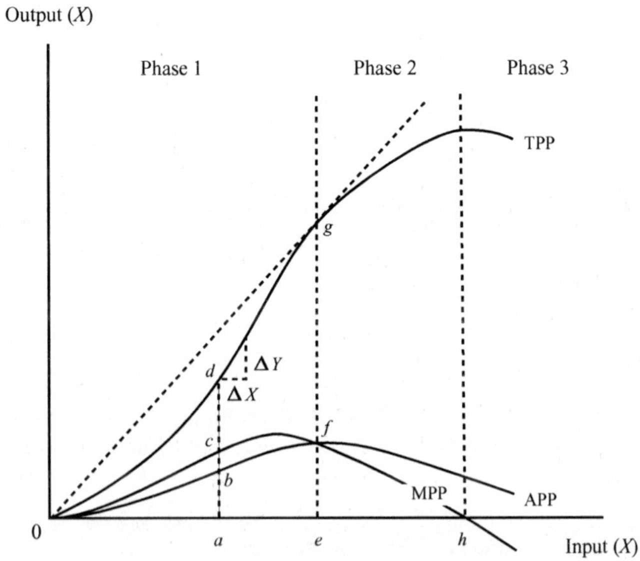

Source: Fukuda, Yori and Mese 1971: 54.

The effect of climatic and other natural conditions

In industrial production, provided that labor and energy are available, and there is a market offering reasonable expectations of profit, capital investment can be made almost anywhere – the climate and other natural conditions of the location are relatively unimportant. For agriculture, however, the amount of available land, its gradient, and weather conditions are of paramount importance. For some products, such as *sake*, soy sauce, and others that require brewing and fermentation, water quality and weather conditions are crucial. It is no accident, therefore, that the production of such goods usually falls under the category of agriculture rather than industrial production.

Managerial structure

Although a variety of business structures can be used for agricultural production, the most common one is the small family business. China and the former Soviet Union introduced collective farms, but these proved inadequate as to need and were soon dissolved. For agricultural production, close observation of plants and animals is vital, and much of the work must be done by hand. Also, the immense effect of natural conditions, such as rainstorms, on the production process often creates a need for continuous work, including nights and weekends. But personal responsibility for these tasks diminishes when production is carried out collectively, especially as under the communist system. As David noted, in the event the owner hires many laborers over a vast territory of land, supervision and evaluation will be extremely hard, which is one of the reasons why farmers usually choose the form of a small family business. This business structure makes it possible to respond to the specific features of agricultural production rather efficiently. Although, the American agricultural system is often referred to as a corporate system, the number of workers employed remains small: in essence, the American farm still operates as a small family business. Thus, the agricultural sector, unlike the industrial sector, operates on a small production scale and is characterized by perfect competition.

Distribution and consumption features of agricultural products

Perishability, susceptibility to damage and other features

Although some agricultural products are used as resources for industrial production, for all other products, freshness is a very important criterion for agricultural goods, especially those sold as raw food or used in the manufacture of processed foods. However, fresh crops are highly perishable and can easily be damaged. Hence, it is imperative that they be transported from farm to market as quickly and as gently as possible. Where quick delivery is not possible – for example, where the physical distance between farm and market entails long transportation, costs associated with conservation, packaging, and storage are important factors. There is thus a special need to develop clear handling standards prior to harvesting, selection and sale. Additionally, compared to industrial products, agricultural goods are heavy and must be transported in bulk quantities of large volume, which results in high transportation costs (Fukuda et al., 1971: 102). Furthermore, not only do these bulk agricultural products need to be divided into smaller portions, but all preparations for transportation, distribution, and sale must be carried out on a relatively small scale, in different places, and through multiple steps. At every stage, seasonal considerations and other features of production come into play, presenting substantial risks. Technological innovation in recent years has seen the development of new cold storage and freezing techniques as well as improved transportation technology, thereby increasing the range and quantity of products that can be supplied to more diverse and distant places. However, these processes entail significant costs, and raise questions about the safety of imported food and the use of chemical preservatives.

Elasticity of supply and demand and price fluctuation

The demand for agricultural products, especially grain, is less elastic than for industrial products. Whereas demand for industrial goods typically increases following either a fall in price or a rise in income, the quantity of grain demanded remains relatively constant, regardless of increases and decreases in income and prices. Furthermore, as previously mentioned, since agricultural products have a fixed, long-term production cycle, it is extremely difficult for producers to quickly respond to changes in demand. In other words, the elasticity of supply is also quite low. At the same time, supply quantities are significantly affected by climatic events that are largely beyond human control.

Agricultural products are absolute necessities for life – people cannot survive without them. In contrast, industrial goods are only relatively necessities – people can and do survive without them. In other words, agricultural goods are indispensable for life itself, whereas industrial goods affect only the quality of life. Consequently, a slight decrease in agricultural production leads to dramatic increases in

Figure 4.4: Supply and demand for agricultural products

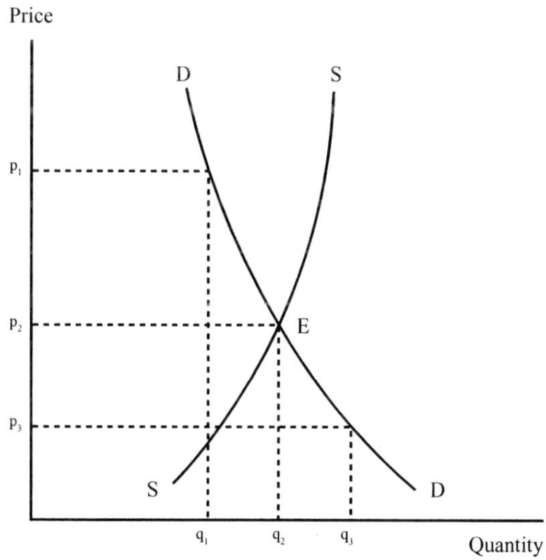

price. Conversely, even a slight increase in production leads to reduced prices, since the total amount of food consumed by the population remains relatively stable. Thus, even slight changes in output produce significant price changes.

Samuelson (1995: 70) explains this supply and demand relationship through a graph, such as Figure 4.4. Inelastic supply and an inelastic demand means that, both supply (S) and demand (D) are nearly vertical. Consequently, any small change in supply or in demand results in a significant change in price.

The long-term fluctuation of the price of agricultural products

Over the long-term, substantial and periodical fluctuations can be observed in the prices of agricultural products. Tamai Torao suggests that global agricultural production is characterized by either shortages or surpluses once every 14 years (Tamai 1980: 112).

Meteorologist Asakura Tadashi observes that twice a decade, in the equatorial Pacific (off the Peruvian coast), the water temperature rises, producing the El Niño effect, which generates extreme climatic conditions in various countries around the world (Kawai 1988: 68). He also notes that occurrence of frost damage to crops in Japan and of droughts in the US every four or five years seems to support this claim. The regional imbalances in agricultural production that result from these climate changes renders international trade indispensable. Yet, when one nation is dependent on agricultural imports from another (or others), if the other(s) suffer(s) from a bad harvest, the prices of these products will rise sharply.

In the 1980s, rice constituted a mere three percent of the total world output of agricultural products traded on the international market, compared with wheat (20 percent) and beans (25 per cent). Rice is therefore particularly vulnerable to price fluctuations, as any small change in total production leads to substantial price changes.

All of the factors and features of agriculture discussed contribute to the susceptibility of agricultural products to sharp price fluctuations. Figure 4.5 indicates trends in agricultural prices over a 20 year period. Prices changed three- to six-fold over this period, with rice being the most severely affected. Especially in 1972, when global agricultural output fell, due to extreme climate conditions, grain prices rose three to fivefold. This is a very sharp price hike, considering that output fell a mere three to seven percent. Under these circumstances, the US prioritized its domestic market demand, and stopped exports of beans, while other countries struggled to ensure that they could secure adequate supplies of grains.

These events not only highlighted the importance of food products for a nation's well-being, but also made apparent the possibility of using

Figure 4.5: Long-term fluctuations of international grain prices

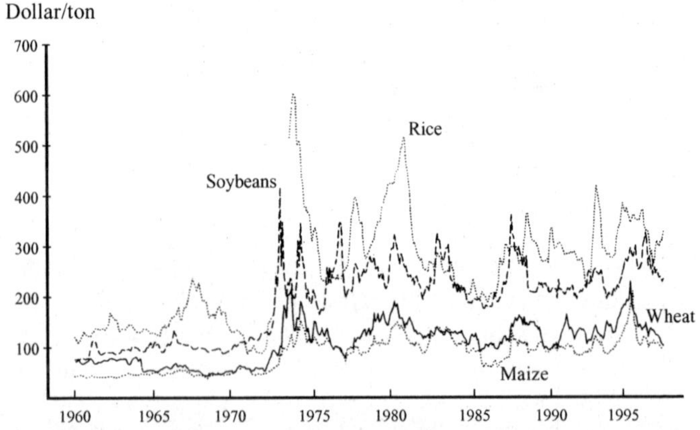

Source: Ministry of Agriculture, *Forestry and Fisheries of Japan*, 1998. Grain prices shown are those on the first Friday except for rice, which are those on the first Wednesday.

food products as a weapon against opposing countries (in the Cold War), on the one hand, and as a foreign aid tool to support its allies, on the other. Today, national food security remains an important issue.

Distinguishing features of fisheries

Agriculture and forestry have been the main focus of the preceding sections. Here, then, I add a few notes on fisheries. Some argue that there are fundamental differences between fisheries and agriculture/forestry. These include, for instance: (1) agriculture and forestry are land-based production activities, while fishery is water-based; (2) some aspects of fishery production, such as the reproduction of living creatures, remain unclear and out of the reach of human control; (3) although exclusive fishing zones are recognized, water is typically considered to be without ownership, and enforcing individual property rights is difficult.

Indeed, many other differences might also be pertinent. Nevertheless, compared to industrial production, there are many similarities between fishery and agricultural production that become apparent, especially if we include coastal fishing and aquaculture farming, which are usually family-operated. These similarities include: (1) engagement in biological production; (2) stable and planned production is impossible, since the size of a catch is dependent upon uncertain natural/climatic conditions (although with the development of freezing technologies it has become easier to control the amount supplied); (3) aquaculture can be seen as comparable to intensive agriculture, where the rules of diminishing marginal returns apply; (4) production in both is dependent upon the well-being of the ecosystem, and therefore preservation of the ecosystem is an important dimension of production.[2]

Capital outflow from agriculture and agricultural protection policies

Some see agriculture as a backward and stagnant industry, pointing to the fact that capital is moving out of the industry to justify this view (the issues canvassed in this section are drawn from Tōbata 1938: No. 18; Kashiwa 1957: 81). The movement of capital investment also indicates that land and labor investments are moving to other industries. This is precisely what happened in Japan during the intensive industrialization that began in the postwar reconstruction period and continues largely unabated today. Capital investment aims to secure maximum profit (entrepreneurial profit), and is therefore overwhelmingly directed towards the sector that has the strongest potential for expanding profits. As we have seen, agriculture is characterized by features that render it incompatible with market principles. Industrial production has a greater capacity to deliver and sustain substantial profits than agriculture, which explains the reduced rate of investment in agriculture.

Allow me here to once more summarize the main distinguishing features of agriculture to segue into discussing some of the possible ways of addressing the issue of declining investment.

Industrial production is conducive to land-intensive production in a wide variety of geographical areas, is not affected by seasonal changes, and its requisite components may be produced in regions remote from where they are ultimately assembled, while its time constraints are few and variable. In contrast, agriculture entails organic production that requires natural resources such as air, water, and sunlight, and its time requirements are relatively fixed. All of this makes it susceptible to climatic (seasonal) change and limited by land constraints. Furthermore, machinery is only useful during particular operations (for short periods of the entire production cycle), the production process is extremely slow and therefore inefficient. In addition, whereas some products, such as rice, are produced in a one year production cycle, others, such as fruit, require about 10 years to get a return on investment, while the production cycle in forestry can be 30 to 50 years. Although the intensity of production can be varied, products and production processes certainly cannot be changed to the same extent as in industrial production.

The supply and demand elasticity of agricultural goods is low compared to industrial products. Although many industrial products have become relative necessities, agricultural products are absolute necessities. Yet, agricultural production is largely dependent on climate conditions and other factors that are beyond human control. Consequently, even small changes in production lead to substantial changes in prices for these products, contributing to the instability of agricultural income.

In Japan, the agricultural products market is competitive with a large number of small family-operated farms. Compared with industrial production, agriculture does not provide much scope for innovation, and the potential for securing and sustaining profits is low. Furthermore, since agriculture is involved in the production of absolute necessities, it is open, public, and does not imply any trade secrets or require any licensing. On the contrary, for public agencies, ensuring the rapid diffusion of new agricultural technologies was the highest priority. Consequently, the benefits for those pioneers who first started working the land were not significant, and the barriers to entry remain very low. In other words, there are not as many restraints to sharing agricultural technology with other farmers as there are in industrial production.

The long-term and fixed nature of production alongside the uncertain/unstable profits leads to the prolongation of the capital turnover cycle and thus decreases in agriculture's overall efficiency while increasing risk. Thus, if agriculture were subject to market principles alone, most capital would migrate towards industry and importing agricultural products, which is easier and rather cost-effective.

I mentioned that the returns on investment from agriculture are relatively small, which leads to capital outflows. But this does not mean that agriculture can be seen, in capitalist terms, simply as an under-developed industry, and therefore unimportant for a nation. It cannot be denied that agriculture plays a paramount role in any national economy. The problem lies in the large gap between notions of national development, human values, and the logic of capitalism. Capital is indifferent to agricultural development and unconcerned by the importance of agriculture for society. The capitalist market economy is based solely on the pursuit of profit and thus sets little value on social stability, social harmony or society's need for agricultural products. This disjuncture contributes significantly to market failures.

For all of these reasons, it appears that this situation can only be ameliorated through government policies and funding aimed at redistributing capital.

If this is correct, then agriculture needs to be viewed from the same perspective as roads, railways, ports, airports, sewage and drainage, welfare facilities, education facilities, land development (including restoring polluted rivers, dam-building, reforestation, etc), and other public works. None of these public works would have been provided through reliance on private funding alone. Although it may be important to encourage private investment in order to improve the efficiency of such projects, attracting individual investment is highly problematic. The absolutely essential role of agriculture in providing food for the nation, and the constraining features outlined above, lends agriculture many of the characteristics of public works projects. Such public goods require social funding and are therefore dependant on state policies governing the redistribution of capital. This is especially true of agricultural infrastructure, such as building and maintaining farm roads, developing farmland, establishing irrigation systems, etc.

In the short-term, investment in industrial production seems to return significant income for the nation. However, to achieve stable long-term growth, it is necessary that this income be redistributed towards agriculture. This is especially true today, with growing political tensions internationally, developing countries facing possible food shortages due to overpopulation and environmental problems becoming ever more acute. Furthermore, as energy problems continue to escalate, the many limitations to the theory of comparative advantage become increasingly apparent.

Distinctive features of agriculture as a multi-value industry

Recognition of external effects in agriculture and forestry

Long-term sustainable agricultural production requires preservation of the environment. But capitalist mechanisms fail to accommodate this requirement, insisting that agriculture is merely an industry producing goods for the market. This has been the predominant view of agriculture in Japan since the beginning of the rapid economic growth era.

Although both environmental pollution and recognition that agriculture is a multi-value industry highlight internal flaws of the market, economic theory tends to regard these issues as externalities. An externality exists when an economic activity affects another economic activity, without any direct payment or transaction being involved. Where this effect is positive, it is called an external economy, while if it is negative, it is referred to as an external diseconomy. Neglecting externalities is considered to be an internal failure of the market.

Marshall was the first to introduce the idea of externalities, and Pigou substantially extended this idea (Marshall 1898; Pigou 1920). They exist in an almost infinite variety of forms, with both narrower and broader effects. Here are some examples. First, where the activities of a firm result in air and water pollution – in the event that there are no regulations to compel the polluter to rectify the damage – it will affect an unspecified number of people in an indiscriminate manner (external effect from the firm to the consumer). Second, the noise produced by one industrial plant may affect the production efficiency of another (external effect from one firm to another). Third, possession and use of the latest (or highest quality) consumer goods by one consumer may affect the sense of contentment of others; or the creation of a garden at one house improves the amenity of the entire neighborhood (external effect from consumer to consumer). Furthermore, in the same way that 'the trees and bushes in a city and its suburbs not only improve the amenity for individual passers-by, but also contributes to cleaner air', the economic and social effects of improved levels of education are beneficial not only to the educated individual, but to society as a whole (Suzuki 1974: 243, 245). Finally, the present global environmental problems generate externalities of the type just outlined not only for our contemporaries but for future generations (Alier 1991: 302).

As we can see, the concept of economic externalities refers to a wide range of phenomena, including the amenities of life, which makes the definition quite ambiguous (Suzuki 1974: 75). Even if we examine the interrelatedness of the occurring phenomena and the difficulties in accurately assessing their effects, the concept remains vague. Nevertheless, recognizing such externalities is crucial for developing a sustainable economic society for the future.

External economies in agriculture and forestry

This section considers the economic and social role of agriculture (here defined broadly to include forestry) – that is, the external economies of agriculture. (I will discuss the diseconomies in the next section.) These can be seen in the following situations: (1) the cultivation of apples, rape seed, and flowers has positive affects for a honey producer (external effect between producers); (2) the quality of water needed by an industrial plant or local community living downstream depends on the actions of an agricultural producer upstream (external effect between a farm and an industrial plant or between producer and consumer). This section focuses primarily on the second case as my primary concern is with the potential for agriculture to affect public welfare.

As discussed in Chapter 3 (see Table 3.3), agriculture is a multi-value industry, having a value for the economy, for sustaining the ecosystem, and for sustaining human life. I will now elaborate on this concept.

Economic value of agriculture includes the following: (1) the adequate protection and development of domestic agriculture provides high-quality reasonable-price food goods in a stable manner, thus serving to relieve the pressures of changing demand or supply and fluctuating prices on the international market; (2) agriculture and the part-time farming systems as in Japan add diversity to the economic structure of the country, which leads to a more vibrant response to changes in the national or regional economy, including industrial activities, and ensures more stable development, while securing food supply in the event of a crisis situation; and (3) promotes regional revitalization and serves to increase employment opportunities for seniors and women.

The ecological value of agriculture can be summarized as follows: (1) agriculture serves to cultivate water supply resources, preserve the soil, helps prevent floods, provides for water reservoirs below and above the ground, thus contributing to sustaining national land; (2) existing agriculture contributes to cleaning water and the air, redresses smell and noise pollution, preserves comfortable landscapes and the land needed for evacuation purposes during natural calamities, thus helping to sustain the environment for daily activities; and (3) agriculture makes it possible to recycle by cooperation with city and make efficient use of regional energy and resources.

The value of agriculture for everyday life (social and cultural value) would include the following aspects: (1) it contributes to prevent social standardization and destabilization, while helping to promote diversity, stability, and continuity; (2) agriculture provides space for social exchange to take place and is a venue for developing perseverance and creativity by enabling workers to become healthy in body and soul; (3) it serves as a welfare function; (4) it serves as an educational function; and (5) urban dwellers can enjoy rustic experiences and relaxing times that enrich their lives.

Among the aforementioned values, agriculture's contribution to the preservation of ecosystems and its social and cultural values can clearly be defined as external economies. The economic value of agriculture can also be thought of as an externality considering that securing food products contributes to social stability and allows for a balanced development of various industries, including agriculture, forestry and fisheries, thereby fostering a stable and simultaneous development of commerce and industry. This final point concurs with Nurkse's theories of developing countries, which argued that industries enjoy a multiplier effect, leading to balanced growth (1953).

External diseconomies in agriculture and forestry

Above, I discussed the active role of agriculture and its external economies. However, agriculture can be a source of environmental pollution just like industrial production. This point certainly needs to be addressed (Kyūma and Soda 1995).

Examples of the external diseconomies of agriculture include: (1) soil and water pollution from the use of chemical fertilizers and of other agricultural chemicals, killing wild plants and animals and damaging the ecosystem; (2) the production of methane gas as a by-product of irrigated rice fields and large livestock rearing, plus the air pollution generated by using fossil-fuels in production; (3) the deteriorating quality of food products from the use of agricultural chemicals and the resulting damage to people's health; (4) on a global scale, growing populations in the developing countries, increasing deforestation for the purpose of expanding the amount of usable agricultural land as well as for harvesting trees for timber and pulpwood for paper, damage caused by cattle, contribution to global warming, salinization from the excessive use of ground water, and landslides owing to the erosion produced by large-scale agricultural activities. Note also that deforestation in developing countries is aggravated by the fact that a significant proportion of tree harvesting for timber and pulpwood – is carried out by giant multi-national corporations with little or no interest in the local area beyond extracting maximum profits. Addressing such diseconomies requires a fundamental paradigm shift and the development of new technologies for sustainability.

Evaluating and addressing external economies

As discussed, agriculture, like industrial production, involves multi-faceted externalities – both economies and diseconomies. It is therefore imperative to elaborate how these issues are to be assessed and how the positive potentials can be realized.

Agriculture plays many roles in society, not all of which are valued adequately. In fact, now that market principles have taken precedence and the international division of labor based on the theory of comparative advantage has become dominant in the Japanese mainstream, agriculture is receding in various senses. Pigou (1920) outlined the failures of market mechanisms and argued that external diseconomies may be dealt with through targeted taxation, while external economies might be promoted through subsidies – both aiming to balance private with public benefits. However, food safety and environmental problems require more than merely the second best solution; they also require an accurate and complete position. There is therefore a need to revise the basic approach that Pigou introduced to economics. Similarly, Röpke's 'framework policies', while basically accepting market principles, argues for the need to supplement them and correct their imperfections. In short, we must carefully reconsider the over-simplified theory of the international division of labor.

Supplementary note

In the 1980s, the Ministry of Agriculture, Forestry, and Fisheries evaluated the public benefits derived from land used for farming and forests and

> reported that the total value – with regard to cultivation of water resources, the prevention of land erosion/degradation, landslides, provision of vacation and health facilities, preservation of wild birds and animals, and cleaning the air – was approximately 36 trillion yen (about $152 billion at the time). This figure would be even higher today if we adjusted to current money values. In 1991, Mitsubishi Research Institute evaluated the external effects of irrigated rice fields to be 12 trillion yen ($89 billion) according to the hedonic price method and 5 trillion yen ($39 billion) according to the substitute cost method. Since the total value of rice itself is only 3 trillion yen ($22 billion), according to both the hedonic price method and the substitute cost method, rice production results in substantial positive externalities. Furthermore, in 1998, the Agricultural Research Institute attempted to estimate the total price value of agriculture and farming villages and came up with a figure of 10.46 trillion yen ($80 billion). The valuation techniques used for these calculations contain numerous vagaries, and further research is required to clarify these. However, regardless of the validity of the valuation methods, agricultural policies taking account of these external economies must be quickly revised as soon as possible.

Although agriculture has many positive externalities, negative externalities give rise to the so-called external diseconomies discussed above. These diseconomies must be addressed quickly, as agriculture is an industry that directly affects the health and wellbeing of citizens.

As mentioned, the external diseconomies of agriculture are closely linked to modern farming methods. We need to change our approach to agriculture towards making full use of the power of nature so as to promote the symbiotic relationship between different crops as well as between animals and crops, and seek to reduce the use of chemical pesticides and herbicides that are harmful to crops and other life-forms through the increasing use of natural remedies, such as deploying the natural enemies of harmful insects. In the EU, various limitations have been imposed on the use of numerous agricultural chemicals, with broad public support for a direct income subsidy system (or direct payment system) to cover the financial losses incurred through reduced productivity, leading the way towards sustainable forms of agriculture that are protected from the failures of market mechanisms.

Inter-generational external effects

As we have seen, the externalities of agriculture affect not only the present generation, but will have a tremendous influence on future generations. It is therefore imperative to remain cognizant of the 'participation of future generations' (Alier 1991: 302). I will now expand on this idea, to include preceding generations and introduce the concept of 'inter-generational externalities'. Preceding generations developed agriculture through long years of coexistence with nature. For example, irrigated rice fields in Japan before high economic growth could be used continuously for many years, while the retaining ridges between them helped to prevent land slides and floods, thus contributing to the preservation of water courses both above and below ground, and thereby helping to preserve the environment. Our ancestors also promoted organic interaction between crops and livestock, thereby encouraging/facilitating a desirable circulation of elements. Rice is particularly suited to the climatic conditions of Japan, is highly nutritious and filling, and is suitable for long-term storage. Technological innovations have further enhanced these properties, providing greater resistance to extreme temperature changes, and developing new higher-yielding varieties that help to ensure more abundant and stable harvests. Modern agriculture, however, has largely ignored the traditional wisdom that has accumulated over generations, arguing that it is not scientific. This attitude and its effects are currently perhaps most apparent in modernizing Africa, where small-scale farming is increasingly neglected. Forests are not simply a means for generating profit, they also represent the heritage that the previous generation prepared for their descendants. In a way, we can see ourselves as simply living off this inheritance.

We need to look respectfully on our inheritance when attempting to solve present problems, as well as considering future generations by developing sustainable production technologies that do not wastefully consume energy and natural resources. It is our responsibility to leave future generations a healthy and vibrant environment. I propose that understanding the scope of inter-generational externalities comprising the past, present, and future generations is necessary if we are to embrace the social role of agricultural production and its long-term effects.

It has become impossible to ignore the environmental problems that have been and continue to be caused by modern agriculture. Resolving these problems through agricultural and trade policy is the tremendous task that lies before us today. To do so, we must recognize agriculture's externalities, regional differences (regionality or placeness), and multiple values.

Regional Features

As discussed in more detail in the next chapter, agriculture, forestry, and fisheries are highly affected by natural and geographical conditions, such as temperature, rainfall, humidity, and the gradient of the land. Those areas that for whatever reason are deemed poor for farming have been labeled 'areas with unfavorable conditions'. In the EU, the direct income subsidy system mentioned earlier includes policy specifications for such areas (Izumi 1989; Yaguchi 1995). In Japan, in 1999, a system involving direct payments to subsidize farming in unfavorable areas, such as mountainous regions, uses the gradient as one important index. This system was further refined and then implemented in 2000 (Ministry of Agriculture, Forestry, and Fisheries 1999).

Compared to other developed countries, Japan's agriculture is typically conducted in areas with unfavorable conditions. The difference is particularly striking if we compare Japan to countries such as the US, Canada, and Australia. For example, in US agriculture, the availability of vast amounts of level and fertile land has enabled the large-scale use of machinery, helping to make the country a producer and exporter of cheap agricultural products from the very beginning. Japan's agricultural landscape, in contrast, is characterized by a high proportion of mountainous valleys and rivers, and thus agriculture has developed historically to be characterized by many small-scale farmers with high costs. In 2000, the average area under agricultural management in the US was about 200 hectares compared to only 1.4 hectares in Japan – a difference of almost 140 times. Yet, if we consider places like Africa and elsewhere, where agriculture is practiced in the savannah or similar very dry climates, we find many places that have even more unfavorable conditions than Japan.

During the Uruguay Round of the GATT, the US argued for an indiscriminate liberalization of trade in both agriculture and industry. Such a position would favor the US, who could take full advantage of its agricultural conditions. However, while the agricultural conditions in Germany and other EU countries are better than Japan's, they are not comparable to the US's. Consequently, Europeans argues that US agriculture is commercial, while European agriculture is peasant-based. Europeans therefore continue to oppose further liberalization of agricultural production. The US position appears to ignore the fact that agriculture is dependent on regionalty or placeness (Soda 1989: 48).

Treating agriculture and industry in the same way, arguing for free competition and free trade on the same terms for both sectors, threatens the very existence of regions, and is therefore clearly problematic.

The US stance on the GATT has been to repress or avoid any discussions of environmental problems, with the aim of obscuring the link between agriculture and the environment and thereby turning the issue of agricultural trade to its advantage.

The economic basis for agricultural trade relies on the principles of comparative advantage mentioned earlier. As discussed, Heckscher's and Ohlin's theory of the international division of labor maintains that the main factors of production, including land, labor, and capital, if distributed in the most efficient manner, will result in increased productivity of various management organizations, thereby delivering higher incomes and improved welfare. Both Japanese and American industrialists refuse to recognize either the general distinctions between agriculture and industry or the regional and geographical differences between the two countries, thereby treating all factors – both positive and negative – as equivalents and assuming the same, uniform approach to calculating costs and benefits in both countries for all 'industry' sectors.

The theory of international division of labor relies on a number of assumptions. Three of the most important for agriculture are: (1) the existence of perfectly free competition; (2) natural conditions that influence agricultural production are considered to be a factor in production; and (3) there are no externalities pertaining to either production or consumption (Kanemitsu 1991: 89).

Supposedly 'free' competition often results in the creation of large oligopolies who can monopolize the market. This is quite different from situations where the price structure is perfectly competitive. Oligopolies emerge as a result of differences in scale of production between firms, whereas agricultural production, having distinctive spatial and time properties, is conducted on small scale, which is quite different from industrial production, especially in Japan and other small-scale farming regions around the world. In addition, with regard to (2), above, in the event of inferior natural conditions, any country that is part of the global economy will be forced to drop agricultural production. As for (3) above, any theory that ignores the environmental degradation or externalities being generated by contemporary economic activities is significantly flawed. Any theory that ignores the differences between regions, states or industries is just a wild argument. Most countries are worried about how to cope with agricultural affairs. Adam Smith's theories of free trade from the perspective of the developed countries and Heckscher and Ohlin's theory of the international division of labor certainly have their strengths, but they also have serious shortcomings that must be addressed.

It is important to note that in the event that imports and exports become even more concentrated in certain countries, the associated environmental problems will only increase. On the one hand, large importers, like Japan, may enjoy some relief from the negative effects of agriculture, but will not be able to take advantage of the positive effects, either. On the other hand, large exporters, like the US, will see their agricultural sector continue to expand at the cost of increasing the negative impact on its environment.

It is therefore absolutely necessary that we develop some degree of public consensus about the need to change direction if we are to achieve a sustainable balance in our agricultural, forestry, and fishery policies

The international economy and agricultural economics

Small government theory and the trend towards deregulation

In the early 1980s, in response to the economic stagnation and the huge financial and trade deficits that the US had accumulated since the 1930s, during the era of 'big government' and the welfare state, the country moved towards 'smaller government' and deregulation. This was the period of 'Reaganomics', paralleled in the UK as 'Thatcherism'. In Japan this new policy approach manifested in the privatization policies of the Nakasone Cabinet. This movement is perhaps most clearly expressed in the Economic Report of the US President and the accompanying Economic White Papers.

These reports argue that big government is very expensive and significantly retards the entrepreneurial spirit. Furthermore, they claim that the positive effects of privatization and liberalization are highly regarded in both Eastern and Western countries. They conclude that 'small government' will remove market rigidities and increase the economic freedom of the private sector (Council 1988: Chapter 8). The US Economic White Paper of 1992 argues for fortifying the competitive market – marking an attitudinal change towards the omnipotence of the market; i.e., a return to Adam Smith and away from Keynesian economics (Sawa 1991: 28).

Despite these efforts, however, government finances and trade deficits continued to grow. Meanwhile environmental policies were demoted or trashed altogether. The consequences of this policy direction starkly highlight its basis in ignorance of the history of market and government failures. While these 'neo-liberal' economic policies dichotomize liberalization and control, and then wholly reject the latter in favor of the former, the German theory of the social market economy has taken into account the lessons of history and advocates a combination of freedom, restrictions, and justice.

Indeed, in Japan, there are many reasons for deregulation and reform (Katō 1994; Tajima 1994; Sawa 1991: 28): (1) the increasing international and regional interdependence known as globalization; (2) the need to adjust international regulations in respect of increasing export-import activities; (3) the continuing expansion and increasing inflexibility of the national bureaucracy, and increasing national deficit; (4) the advent of various problems associated with corporate downsizing and other efforts to sustain competitiveness in the face of a prolonged economic recession; (5) the transformation of information resources from basic public goods to commodities with far-reaching influence; and (6) the 'coming of age' of industries that were born long after their counterparts in other countries, and have now developed not only to the point of independence, but where they can challenge the mightiest international competitors. Most prominently, the Japanese automobile and aviation industries were developed with significant government protection. Deregulation often involves decentralization of power, which depending on the situation, often has great merit.

Deregulation can deliver significant benefits to a maturing economy. But not all economic sectors can adjust to such changes in the same way. Agriculture is prominent amongst these, due to the various distinguishing characteristics that I have been discussing in this chapter. Social market economic theory argues for a productive synthesis of freedom and regulation, clearly acknowledging these sectoral differences, maintaining that the free market economy requires some degree of government regulation if it is to develop adequately.

Corrections of market principles: social market economy

As we have seen, in the 20th century, capitalism underwent two very sharp transformations. First, in response to the Great Depression, following Keynes, there was a transformation from *laissez faire* free market capitalism to strong government intervention intended to regulate and correct the inherent excesses of capitalism. Then, after the oil crises of the 1970s, with its accompanying stagflation, and in the face of mounting government debt through the big government spending of the Keynesian welfare state, there was a transformation back towards *laissez faire* market determination under the auspices of Reaganomics and Thatcherism – or more generally, neo-liberalism. However, neither of these transformations adequately redressed the problems of the market and government failures that prompted the changes of direction; in fact, both resulted in the creation of many new problems, as discussed above.

Here then I want to explore a more marginal, alternative theory, the social market economy – *die soziale Marktwirtschaft* – which became mainstream in West Germany after World War II when that country tried to find its own 'third way' (between *laissez-faire* capitalism and Marxist-Leninist communism). This 'third way' was born of the experience of skyrocketing inflation after World War I – when prices rose one trillion times over three years – and remorse for the excesses of the Nazi's efforts to 'correct' this through extreme government intervention and militarization. With great effort, the postwar West Germans developed the social market economy, which theoretically rejects even Keynesian-style government intervention – primarily because Keynesian theory welcomes inflation.

Figure 4.6 depicts the basics of the German position. It indicates that at one extreme lies economic freedom of staunch individualism and economic anarchy, while at the other extreme lies strong collectivism and communism. It also indicates seven intermediate stages between these two extremes, with the social market economy located at the very center of the continuum. A social market economy thus represents an attempt to forge a productive synthesis of opposing principles: freedom–regulation, individualism–collectivism, liberalism–communism. As such, it truly represents 'a third way'. Analyzing this situation, Pilz argued that the social market economy is a concept that combines economic freedom guaranteed by the state with social stability and social justice (Pilz 1974: 37). This theoretical framework was developed by E. Eucken, Müller-Armack, and W. Röpke, while the political policy was implemented by Chancellor Ludwig Erhard. Under this system, West Germany achieved 'miraculous growth' and responded flexibly to new challenges (Röpke 1944).

Although the social market economy was developed by the Christian Democratic Union (CDU), the Social Democratic Party

Figure 4.6: Weddigen's positioning of social market economy

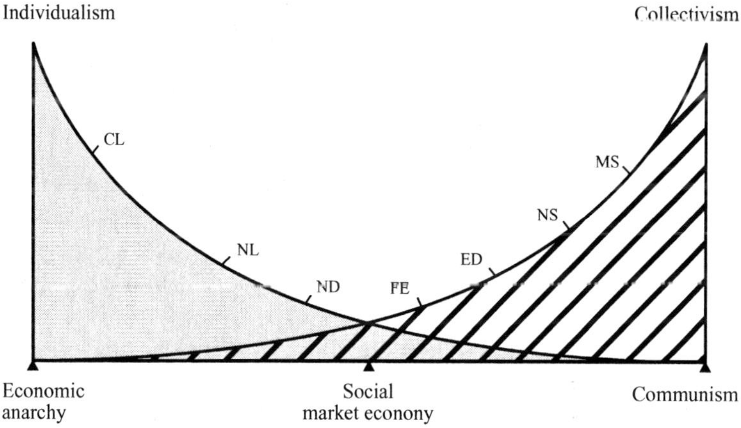

CL = classical liberalism
NL = neoliberalism
ND = New Deal liberalism

FE = functional economy
ED = economic democracy
NS = neosocialism
MS = Marxist socialism

Source: Weddigen 1966/1979: 48.

(SPD) continued to support it when it came to power in the 1970s, abandoning the ideas of Marxism and socialist revolution. CDU and SPD share basically the same economic policy/theory, though the latter might be, only relatively speaking, more in favor of social justice and welfare and close diplomatic relations between capitalist and socialist economies. The social market economy implies a 'productive synthesis of freedom and regulation' (Weddigen 1966).[3] Freedom as used here implies prioritizing market principles, while regulation implies the social measures aimed at preserving social equilibrium in a state that has overcome interventionist policies. In other words, regulation here refers to a policy framework that consists of measures aimed at those areas where the market mechanism alone is expected to fail. Specifically, these measures would relate to financial, welfare, social, spatial planning, education, environmental, and agricultural policies. The objective of productive synthesis recognizes the fact that some degree of regulation is required to maintain freedom.

After coming into office, the SPD, with Karl Schiller as Minister of Economic Affairs, adjusted the economic process policies, affecting aggregate demand, financial policies, and distribution policies, and bringing the social market policies closer to Keynesian economic principles (Ōnishi 1992: 24), but without abandoning the principles of the social market economy. The principal differences between a social market economy and Keynesian economics are as follows: (1) inflation is viewed differently, with the social market economy prioritizing long-term price stability; (2) accounting for regionality (regional differences/particularities) in the formulation of land and regional policies; (3) the social market economy prioritizes maintaining quality of life, justice, and equality, primarily focusing on the middle class; (4) the social market economy combines respect for market principles with the ability to correct failures of the market mechanism.

By directly addressing the problem of inflation, West Germany managed to stabilize it at around 3 percent (from the 1950s until reunification). This is in sharp contrast to the 7–8 percent inflation in Japan during the period of rapid economic growth (when economic growth was regarded as the highest priority). Maintaining distance from the Keynesian approach that regards economic growth as the most important criterion and sees inflation as an inevitable consequence of such growth, Germany managed to achieve miraculous growth while sustaining relatively low levels of unemployment, compared to other European countries and the US.

Additionally, as discussed in more detail in Chapter 8, Germany pursued a regional-land policy that promoted a decentralized (or multi-centered) social structure, including efforts to limit the growth of large cities, and the integration of small or medium cities and country areas, while striving to achieve regional balance with regard to welfare and other aspects of social justice, etc. In this respect, Germany is quite different from larger agrarian countries, like the US for example, where there is an abundance of fertile land but there is no village society requiring careful spatial planning. The German approach is radically different from that adopted in Japan, where the problem of urban sprawl, although acknowledged as problematic, was accepted as inevitable for efficient economic development. As Yoshihirō Tamanoi notes (1974), the German economy follows economic principles that continuously feature regionalism. Röpke's economic theory (1994: 159), which, as discussed, formed the basis for the economic policies of West Germany, includes conceptions of dispersed land policies and spatial development, and emphasizes the natural environment and agriculture in the everyday lifeworld. He refers to the 'magnetism of the garden (Kleingarten)' to denote this idea.

This particular social policy approach was embraced very early in West Germany's reconstruction period, and there is very widespread public support for policies that prioritize a stable society with the middle class located at the center of the social market system. This public support base has enabled the country to pursue social justice as an economic policy priority. The qualifier 'social' in the description of the economic policy is a reflection of this priority.

More recently, however, a new dimension has begun to be recognized in the already synthetic German postwar economic system, as the Eco Social Market Economy has come into focus (Nojiri 1995: 109–155). In this vein, Stratmann-Mertens and Thiede have argued that the economic structure has three important dimensions – the economy, society, and ecology (Stratmann-Mertens 1991; Thiede 1988).

As we have seen, the social market theory includes important elements of Keynesian economics while rejecting its short-term approach. It values balanced and stable growth, but also emphasizes social justice to a greater extent. Upholding market principles, it understands freedom and regulation not as opposed, but as complementary values, and strives for a flexible synthesis in one productive theory that prioritizes stable growth. It must be recognized that these principles were seriously challenged after the collapse of the Soviet Bloc and the reunification of East and West Germany, as the new Germany found itself facing many new problems. But today it appears that they have succeeded (to a great extent, if not entirely) in integrating the two parts into a revitalized social market economy.

For our purposes, then, considering marginal agriculture and other issues affected by the tensions between freedom and regulation, the principles of the social market economy appear to be best able to addressing them in a satisfactory and flexible manner.

Agricultural technology and agricultural economics

So far, this chapter has focused on trends in general economic theory. Before we conclude, then, we must more specifically consider the trends in agricultural economics.

In *Seimeikei no Economī* (Living Economics), Tamanoi Yoshihirō argues that agricultural economics should be the starting point for economic theory, or rather, agriculture should provide the basis for rethinking economics.

Unfortunately, though, agriculture has been subject to processes of industrialization for at least half a century – processes that have generated numerous problems. Since WWII, Japanese agriculture has been modernized, which means increasing mechanization, widespread (and excessive) use of chemicals (herbicides, pesticides and fertilizers), increasing professionalization of on-site farm management, and a continuing trend toward monoculture cropping. These developments have freed farmers from heavy labor, have largely overcome the problems of food shortages, stabilized supply, and satisfied the ever-growing market demand for agricultural products.

But the cost of these impressive achievements has been enormous. Modern agricultural techniques consume huge amounts of fossil fuels, which, combined with the widespread use of chemicals, has heavily polluted the land, water, and air, leading to the disappearance of innumerable plants and animals. These large quantities of chemicals have also affected the health of the farmers who handle them, and the people who consume their contaminated products. Professionalized management and monocultural farming have undermined the organic interaction between crops, and treated the by-products of farming as waste, further contributing to pollution. Monoculture farming continually plants the same crops in the same fields, thus stripping essential nutrients from the soil, leading to even heavier applications of fertilizers, which then demands further pesticides, etc – creating a horribly vicious circle. In developing countries, population growth and reclamation of forests for agriculture combined with the processes of modernization just described have created unprecedented levels of environmental degradation. The US has responded to global environmental destruction by developing Low Investment Sustainable Agriculture (LISA) practices, while Europe likewise endeavors to develop sustainable agricultural practices.

As discussed earlier, ideally agriculture contributes to land and environment preservation, nurtures various plants and animals, and fulfills countless social and cultural roles. That is, it has many non-economic values. But the market economy has ignored all non-market values, making it almost impossible today to accurately assess the economic externalities of agriculture, while the external diseconomies of agriculture highlight market failures.

As Tamanoi points out, agriculture is necessary for sustaining human life, and agriculture and agricultural economics are fields that can be life-promoting. A comprehensive science incorporating both agriculture and agricultural economics could provide important guidance to many aspects of industrial production and general economics. Modern science is anthropocentric, grounded on Newtonian physics and mechanics. Following Descartes, modern science has embraced and pursued rationality (which in the late 20th century has come to mean especially economic rationality), seeking to subjugate nature to maximize human efficiency and comfort. But this anthropocentric approach has driven us to push nature beyond its absorptive and restorative capabilities. Under these circumstances, a new holistic view of nature that centers on biology and ecology is essential. Rather than pursuing new techno-scientific solutions, we must look to agriculture as a comprehensive and life-promoting science for a new direction forward. The mission of a new era awaits agricultural economics.

5 Agriculture and the Ecological Environment

Environmental problems are considered to be among the most pressing issues of the 21st century. The mishandling of the environment threatens the survival of humankind itself. Agriculture – which for the purposes of this chapter includes forestry – is concerned with nurturing and exploiting nature, particularly living organisms. In the process, it burdens the ecological environment in a variety of ways, even while, at the same time, it is affected by changes in the ecological environment. The long-term preservation of the environment is therefore essential for agriculture. This chapter examines the overall relationship between agriculture and the ecological environment. First, I consider the ways in which the ecological environment regulates agriculture. Second, I look at the need for increased food production brought about by global population growth. Third, I discuss the deterioration of the environment and agriculture. Fourth, I consider the affect of agriculture on ecology and its meaning in sustaining domestic agriculture. Finally, I outline some important issues for the future relationship between agriculture and the ecological environment.

Ecological regulation of agriculture

Agriculture is inseparable from ecology. Meteorological and geographic conditions such as temperature, precipitation, soil quality, humidity, sunlight, gradient, and wind velocity directly affect the forms that agriculture takes. It is also inseparable from broad social trends, and hence, with the development of industry, agriculture has tended to expand in scale. The 'industrialization of agriculture' has affected significant changes to water, air, soil, temperature and other environmental conditions – and in a circular relationship those changes, in turn, affect agriculture. The relationship between agriculture and the ecological environment is indeed quite complicated.

Here the term 'ecological environment' denotes the entire ecosystem, which is essential for agriculture (including forestry and fisheries),

Figure 5.1: Köppen's classification of world climate

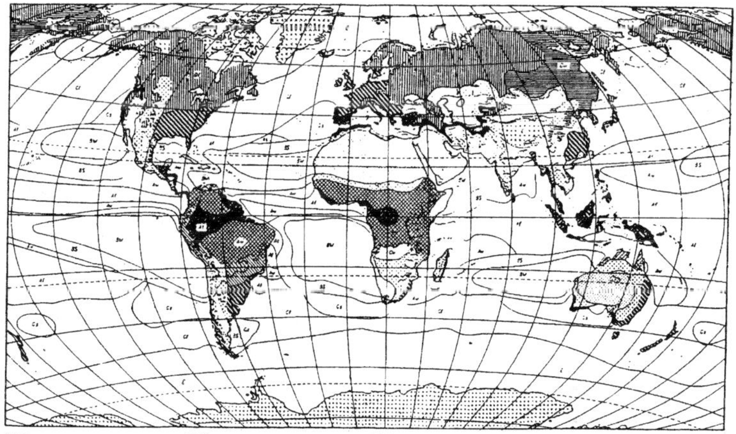

AF: tropical rainforest climate
AW: savannah climate
BS: desert climate
BW: steppe climate
CS: temperate rainy climate with dry summer
CF: temperate rainy climate
CW: temperate rainy climate with dry winter
DF: boreal rainy climate
DW: boreal rainy climate with dry winter
E: polar climate

Source: *Sinpan: Kishō handobukku* (New edition: Climate handbook), 1983: 369.

industry and all other activities that sustain human life. It should go without saying that humans and their activities are part of this ecosystem. However, rejecting any Cartesian dichotomy that treats subjective existence and the material/external world as distinct from one another, I shall consider the mutual relationship between the ecological environment on one hand, and agriculture, an activity that primarily treats the human as both its subject and end, on the other. The full ramifications of this relationship are more apparent if we consider it in a global light.

Figure 5.1 depicts the world's major climate zones. With differences in precipitation and temperature, landscapes vary from sparsely vegetated deserts to lush tropical rain forests, and from perpetually frozen continents to the permanently hot tropical areas. Topographically, these differences manifest in assorted patterns such as mountains, oceans, coasts, and inland areas. Agriculture is invariably affected by these climate zones.

This of course does not mean that within a particular climate zone, the weather conditions are unchanging. Subtle changes in environmental conditions in one zone can produce drastic changes in another part of the world. One of the better-understood instances of this is the El Niño effect, in which warming in the eastern tropical area of the Pacific Ocean alters trade winds and generates unusual climatic conditions all around the Pacific Rim. The El Niño effect has occurred every four to five years.

In general terms, there are twelve different soil patterns on Earth: tundra, podzol soil, brown soil, prairie soil, chernozem, chestnut soil, desert soil, reddish-brown soil, red-yellow podzolic soil, laterite soil, mountain soil and so on. It has been pointed out that approximately 70 percent of soil has some problems for agricultural production and only 11 percent of that can be improved to function as agricultural land (Uemura 1997: 78).

Integrating the 10 climate zones from Figure 5.1 with the 12 soil patterns, the Earth can be divided into eight different biomes with varied vegetation: tundra, northern conifer forest, temperate deciduous forest or rainforest, temperate prairie, desert and sparse woods, tropical rain forest and deciduous forest, tropical copse–grassland, savannah, and mountains.

Despite this great variety of biomes, human beings have adapted to most of the different climates and soil patterns and developed corresponding life-styles and agricultural forms.

Figure 5.2 displays the global distribution map of the patterns of agriculture and forestry. It was initially designed by Whitley and revised by Grigg and Hamroch. Although a variety of agricultural patterns exist in various regions, the figure demonstrates the most characteristic and predominant forms of agriculture, which reflect each region's climate and natural features. The figure exhibits ten types.

Figure 5.3 shows the northern or southern latitudinal limits of crops in the world, which reflect temperature differences around the globe. While agriculture and agricultural science make it a general principle to cultivate the right crop for the land, they have overcome natural limitations by the improvement of species as well as that of cultivation techniques, a development brought about by modern human civilization.

In hunting and gathering societies, humans simply gathered the fruits of nature in forests, rivers and the sea. With the advent of agriculture, they started burning forests and cultivating land and engaged in swidden agriculture, dry field farming and paddy field agriculture. Subsequently, the human population gradually expanded with the development of agricultural methods and techniques, which led to the accumulation of capital. This gave rise to industrialization

Figure 5.2: Distribution of different types of agriculture and forestry in the world

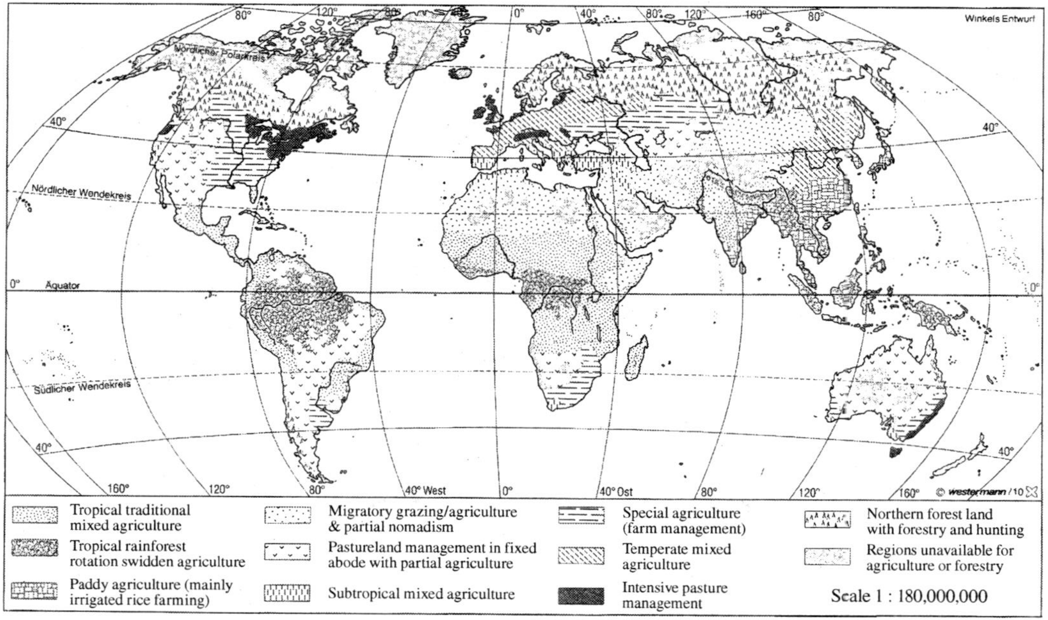

Source: Sick 1993: 161.

Figure 5.3: Latitudinal limits of distribution (northern and southern limits of cultivation) of main crops in the Northern Hemisphere

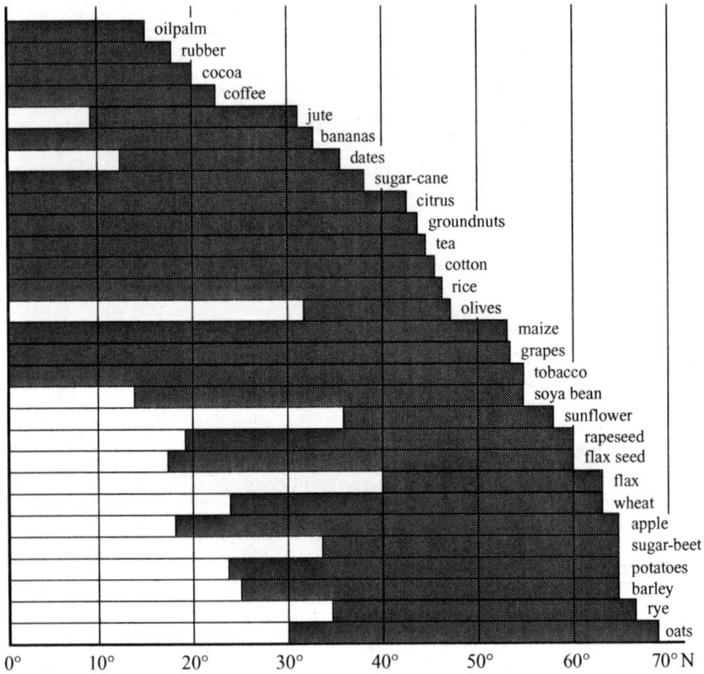

Source: Grigg 1995: 22.

and urbanization subsequent to the Industrial Revolution. The industrialization of agriculture dramatically expanded the productivity of agriculture and contributed to a population explosion, a process that was accompanied by large-scale exploitation of forests and environmental destruction. Some observers go so far as to claim that agriculture is the main culprit of environmental destruction. However, human beings must eat. If we follow the argument that we should not pin any hopes on the development of technology for the expansion of agricultural production, we would have to deny human existence itself or at least drastically reduce the human population. I would maintain that we should find future solutions of these issues by facing the realities of human beings today, which such arguments fail to address.

Food, agriculture and the 'population explosion'

Agriculture in the 21st century will undoubtedly impose further pressure on the ecological environment. Population growth and changes in eating habits will be the major contributors to this pressure. It is well anticipated that with such problems as environment deterioration drawing more and more of our concern, how to develop self-awareness and self-restraint is an ethical question that confronts all of us. Given that humankind is currently moving towards a satisfaction of spiritual desires, this fact means that we must shoulder a new, most difficult and noble task.

In 1999, the world population reached six billion. As Figure 5.4 indicates, the world population has grown dramatically since the Industrial Revolution. While it took 123 years for the global population to grow from one to two billion people, it took only 23 more years to reach three billion, a further 14 years to reach four billion and only 13 more years to arrive at five billion. At present, however, developed countries are experiencing declining birthrates. Nevertheless, the world's population is expected to reach nine billion in the middle of

Figure 5.4: Basic population curves (world and agricultural populations)

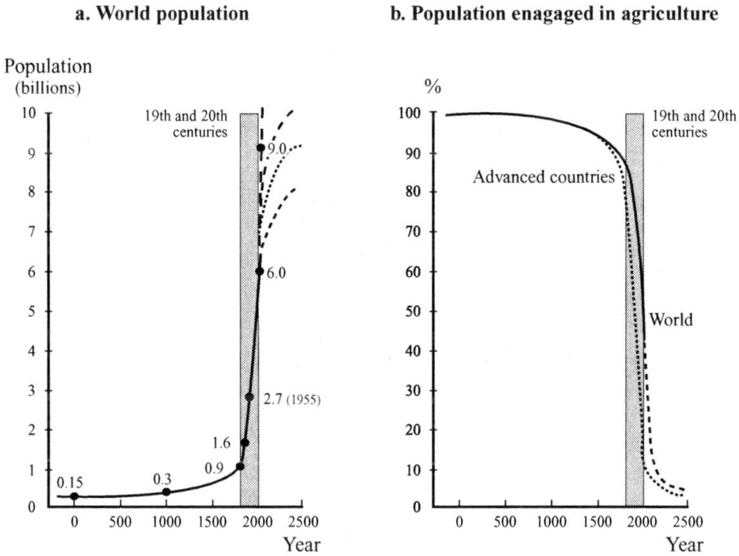

Source: Haushofer 1973: 8, with some modifications.

the 21st century. This accelerating growth trend came to be known as a 'population explosion'.

In contrast, the proportion of the population employed in agriculture has fallen dramatically over the same period. At the time of the Industrial Revolution, more than half of the world's population was engaged in agricultural production. Today, however, in the developed world, only between two and five percent of the population are thus employed. Agricultural productivity – concomitant with rapid industrialization, urbanization and the development of market economies – has clearly increased even faster than population.

In *An Essay on the Principle of Population* (1798), Malthus argued that while population grows geometrically, due to natural constraints, food production can only increase arithmetically. It follows that population growth would soon outstrip food supplies, leading to social disruptions such as poverty, immorality, war and starvation. For Malthus, then, the remedy was not to be found in charity or philanthropy, but in state policies to promote and protect agriculture and control birth. History, however, has not borne out these dire predictions; although there are certainly numerous destitute countries in the world today, this is clearly not because agricultural production is unable to keep up with population growth. In fact, it appears that the problems predicted by Malthus have been 'successfully' overcome by modern agriculture (Soda 1998: chapter I-1).

After this population explosion, consumerism was born – an explosion in material desire to purchase and possess assorted industrial products. Once their needs for food had been met, people began to covet the copious quantities of commodities being produced by the continually expanding non-agricultural population. The maxim that 'the more one possesses the more one desires' prompted the production of new industrial products, leading to the emergence of a highly industrialized society. The world appears to have risen above the problem of food shortages and transformed into a commodity-hungry creature, which must continually try to mollify its insatiable cravings for more-and-more things through mass production and mass consumption. In the process, we appear to have become mired in a bottomless pit of avarice, gradually falling prey to our own cupidity.

In mass-consumption society, food is no longer merely a life-sustaining necessity, having been transformed into yet one more commodity for unlimited consumption. These changing consumption patterns have led to at least two major changes in agricultural production: (1) a shift from grain consumption to meat consumption;

and (2) a shift from mass production and mass consumption to mass waste.

Figure 5.5 charts the world's aggregate meat production from 1950–1993. As is clearly shown, the consumption of pork, beef and chicken has grown rapidly over this period. According to FAO statistics, as of 1990, animal products account for approximately 30 percent of food consumed in the developed countries and 10 percent in developing countries. This implies a positive correlation between the level of industrialization and average income on one hand, and meat consumption on the other. However, the production of 1 kg of pork requires 6 kg of corn. Beef and chicken require 8 kg and 2 kg respectively. In order to produce 100 kcal of pork, 1050 kcal (or 10 times as much) of corn is needed, while the figures for beef and chicken are 1400 (14 times) and 420 (4.2 times). For grazing (i.e., grass-fed) cattle, 1 kg of beef requires 30 kg of grass.[1] Either way, vast amounts of farmland are required to satisfy these modern meat consumption habits.

Furthermore, even in a country such as Japan, which is heavily dependent on imports to meet its demand for foodstuffs, it has been estimated that approximately 28 percent of foods (measured in calories) is thrown out as garbage (Tamura 1997: 51). This pattern of mass consumption/mass waste can largely be attributed to four factors: (1) a growing tendency to dine out rather than eat at home; (2) the shift

Figure 5.5: World meat production by type (1950–1993)

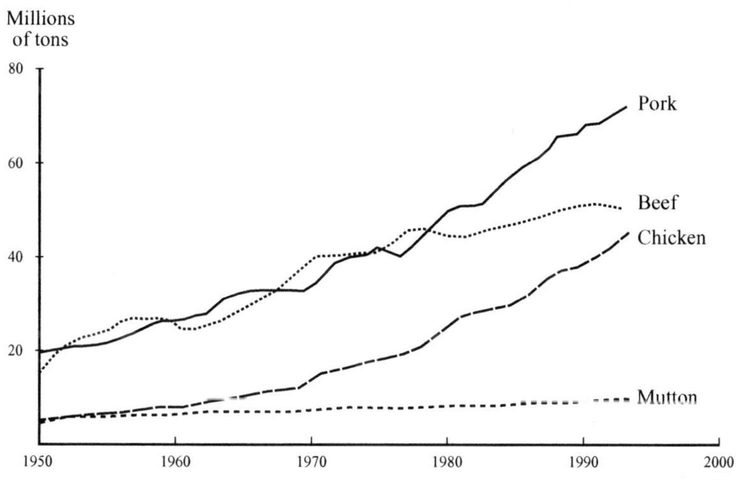

Source: Brown 1995: 53 (FAO data).

from food self-sufficiency to import dependency; (3) the loss of a sense of ethics in eating habits; (4) the absence of agricultural and food education in school. Added to this, modern consumers demand tasty food, rather than nutritious or sustaining food, which generates energy loss. And the cultivation, refrigeration and preservation technologies that enable us to grow and consume fresh fruit and vegetables out of season, consume large quantities of fossil fuels (which of course produce large quantities of 'greenhouse' gases).

In short, the emergence of highly industrialized society has not only entailed expanding covetousness for material goods but also a transformation of dietary habits from grain to meat consumption and new demands for more palatable, high-class agricultural products. It could be said that while hunger for food is a problem that puts our basic survival at stake, hunger for commodities is one that concerns our daily life level.

Figure 5.6 charts the relationship between the increasing population and crop production. Although the last three decades of the twentieth

Figure 5.6: World population and crop production (augmentation index)

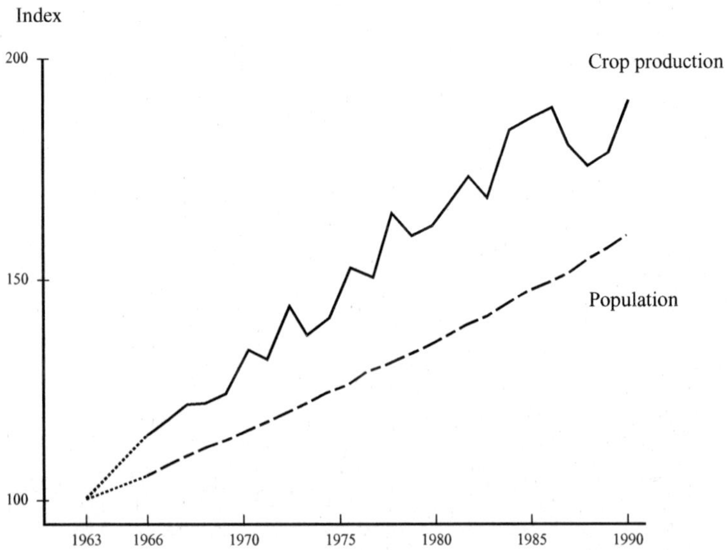

Source: Egaitsu 1994: 9. The population data is based on *Production Yearbook* by FAO, and the crop production data is based on *World Grain Situation and Outlook* by the U.S. Department of Agriculture. (Data for 1961–1965 = 100)

century were marked by unprecedented growth in the human population, the growth rate in the production of food crops nevertheless outpaced population growth. And this is despite the fact that in many Western countries production was artificially dampened in order to protect the financial interests of the producers.

Owing partly to continued expansion of the area of land under cultivation around the world, this amazing achievement is primarily the fruit of efforts by farmers and agricultural researchers that greatly enhanced productivity. Since the time of A.D. Thaer, the founder of agricultural science, the field has enjoyed revolutionary developments such as the invention and subsequent use of fertilizers and the adoption of improved management techniques. Mendel's discovery of the laws of heredity facilitated the invention of multi-yield breeds. The development of disease- and pest-resistant breeds has made it possible to expand the acreage under cultivation (by overcoming crops' northern and southern cultivation limits). The introduction of agricultural machinery and facilities, the expansion of irrigation systems and improvements in water management techniques, and the discovery of agrichemicals are but a few examples of the numerous developments that have enhanced agricultural productivity. Other developments of significance in this regard include: increasing specialization, the movement to monoculture and continuous cropping, chemicalization, mechanization and facilitation (for example the use of greenhouses). The sum total of these developments is referred to as the modernization of agriculture, through which the mass production of agricultural goods has become possible. These advances also contributed to mitigating the impact of meteorological changes on agriculture, thus stabilizing annual food supplies. Mechanization and facilitation have made it possible for a small agricultural workforce to produce enormous amounts of food.

Figures 5.7 (production of agrichemicals) and 5.8 (use of machinery in agriculture), indicate that since the 1960s, Japan has experienced a very rapid 'industrialization of agriculture'. In the final 40 years of the 20th century, the average crop harvest (harvest per unit of land) increased 2.4 times. When agrichemicals were first discovered, they were hailed as a miracle that would release farmers from onerous labor and deliver humankind from hunger.

The unintended consequences of this miracle, however, are that agriculture has contributed in many ways to the environmental problems that confront us today. In the following section, I examine the connection between agriculture and some of these major environmental issues.

Figure 5.7: Changes in the amount of agrochemical production (Japan, 1934–1987)

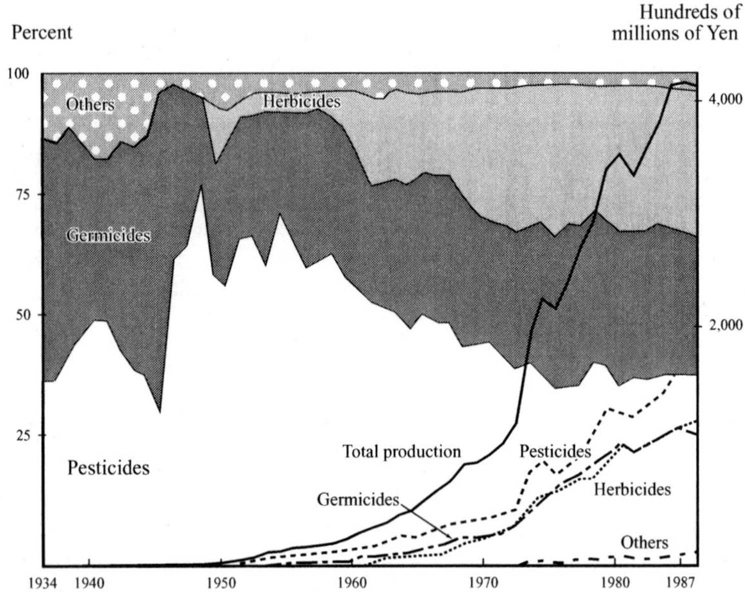

Source: *Me de miru Shōwa Nōgyōshi* (The visual history of agriculture in Shōwa period), 1991: 109. (Data for 1934–1949, Plant Protection Division, Ministry of Agriculture, Forestry and Fisheries (ed.), *Nōyaku no ayumi* (Development of agrochemicals); data for 1950–1969, Shokubutsu Bōeki Jigyō 30–shūnen Kinen-kai (ed.), (Plant Protection Service 30th Anniversary Memorial Society), *Shokubutsu Bōeki Jigyō 30–shūnen Kinen-shi* (Plant Protection Service 30th Anniversary Memorial Book); and data for 1970 onward, Plant Protection Division, Ministry of Agriculture, Forestry and Fisheries (ed.), *Nōyaku yōran* (A general survey of agrochemicals).)

Environmental deterioration and agriculture

The environmental problems that we face have numerous dimensions. Agriculture, while being a major contributor to these problems, is also heavily burdened by the accompanying changes to the environment.

Global warming

Figure 5.9 graphs changes in the average winter temperatures in the central mountains of Japan for the past 800 years, calculated from the

Figure 5.8: Changes in popularity (the number of machines owned by farms) of main agricultural machines

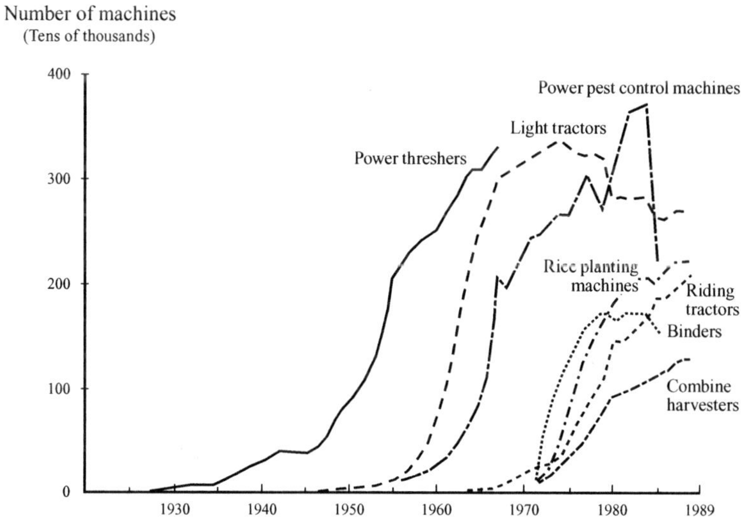

Note: No statistics are available for power threshers from 1965 onward.

Source: Teruoka 1996: 246.

Figure 5.9: Average Winter temperatures in central Japan's mountainous region (800 years)

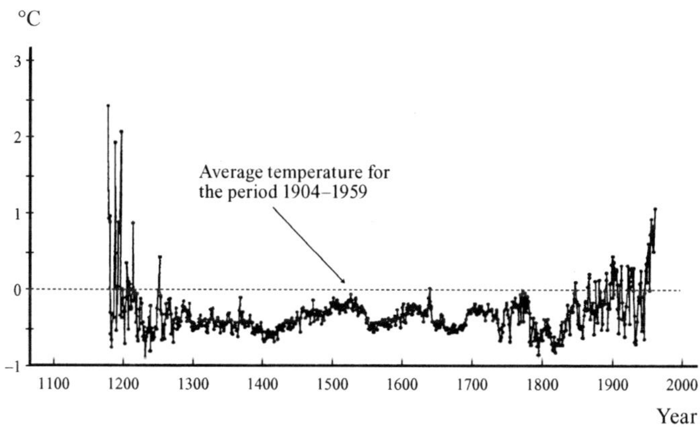

Note: Temperatures calculated from the growth rings of Japanese cedar trees.

Source: Kyūma and Soda (eds), 1995: 152.

annual growth rings of cypress trees in the Kiso Valley. The results indicate that since the Industrial Revolution of the 18th century, the average temperature has risen approximately 1.5 degrees Celsius. Most scientists have reached a consensus that this temperature rise is not limited to these mountains but is occurring globally.

The primary factors that have contributed to global warming are believed to be the emission of carbon dioxide and other waste gases containing methane, nitrous oxide and chlorofluorocarbons. The effects of global warming include an accelerating loss of coastlines, in part from damage by tidal waves, changes in underground water contaminated by inflowing seawater, coral reefs dying due to higher water temperatures, deteriorating living conditions on islands leading to increased emigration, changing sea currents and exacerbated living conditions in populated areas (Yonemoto 1994: 116). An indirect effect is the increased use of non-renewable energy for air-conditioning to create comfortable working and living environments in response to rising temperatures – which has the circular effect of pumping even more greenhouse gases into the atmosphere, and thereby exacerbating global warming.

While it is still not certain what will become of agriculture as a result, the effect is likely to be twofold. While it will have a negative impact on the primary areas of current food production, global warming may make the now frigid lands of high latitudes potential sources for agricultural production (Kyūma 1997: 25). Increased temperatures are expected to melt glaciers and raise sea levels, flooding coastal farmlands and causing salt damage. Whatever the particulars, the entire ecosystem is likely to undergo great changes.

Moreover, as the developing countries progress with their efforts to industrialize, emissions of greenhouse gases will increase rapidly. While the average temperature increase is expected to be between 1.9 and 5.2 degrees Celsius, in the higher latitudes, especially in the northern hemisphere, increases of up to 16 degrees Celsius are predicted (Kyūma and Soda 1997: 154–5). If we cannot drastically reduce our greenhouse gas emissions, the Earth, its atmosphere, and its life-forms will face an acute crisis.

Deforestation

As of 1995, forests occupied about 27 percent of the Earth's total landmass (3.45 billion hectares out of 13 billion hectares) (Takeuchi and Tanaka 1998: 30). The total forest area was reduced by 0.32 percent between 1990 and 1995. If this rate of reduction is allowed to continue,

Table 5.1: Forested area (percentage of total land) for ten countries

Nations	Percentage
Britain	9
France	24
Germany	30
Switzerland	28
Italy	22
Europe (subtotal)	27
Soviet Union	35
USA	23
Japan	66
China	11
Australia	5
Industrialized countries (total)	27

Adapted from Maruzen 1996: 703.

the total forest area will decrease by one third in the next 100 years. Rain forests, which make up approximately half (1.7 billion hectares) of the entire forested area, have been declining rapidly. According to the FAO, if the decline continues at the current rate, there will be none left in 110 years. Table 5.1 tabulates the forested area (percentage of total land) for ten countries. Japan appears to be quite exceptional in this regard, with forests covering about 66 percent of the country. In contrast, for most Western countries the proportion is at best 20 to 30 percent. In most of those countries, the natural forests were destroyed to procure arable land for agriculture. This massive deforestation goes some way to explaining the US's, Australia's and Canada's 'advantages' in large-scale, low cost agriculture.

In China, home to one-fifth of the world's population, 'since the communists took over in 1949, forest cover has declined to less than 11 percent, despite massive attempts at reforestation, and it will fall to 8.3 percent by 2000' (Ehrlich and Ehrlich 1990: 116). The total population of China is expected to reach 1.7 billion in the near future, and as it continues to embrace consumerism its total meat consumption is expected to move closer to the levels currently seen in Western countries. As these two factors combine, China will most likely be confronted with severe food shortages.

This rate of deforestation is also occurring in other developing countries. Between 1980 and 1990, the total area of tropical forests declined from 1.91 billion hectares to 1.756 billion hectares at an annual rate of 0.8 percent or 15.4 million hectares (three times the total cultivated acreage in Japan). The causes include: those countries' efforts to secure arable land to meet the needs of a booming population; timber harvesting to meet both domestic demand for wood and charcoal, and export demand for lumber and agricultural products to developed countries such as Japan; the shortened rotation cycle for slash-and-burn (swidden) farming; over-cultivation and overgrazing. Deforestation thus contributes to the deterioration of the natural environment.

Furthermore, the fossil fuels consumed in the process of deforestation and the industries that transform the raw materials of the forest into consumer products emit more carbon dioxide, nitrogen oxides, sulfur oxides and hydrocarbons. The consequent atmospheric pollution is the primary cause of global warming and acid rain, bringing pernicious effects to man and plants. The world's first reported case of the acidification of a lake from acid rain and the subsequent extinction of living organisms was reported in Sweden. Since 1982, the UN European Economic Committee has kept annual records of the damage caused by acid rain. In 1988, acid rain appeared to have been responsible for the withering of 35 percent of the total European forest.

Desertification, salinization and land degradation

While desertification literally denotes the expansion of desert, it is now usually defined as 'land degradation in arid, semi-arid and dry sub-humid areas resulting mainly from the impact of unsuitable human activities' (GLASOD, Global Assessment of Human-induced Soil Degradation, 1990; Takeuchi and Tanaka 1998: 62). The major causes of desertification are deforestation for fuel (firewood and charcoal) for everyday living and iron manufacturing, overgrazing, and shorter fallow periods in swidden agriculture. Overgrazing is usually attributable to efforts to unduly increasing the number of head of livestock in a given grassland area, either in pursuit of higher cash incomes in order to purchase the commodities necessary to adapt to civilized society, or in response to a surge in population.

The Aral Sea in Kazakhstan is a well known site of salinization. During the 1950s, by exploiting the Syr-Darya River, farmland was expanded, and irrigation introduced to grow paddy rice. However, some of the newly developed farmland did not manage to get enough water and gradually the soil became salinized. Eventually, the farmlands had to be abandoned. At the same time, extracting water from the Syr-Darya River for irrigation reduced the amount of water flowing the course of the river, with the downstream effect of lowering the level of the Aral Sea, thus causing a receding shoreline. At the same time, the proportion of salt in the water soared and the Sea became a fishless pond, depriving the locals of both fishing income and dietary protein.

Similar processes of salinization in river basins and underground waterways have been seen all over the world. The ill-fated irrigation project in Pakistan's Punjab region and the inundation in the Nile Basin following the construction of the Aswan Dam are just two examples.

Soil has been washed away as a result of land becoming bare and exposed to rain and wind in the areas of large-scale agriculture and monoculture. The situation is serious and cannot be ignored in the grain fields of the United States. A similar condition exists in the developing countries of Africa and Latin America in which the commercialization of agriculture has shortened the fallow period, fractionated land and expanded production into sloping areas (Kyūma and Soda 1997: 48–9).

Lives of human beings and other living beings in crisis

The development projects and productive activities of human beings have decreased the number of wild life species and driven many species to the verge of extinction by the destruction of natural habitats and excessive hunting. The International Union for Conservation of Nature and Natural Resources conducted a survey of 240,000 species in 1997 and found that one of eight species was threatened with extinction (Brown 1999: 97–8). Agriculture has undoubtedly contributed to the demise of many living species. Agricultural chemicals and chemical fertilizers are extensively used, and rivers and waterways are paved with concrete, with the consequence that many species, including some dragonflies, killifish, mud snails and fireflies, have become extinct. This has reduced the diversity of living beings, led to the destruction of the ecosystem and brought about a decline in hereditary capital.

We should also be concerned about the safety of products of agriculture, forestry and fisheries – the foods that humans eat. Human beings are located at the top of the hierarchy of the food chain of plants and animals and adversely affected by the extent to which agricultural chemicals are used, harmful substances derived from factory wastes are accumulated in the bodies of plants and animals, and post-harvest crop-spraying for long-term maintenance is conducted. Moreover, food additives – either natural or chemical – are extensively used to manufacture, process and maintain foodstuff. There are 347 kinds of chemical agents in use, including preservatives, disinfectants, antioxidants, adhesives, emulsifiers, food manufacturing agents, inflating agents, reinforcing agents, seasonings, sweeteners, acids, flavoring agents, coloring agents and bleaching agents. There are also 724 natural agents in use (Kurosaka 1990). These agents are approved for use once they satisfy the set safety criteria. However, no firm confirmation has been made as to the complex interactions between these agents. We cannot be certain that they have no effects on human bodies (Hirasawa 1996).

In the process of raising cattle – especially in high density cattle farming in large-scale farms – antibiotics that protect animals from disease and growth promoters are mixed with feed. This weakens human capacities to resist disease as well as the effects of antibiotics and affects human reproductive functions, an issue that is a source of friction between trading countries. More recently, gene manipulation has contributed to improving crop production on one hand, while concerns have been expressed over safety on the other.

Thus, we should carefully study the issues of agricultural chemicals in the production and distribution processes and additives and chemical agents in food processing.[2]

The positive roles of agriculture

Preserving the ecological environment

So far, I have discussed only the negative impact of human productive and lifestyle activities on the natural environment, beginning with agricultural and industrial production. As discussed in Chapter 2, agriculture's roles in preserving the environment, that is, its positive impact, also deserves attention.

Humans must eat to survive. Therefore, whatever the pros and cons, they must exploit the environment for their advantage. To this end, we have invented tools and machinery to convert nature, building artificial constructions such as paddy fields, canals, roads and buildings. We have managed to conquer the world of plants and animals by growing crops and domesticating livestock. Forests have become objects of forestry and the oceans sources of fish and other resources. In a sense, there is no longer any natural environment, but everywhere a synthetic or 'man-made' construction, irreversibly modified through human intervention. Even where industry primarily utilizes and processes inorganic materials, its by-products (called industrial waste) have polluted waterways, soil and atmosphere, inflicting damage on flora, fauna and humankind. The challenge currently facing industry is how to minimize (down to the sustainable level) the negative impacts of its by-products.

The problem facing agriculture is quite different, however. Since it is directly involved in the production of living organisms, its processes are inevitably affected by problems with water, soil and the atmosphere. Hence, while it certainly may have a negative impact on the environment, it can at the same time help to, for example, store and purify water and minimize floods. Moreover, it can also serve to preserve soil quality, guard the natural environment and clean the atmosphere. In sum, in order to ensure sustainability, the negative impacts of agriculture must always be kept to a minimum within the production process (Yaguchi 1995: 19). The fruit such efforts bear would not only benefit the environment, but would also bring positive externalities (external economies) from social and cultural perspectives, to the human environment. To conclude, for industries in which most of the production processes are confined to the limited spaces of factories, effort need only aim toward minimizing negative effects (external diseconomies), while for agriculture, we must endeavor both to minimize the negative externalities and to maximize the positive externalities.

What follows is a summary of some of the externalities of the natural environment that are a cause for immediate concern (Soda 1989; Sugaware 1993; Honda 1982; Sekiya 1992; International Farm Management Association 1980).

Storing and purifying water, and preserving the land

Forests and rice paddies serve to store and clean water. Although trees themselves serve a water-storage function, the thick, aerated soil also serves this purpose. Forests thus function as effective buffering mechanisms to store water from precipitation (rainwater and snow-melt) underground, dispersing the peak of the flow at the lower level. In effect, they are water dams that help to alleviate the pressure of droughts and floods and prevent sediment runoff, thus preserving the soil. At the same time, the soil serves as a filter that helps to purify water.

In the case of rice paddies, water is stored as in the forests, but also as in ponds or dams. Of course, when these fields are abandoned, they cease to serve this function. In contrast to wheat fields, the hallmark of Western agriculture, which are constantly beset by the risk of top-soil loss from droughts and floods, rice paddies are a more sustainable mode of production which helps to stabilize water cycles.

Purifying the atmosphere

In the processes of photosynthesis, green plants absorb carbon dioxide and release oxygen into the atmosphere. By growing green plants, agriculture functions to purify the atmosphere. The oxygen is in turn inhaled by human beings and other living organisms, or consumed by industrial production and the combustion of fossil fuels. But to satisfy the oxygen consumption requirements of Tokyo's 23 districts, a forested area 20 times the size of the metropolitan area is necessary (Fukuoka 1987; Sugawara 1993: 129). According to some estimates, between 1978 and 2000, deforestation reduced the oxygen producing capacity of the Earth by 34.5 percent.

Contributions to national health and recreation

From a socio-cultural perspective, agricultural and mountain villages and the pastoral landscapes they create deliver external economies that promote national health, recreation and education. I therefore define agriculture and forestry as multi-value industries, and agricultural villages as multifunctional spaces. In order to develop a sustainable agriculture that imposes a zero burden on the environment, the multiple roles, functions and values of agriculture and villages have to be fully recognized.

On food self-sufficiency: *agri-minimum*

This section briefly summarizes the extent to which Japan is self-sufficient in agricultural products, timbers and marine goods in the late 20th century.

Figure 5.10 graphs the calorie/food self-sufficiency rates of selected major countries. During its period of rapid economic development, Japan's self-sufficiency rate has dropped to slightly over 40 percent. In 1992, the poor harvest reduced it to a mere 30 percent. When we consider the self-sufficiency rates of individual items – except rice, which is entirely supplied domestically – the numbers are shocking: wheat 7 percent, vegetables 85 percent, dairy products 72 percent, meat 57 percent. For soybeans, the main ingredient of Japanese culinary staples such as miso paste, soy sauce and tofu, only 2 percent is supplied by domestic agriculture. This is in sharp contrast to the European countries, where once low self-sufficiency rates have been significantly improved through the recognition of agriculture's multiple values and the subsequent adoption of policies to subsidize domestic agriculture. Those countries are now characterized by high self-sufficiency rates.

Shifting to lumber: in 1995, 20.5 percent of Japan's lumber was supplied by the domestic forest industry; as argued above, this large reliance on imported timber is a significant factor contributing to the rapid deforestation of developing countries.

Furthermore, Japan also has high demand for marine products. In 1995, the production of seafood in Japan dropped from a peak of 13 million tons to 7.5 million tons. The self-sufficiency rate for edible and non-edible products are 57 percent and 34 percent respectively. With the two combined, the self-sufficiency rate of seafood is 49 percent. These rates are very low by world standards.

As we can see, as its economy boomed, Japan's self-sufficiency in agricultural products has declined remarkably. Japan has historically

Figure 5.10: Changes in food self-sufficiency rates in various countries

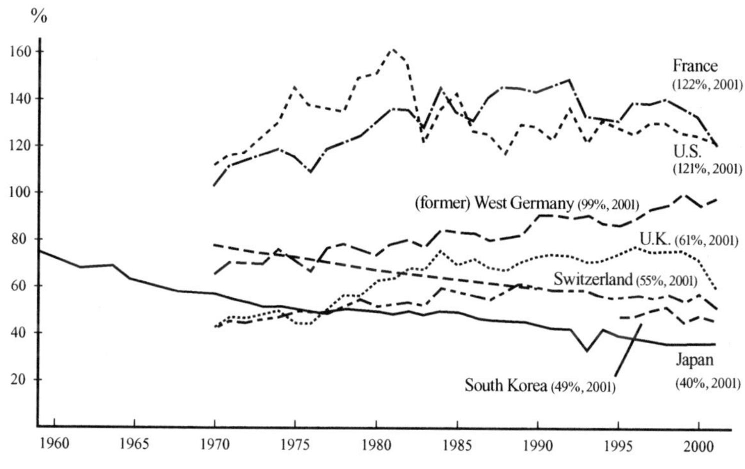

Source: Ministry of Agriculture, Forestry and Fisheries, *Shokuryō jukyū hyō* (Food supply and demand tables). Self-sufficiency rates on a supplied caloric basis.

been, and today remains, highly dependent on foreign imports for foodstuffs and other natural resources.

As a result, Japan is deeply implicated in many of the world's problems, such as the expansion of cultivated areas, deforestation, diminishing sea resources, and declining energy supplies, all of which contribute to pollution of the atmosphere, water and the loss of soil. In order to support a world population of 9–10 billion in 2050, a minimum area of 3–4 billion hectares of cultivated land will be required, which is 2.7 times the 1.5 billion hectares currently under cultivation. The expansion of meat consumption necessitates vast land for livestock feed production. This expansion will have to increasingly use marginal lands (the producing capacity of which is about 70 percent of that of normal arable lands). Even if this could provide a further 1.1 billion hectares of arable land, it remains far short of the required 3–4 billion (Ministry of Environment 1996).

At the same time, overgrazing and other forms of land overuse, combined with salinization and other forms of erosion, is turning approximately 6 million hectares of arable land into desert each year. This includes 3.2 million hectares of pastoral fields, 2.5 million of rain-irrigated fields, and 0.125 million of irrigated fields and so on. Furthermore, a further 3.5 billion hectares, mainly pastoral fields, are expected to become desert over time. In this era of global warming, it is unclear whether we will continue to be able to secure sufficient food supplies to sustain the growing human population in the near future. As previously suggested, if we continue on the current path of trying to force our way out of this predicament, we will only accelerate the deterioration of the environment. Even if we manage to secure enough food, a death sentence will have been handed down. If this situation continues to prevail, food security and ecological conservation will have to be traded-off.

As we have seen, agriculture can play an important role in preserving and sustaining the natural environment. It has expanded to meet the needs of six billion people and will have to continue to expand if it is to meet the needs of nine or ten billion people in the years to come. However, as argued, to do so we must minimize the negative impact on the environment and maximize agriculture's capacity to preserve ecological and life values. From this perspective, in this era of globalization, it is necessary for countries to consider not only the economic benefits or the comparative prices of agricultural products, but how to fully and sustainably utilize natural resources – which necessarily means refraining from consuming products that come from deforestation, over-farming, over-grazing or any of the other ways that we over-exploit our natural environment.

Japan's 1998 food self-sufficiency rate is abnormally low at 40 percent. This is not only a cause for concern about the food security of the Japanese people; more importantly, it indicates that Japan is responsible for extensive environmental destruction in other countries. Having given up the development of its own national resources and multiple functions of agriculture and forestry, it has significantly contributed to many major environmental problems that put the survival of the Japanese and all of the world's people at stake.

It is for this reason that I have long insisted that all countries should set minimum limits above which we must secure agriculture, and an upper limit below which we must limit industrial development. In short, we must set 'agricultural-minimums and industrial-maximums' (informally, *agri-minimum* and *indus-maximum*) beyond which we should not venture. The principle of agri-minimum is based on three components: (1) that all nations should mutually accept policies that enable them to sustain food self-sufficiency of at least fifty percent, while keeping free trade as the basic rule; (2) that, in WTO agricultural negotiations, each nation should decide its quantities of imports and exports of agricultural commodities in total terms rather than debating on particular item-by-item self-sufficiency rates, since agriculture is governed by placeness and regionality; and (3) that we should consider the proper regional arrangements of agriculture on the global scale. Only by following this precept might we finally achieve balanced development and a harmonious realization of the comprehensive values (economic, ecological and life values) that form the core argument of this book (Soda 1989).

Agriculture and the ecological environment

As mentioned, there are a wide variety of diverse and differing opinions regarding the relationship between the ecological environment and agriculture. No one would now deny the existence of environmental problems. What is in dispute is how to cope with those problems. I discern three dissimilar points of view. The first sees no intrinsic conflict between the current path of continuing economic development and the environment. As long as economic growth continues, those who hold this view maintain that they will find ways to tackle those environmental problems. This perspective might be called growthism, or developmentalism. The second position is that animals, plants and humans are all parts of nature and have equal value. Proponents of this view argue that human beings do not have any intrinsic priorities over other life-forms. This line of argument, known as 'deep ecology', often advocates animal and plant rights and condemns scientific technology in general, believing that human beings should refrain from doing any further harm to the environment. The third approach adopts a more eclectic approach, maintaining that while it is important to try our best to protect the environment, it is not possible to deny the developing countries opportunities to pursue industrialization in their efforts to resolve social problems and satisfy the growing demands of their burgeoning populations. From this perspective, we should seek better scientific technologies to preserve natural resources and resolve our environmental problems.

The first viewpoint is best illustrated by Figure 5.11, 'The environmental Kuznets curve', which suggests that once the GDP per capita reaches a certain level, pressure on the environment will be greatly relieved. To pass that specific threshold, economic growth is necessary. It is therefore important to promote particular driving forces for economic growth, such as the expansion of international trade through further trade liberalization. Proponents of this view, such as Meadows and Randers, who first introduced environmental problems to economic theory (1972, 1992), argue that the current era of economic growth has not yet reached its limits.

Figure 5.11: The environmental Kuznets curve

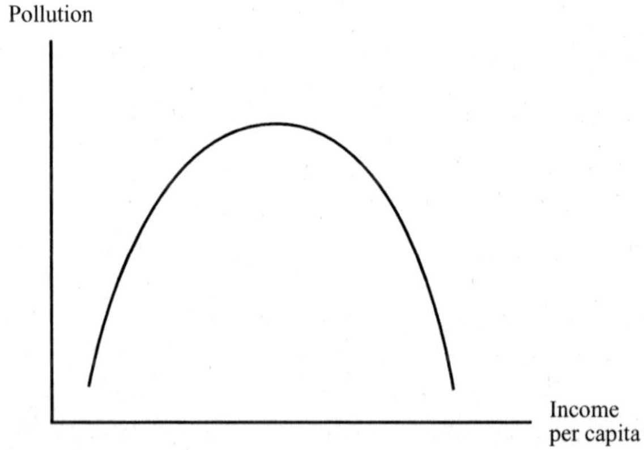

Source: World Bank 1992: 81.

They are not ignoring the current environmental problems. What they advocate is seeking resolutions by pursuing further economic growth. If there is still a reserve capacity for development and growth, they ask, why should we not sustain our current rate of economic growth? (Yonemoto 1994: 5–6) This is a revised version of the original developmentalist view.

The second perspective, deep ecology, positions itself as the polar opposite to developmentalism, seeing the biosphere as the only subject that has real rights. This line of argument is grounded, in simple terms, in the idea that, to quote Luc Ferry, 'the natural world is worthy of respect in and of itself independent of all human considerations' (1995: 98). The existence of nature itself is what matters most. The proponents of this argument excoriate Cartesian philosophy, regarding it as the source of human-centered modern science and economics, both utilitarian and developmentalist. Cartesian philosophy separates man from nature, identifying the former as superior. By deeming animals as simple mechanisms, it is used to justify the human conquest of the ecological environment. Utilitarianism and developmentalism aim at maximizing the exploitation of nature to satiate human wants (Ferry 1995: chapter 4). Deep ecology rejects these views, maintaining that animal rights and plant rights are as important as human rights. It is therefore highly critical of science and development. They are equally as concerned about the rights of those 'harmful animals' that impose a burden on agriculture as they address the question 'to what extent does agriculture inflict harm on the environment?' If taken too far, though, defending the 'rights of the biosphere' and the 'rights of plants and animals' above human well-being poses a threat to human survival itself. At the extreme, some have suggested that it might be best for planet Earth if the human species was extinct. Only slightly less extreme, some have calculated that the Earth might sustain a population of about 100 million (Arne Naess) or 500 million (James Lovelock cited in Asahi Shinbunsha 1994: 21, 34) at current living standards; but this remains a scenario of near-extinction (Ferry 1995: 75).

From a position between these two perspectives, we might describe their disagreement as the conflict between anthropocentrism and environment-centrism. On the one hand, with the human population already in excess of six billion and expected to surpass nine or ten billion in the future, the extreme environment-centrist path appears to be unrealistic. On the other hand, the limits to anthropocentrism are painfully obvious. We appear to have no alternative than to pursue a middle-path between anthropocentrism and environment-centrism, from which we can hasten our efforts both scientifically and ethically. Agricultural science can not accept any premise that denounces humankind or proposes its near-extinction.

In reality, six billion people means six billion mouths to feed and six billion pairs of hands to work out ways of survival. The pressing problem is how to protect the environment without starving. To this end, we need to develop new views and ethics of nature, including prohibiting mass waste and promoting recycling, as well as developing the required technologies.

Developing new views and ethics of nature

Coexistence and circulation

Concerning the solution to the environmental problems of our time, a few opinions have been suggested so far. Among them, the idea of 'coexistence and circulation' is one that is most widely accepted and adopted. This is due to academic achievements in the fields of biology and ecology.

The most widely accepted theory from the academic fields of biology and ecology for understanding our planet in a way that might assist in resolving the environmental problems of our time is the idea of 'coexistence and circulation'. The Earth's ecology is not only a 'biosphere' inhabited by humans, animals, plants and microorganisms etc, but also a 'synbiosphere' in which both competition and interdependence are seen among the species, for each of whom their very existence is dependent upon others (Matsumoto 1993: 2). This idea is central to Tansley's concept of an 'ecosystem' (1935), where living organisms and the inorganic environment in which they live are understood as a unified whole. 'In this ecosystem, diverse activities of the living organisms are observed along with the circulation of sundry materials' (Matsumoto 1993: 26). According to Thienemann's taxonomy of the circulation process, living organisms can be classified by their functions. The process follows a chain that starts from the photosynthetic green plants (the producers), to the animals (the consumers) that feed on green plants and finally the microorganisms or bacteria that decompose and consume dead bodies and animal excrement (the decomposers or deoxidizers).

The major elements in circulation are carbon, oxygen, hydrogen and nitrogen. These four elements make up 99 percent of trees and 96 percent of animals, especially mammals. The perpetuation of the ecosystem depends on the diverse living organisms within it, in a word, its biodiversity. The stability of the ecosystem is therefore dependent upon the dynamism of the material circulation within the biosphere. The population explosion and uncontrolled economic growth have disrupted the ecosystem, leading to a loss of biodiversity, the sustained force for all life on Earth. Consequently, we have thrown ourselves into a predicament in which our very survival is at stake.

Hence we must once again consider our proper position in the circle of life within this sphere of coexistence.

Limits of the Earth

The Earth is limited in two aspects: it has finite natural resources; and a limited cleaning (self-repairing) capacity. Boulding's 'Spaceship Earth', first raised the alarm about natural limits and possible exhaustion of resources such as oil and minerals. Initially these were considered merely as problems that might hinder further economic growth; their associated environmental problems were not then recognized. It was in this context that the issue of generation was raised – the exhaustion of limited resources concerns not only the current generation but also those yet to come. This is what Alier referred to with the idea of 'participation from future generations'.

Sustainable development

This understanding of the natural environment and the Earth's limitations has given rise to several lines of thought, including:

1. Living economics: attempts to introduce patterns of coexistence and the circulation system as regulatory models into industrial production systems (Elkins 1986; Polanyi/Tamanoi 1978).
2. The 'Gaia hypothesis': the Earth functions like a single living self-sufficient organism (the ecosystem). If human beings break the self-adjusting mechanism of Earth, Gaia will in turn erase them from the planet (Lovelock 1984).
3. The 'steady-state economy': the stock of population and resources should be maintained at a steady state with low cost (the minimum cost necessary to maintain the specified level). However, this 'steady-state' is subject to changes in the level of technology and human values (morality). Because of these changes, occasional shifts in the form of the steady state are inevitable (Daly 1977).

4. It is neither the most-advanced most-expensive technology, nor the most-simple least-expensive craft, but moderately-priced intermediate-level technology that serves humankind best, stimulating interest in productive work and life while at the same time minimizing unemployment. This idea of the intermediate most appropriate technology is expressed in the conviction that 'small is beautiful' (Schumacher 1974), which has found fervent supporters, especially among the developing countries.

Since the 1970s these ideas have converged to form the concept of sustainable development. Sustainable development denotes the form of society in which economic growth is pursued according to the principle of coexistence and circulation with due consideration for limits to the Earth's resources and self-correcting capacity. In such a society, environmental impact is minimized and biodiversity is maintained. This model was first outlined during the World Commission on Environment and Development (WCED) in 1987.

The notion of a balanced approach to environment and development might seem simple, but it in fact covers a wide range of issues. Take agriculture for instance. The population explosion and the increase in meat consumption have made the production of more crops not only required but exigent. In this case, it is impossible to advocate zero growth in agriculture, making it necessary to consider sustainable development. However, sustainable development cannot be achieved simply by scientific knowledge and technology alone. Modern industrialization and urbanization has wrought terrible damage to the natural environment, largely through the drive to satisfy insatiable desires. As society has become more and more avaricious, it has created an everyday lifestyle characterized by mass consumption and mass waste. It is fair to say that the people of highly industrialized society are caught in a huge economic spider web.

Throughout most of human history, self-restraint in the relationship between humans and the environment has usually been considered a virtue. More specific questions about this relationship and its implication for our lifestyles are discussed in Chapter 6, but here it is worth recalling some significant modern contributions to our understanding of this need for self-restraint: Charles Wagner and his 'La Vie Simple' (simple life); Henry David Thoreau, who, mesmerized by nature, idealized it in his *Walden, or Life in the Woods*; William Leiss, who advocated that a more fulfilling life must be sought beyond the sheer satiation of commodity and economic benefit in *The Limits to Satisfaction*; and Karl Polanyi, who urges us to return to 'society' in its wider meanings from one dominated by economic determinism. These inspiring authors might find sympathetic tones in various Buddhist dictums too, such as 'learn what is enough' and 'mottainai (what a waste)', which decrees the self-restraint of human beings. More recently environmental ethics has emerged as an independent academic discipline (Katou 1991, 1998; Hasebe 1999; Thompson *et al.*, 1994).

Forming a sustainable agriculture

What shall become of agriculture if we try to reform it according to the sustainable development model discussed above? The field of agricultural science has begun to formulate sustainable practices, introducing concepts such as maximum sustainable yields in fisheries and maximum allowable cut-rates in forestry; and these ideas have already been applied to the preservation, management and utilization of sea resources and forests. While there is undoubtedly still much work to be done in these areas, arguably agriculture faces greater obstacles to reform. Admittedly, in order to meet the needs of a continually growing population and its ever-increasing demand for food, meat and forage products, through converting forests into arable land, improving livestock breeds, prolific use of agrichemicals and fertilizers, mechanization and other forms of industrialization, we have achieved stable yields and a high rate of agricultural production. Yet following the industrial model, agriculture has also developed in such a way that it imposes grave burdens on the environment.

It is now essential that we endeavor to minimize those burdens by limiting the use of oil energy, chemical fertilizers and agrichemicals. Only by doing so might we establish a sustainable agriculture in which biodiversity and food security, along with the future of farmers, are ensured. To this end, every discipline within agricultural science should strive to develop farming technologies that help to preserve – and where necessary repair – the environment. But this is not simply a technological problem. Cultural change is required to, if we are to avoid trying to meet the ever-increasing demand for food through deforestation, over-farming and over-grazing. Achieving sustainable agriculture is a momentous task facing agricultural science and society in general. How to balance the preservation-needs of the environment with the need to increase agricultural production is a question that will become harder to ignore in the 21st century.

Conditions for forming a sustainable agriculture

Despite formidable difficulties, a sustainable agriculture must be realized as soon as possible. Here, I shall address four conditions that may make the task slightly easier.

Creating low-input agricultural technologies

First, we must focus on creating low-input environmentally-friendly agricultural technologies *in addition to* continuing our pursuit of efficiency and productivity. This suggests the development of energy-saving agricultural machinery and facilities, and live agricultural fertilizers and insecticides to reduce the use of synthetic agrichemicals. We must also institute a comprehensive prevention system, to greatly reduce the rates of air pollution, water pollution, soil pollution, the extinction of plants and animals, and the spread of human diseases.

Great efforts have already been made to formulate suitable agricultural practices – to name but a few: Sustainable Agriculture (farming system), Alternative Agriculture, Low Input Agriculture, Organic Agriculture, Biological Agriculture, Radical Agriculture, Ecological Agriculture or Eco-Agriculture, Natural Agriculture, Permaculture and so on (Grigg 1995: 209). These alternatives to mainstream modern industrialized agriculture all share the characteristic of being environmentally-friendly. Whether it is sustainable agriculture, environmentally-friendly agriculture or alternative agriculture, the terms may differ but they all mean the same thing. According to Malten, sustainability in its wider sense connotes natural adaptability and environmental-friendliness. Therefore, the foundation of an agricultural system should be sought in productivity, stability and sustainability (Loomis and Connor 1992: 4). But they differ greatly in their actual recommendations about *how* we should formulate a sustainable agricultural system – from approaches to minimize the consumption of oil energy, chemical fertilizers and agrichemicals, to ones that still largely relies almost exclusively on nature's self-correcting capacities.

Low Input Sustainable Agriculture (LISA) has been proposed by the US Department of Agriculture as an alternative agricultural system. It has the following features: (1) accommodating planting methods, potential farmland productivity and natural conditions; (2) low input production to ensure preservation of natural resources; (3) nitrogen fixation in the air, utilization of natural food chain relations such as that between pests and their predators; and (4) use of the genetic information of plants and animals.

The idea of Organic Agriculture was first proposed by the English agricultural scientist Albert Howard in *An Agricultural Testament* (1940). It later spread to the US through the efforts of people like Rodail. When the Japan Organic Agriculture Association was first founded in 1971, Organic Agriculture was defined as an agricultural system that serves the purpose of nurturing rather than contaminating the environment and aims to develop salubrious and palatable food (Kyūma 1997: 114–18). Ecological Agriculture, Biological Agriculture and Radical Agriculture were developed in various European countries, all arguably derivatives of the Organic Agriculture movement.

Masanobu Fukuoka's Natural Agriculture is a special case, however. It rests on the philosophical principles of 'not acting' and 'natural life', advocating agriculture without plowing, fertilizer, agrichemicals or weeding (Fukuoka 2004).

All of these approaches must continue to be understood as 'works-in-progress', which still require great effort in research and development.

Multiple farming

For the purpose of a sustainable agriculture, the concept of 'multiple' should be emphasized for the following reasons: (1) since planting crops is typically confined to particular seasons and resources like land, capital (machinery, facilities and raw materials) and labor are not fully utilized all year-round, a number of different crops should be combined to maximize utility; (2) in order to gain easier access to organic fertilizer made from stock excretion and by-products like rice straw, stockbreeding and crop planting should be combined (complementary relation); (3) to utilize the nitrogen fixation function of legume; (4) to overcome the problem of soil sickness (replant failure): (5) the distribution of risk as a precaution for drastic changes in the market or natural conditions; (6) to prevent soil loss; and (7) to prevent pests and grow healthy crops. Agriculture is a human craft to nurture organic lives. It is necessary that the form of agriculture should account for this.

Composite agriculture should not only combine different management departments but also encompass processing, retail businesses and so-called green tourism businesses, such as bed & breakfasts, holiday farms, picnic grounds and barbeque facilities. For instance, at the Funakata Integrated Farm in Yamaguchi prefecture, a visitor can tour the five traditional production departments of rice, dairy products, cows, flowers and grass as well as the yoghurt processing and sales department. In addition, part of the grassland area is set aside for children, pets and family barbeques to promote tourism. By hiring a few dozen employees, the farm is running an impressive composite agricultural business.

The establishment of regional multiple farming

In many old world societies, like Japan and much of Asia, farming is typically conducted on a relatively small-scale. Here it is not practicable for small independent farmers to develop multiple farming as described in the previous section. In such cases, the multiplicity needs to center on a village, a group of several villages or some other large traditional social unit, to establish regional multiple farming. For example, farms that specialize in flower or vegetable cropping could either simply combine different genera or integrate various technologies – such as outdoor cropping, tunnel cropping and greenhouse cropping. Such an approach would overcome some of the problems with changing seasons, thus maximizing the efficiency of land and facilities usage. Where stockbreeding, vegetable-cropping and flower-cropping are combined in the one region, the stock-breeders' excretion disposal problems can be resolved in the same process that fixes the croppers' soil fertility problems. It is necessary to create mutually-supporting and complementary relationships through intra-regional conglomerations of farms (Sakai 1981; Chiiki Nourin Keizai Gakkai 1999).

Constructing a total recycling social system

Because agriculture involves a great many different social actors, it will not suffice to focus solely on amending the relationships within agriculture itself, between production sectors, managerial departments and regions. More importantly, we must strive to develop better ways of improving the 'coexistence and circulation' between cities and villages, and between agriculture and industry. I call such a scheme 'a total recycling social system'.

Only by combining our knowledge and experience and ingeniously recreating the wide array of interrelations discussed can we ever institute a total recycling social system that might resolve the grave environmental problems that we face in the 21st century. What is at stake here is nothing less than our own survival. Our success will require a paradigm change, a new revolution in the history of humanity.

6 Agriculture, Farming Villages and Everyday Life

Rural communities and farmers change over time. Agricultural villages in Japan and the developing countries of Asia, where small farmers form an overwhelming majority, are undergoing rapid changes in the face of urbanization and industrialization. However, regardless of the differences, certain fundamental characteristics transcend time and region. In this chapter, I investigate the value of agriculture and farming villages from an everyday life perspective, focusing on what remains fundamental and what changes. First, I examine the essential nature of rural communities and farmers. Second, I analyze the social significance of agriculture, farming villages, and especially agricultural work, along with the significance of the cultural activities in farming villages from the everyday life perspective. Third, I describe new trends and meanings in agriculture and farming villages.

Rural communities and the farm household economy

The structure of rural villages

A rural village in Japanese is often called *mura* or *shūraku*. Farming villages (*shūraku*) comprise the most basic units of social organization in Japan. In 1995, there were about 140,000 villages in Japan, with an average of about 30 households per village. In 80 percent of these villages more than 50 percent of households were farmhouses.

Supplementary note

Figure 6.1 depicts the spatial composition of an average village with about 60 households. This village (*ooaza*), situated in the San-in area of Japan occupies two valleys, with the houses divided into two small clusters (*koaza*), one in each valley. The travel time by car to the nearest neighboring town (population 30,000) is 15 minutes, and 25 minutes to the nearest city (120,000 people). Of the 60 households, 50 are farming households (roughly 80 percent). Almost all of the farmers are part-timers who also have non-agricultural jobs, working

Figure 6.1: Spatial composition of an average village

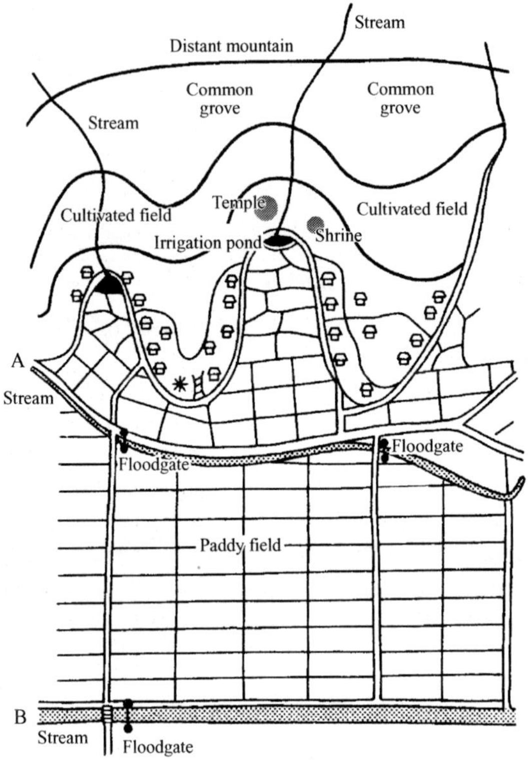

in factories, government offices, farmers' cooperatives, and shops in nearby small-to-medium-sized towns. The average income of farming households in 2000 was nine million yen, of which about seven million yen was from non-farming activities, one million yen from farming activities, and the remainder from gifts, annuities and other sources. The village's cultivated land comprises 60 ha of rice paddies and six ha of upland fields. The average farming household has 1.2 ha (3 acre) of rice paddies and 0.12 ha (0.3 acre) of upland field. These fields are on a slight gradient that extends from the village towards a neighboring mountain. Behind these fields lie *satoyama* (a forest held in common near a village), an undeveloped woodland on the lower slopes of the mountain, and further beyond this lies an *okuyama* – literally the 'deep mountain' – an undeveloped woodland on the upper slopes and mountain recesses (a privately owned forest area).

There are irrigation ponds at the far end of each valley, with a brook from each pond meeting in stream A. There is also a slightly larger stream

B. These ponds and streams are sufficient to irrigate the 60 ha of rice paddies (both the terraced paddies and the flat areas). Thus rice can be grown through managing the watergates. A shrine to the local deity and a temple are situated at one end of the village (separate from each other), and are the principal locations for village festivals or ceremonies. The village thus has a social and spatial unity, providing a place for production and everyday life for the residents. This arrangement or something very similar is typical of the majority of villages, and can therefore be seen as a typical Japanese village.

Cooperation in rural villages

The dominant social characteristic of the village is its highly cooperative nature, which is not found in either urban communities or urban neighborhood associations. Even today, villagers have powerful governing organizations (such as *burakukai* or *shūrakukai*) that treat the whole village as a single unit. These organizations hold periodical assemblies (*jōkai*), where a wide range of matters are discussed and executed, for example: construction, maintenance and cleaning of farm roads and water canals; management of common forests, shrines and temples; conducting traditional events; periodical repair and renovation of temples; management of common facilities such as dams and irrigation ponds; fire fighting organizations, coordinating emergency activities for various disasters like fire, earthquake and floods; providing mutual aid in ceremonial occasions; management and use of community centers and the cultural and sporting activities held in these centers; regulating the buying and selling of farm land; exchange of labor during the busiest farming seasons; organizing joint farm work and village-operated farms; creating group associations and various mutual aid organizations; and, finally, dispute resolution.

Before WWII, there were extensive multi-layered mutual aid relationships in farming villages, based largely on blood ties and neighborhood relations within a community, but also including a wide variety of other ties. These relationships were typically regulated in a highly ordered fashion by a similar group of all households within the community. One example is provided in Table 6.1. During the postwar period of high economic growth, many farmers began to farm part-time, farm management diversified, and nuclear families increasingly became the norm, all all of which weakened the framework of mutual aid relationships. Nevertheless, compared to the cities, there were still far more mutual aid relationships in the village (Hasegawa 1986). Some

Table 6.1: Pre-war mutual aid practices (Tokushima prefecture, Kamiya town, Jinryo district)

Name	Content	Main institution
Labor exchange (*temagae*)	Mutual assistance in peak production periods	neighbors, relatives
Help (*tetsudai*)	Labor assistance for households that need it for planting and harvesting	neighbors, relatives
Emergency aid (*yakebushin*)	Practical assistance during fires, floods, and landslides, etc; donation of money and goods; recovery assistance	Head of Household Association
River accident	Search and rescue operations if someone falls in the river or is lost in the mountains	Head of Household Association
Livestock association (*manninkou*)	Indemnification of dead beasts/livestock	Head of Household Association
Caretaking association (*senninkou*)	Caring for the sick	Head of Household Association
Funeral association (*shindan*)	Help with and execution of funerals	Mutual Help Association
Economic help association (*tanomoshikou*)	Monetary help for economic relief	Mutual Help Association, neighbors
Economic help association (*masumonokou*)	Non-monetary aid (particularly the provision of rice) for economic relief	Mutual Help Association, neighbors
Hermit houses (*anbō*)	A type of welfare support for the elderly	Mutual Help Association, Head of Household Association
Cemetery for pilgrims (*henro bochi*)	Providing graves for people who died on pilgrimage	Head of Household Association
Accommodation (*zankon yado*)	Pilgrims' (wanderers) accommodation and entertainment	individuals
Wedding help (*yomeiri tetsudai*)	Catering for weddings and transporting wedding apparel	neighbors, relatives
House-raising association (*mune age*)	Construction help, roof building	neighbors, relatives
Roof-making association (*yanefuki*)	Roof changing	neighbors, relatives
Making rice cakes (*mochi tsuki*)	Making rice cakes	neighbors
Bath (*morai buro*)	Rotational use of each other's bath	neighbors

Source: Hasegawa 1986: 60.

consider these relationships to be overly restrictive, but considering the unpredictability of nature, the seasonal intensity of farm work, and the particular difficulties and risks associated with small family-based management units, it is clear that such mutual aid relationships are necessary.

Supplementary note

Since WWII there have been two major approaches to defining village society as a village community. First, in the 1950s, in a rejection of the militarization of prewar Japanese society, land ownership by a small number of landlords came to be seen as a backward 'semi feudal' or 'pre-modern' stage of society. It was argued that the village community at the foundation of this system needed to be broken up in order to realize a genuinely modern civil society (Ōtsuka 1955). This perspective, which regarded the democratization of Japanese society to be an important issue, was theorized by Ōtsuka Hisao, Maruyama Masao and Kawashima Takeyoshi.

Second, when the negative affects of the pursuit of rapid economic growth became apparent in the 1970s, critics began to descry the idea that GNP growth had to be pursued at all costs, an idea that posited economic values as the central virtue. These critics pointed to the limits of market principles, and regarded the village society as a communal and non-market organization that had successfully preserved its mutual aid characteristics. Writers who presented this social framework in glowing terms include Hirata Kiyoaki, Hiromatsu Wataru, and, in the field of agricultural economics, Isobe Toshihiko, Kawaguchi Akira and Morita Shiro. Kitahara Atsushi refers to this perspective as form of Japanese 'communitarianism', which generally emphasizes the effectiveness of social criticism – not from a theoretical perspective but from one of ideals and strategy (Kitahara 1996: 52). Between these two positions we find the views of authors such as Watanabe Hyōriki, Kawamoto Akira, Hasegawa Akihiko and Aikawa Yoshihiko, all of whom attempt to demonstrate empirically what village society actually is by clarifying the positive meanings of and roles played by the village.

But the village community concept is originally based on the medieval landlord-peasant relationship, which formed around particular land areas, and was accompanied by coerced cultivation and non-economic coercion (the right to judge, for example) and thus took as an ideal type a closed and self-sufficient village community. To apply this concept to villages in a capitalist society is in and of itself logically self-defeating. There is a gap between theories of the community on the one hand, and on the other, the fact that agriculture and farming villages differ from urban societies, cannot be wholly reduced to market principles, and must adopt cooperative strategies.

Communal characteristics of agricultural land

The forest commonly owned by the village is, naturally, the object of common management. A similar idea exists with regard to the farming land. Following the Meiji Restoration, farming land was privatized and traded. Today, as mentioned, Japanese farming households own an average of 1.4 ha of land. However, when a family wants to sell farmland, it attempts to find a buyer within the village. More importantly, the village itself, through community officers or agricultural committee members, mediates to ensure that the transaction takes place within the village. The reason is that the village is concerned that if an outsider purchases village land, there will be various problems with, for instance, the fulfillment of water management, road construction, watergate management, and other community obligations as previously outlined.

As Kawamoto Akira explains: 'the motherland that is the common property of the village is called territory… This territory has the same meaning for the village as the state borders of national territory, and the villagers are conscious of where their boundaries lie' (Kawamoto 1972: 138). Following the villagers' first New Year's meeting, their representatives travel with the youth along the borders, inspecting their territory. This is called 'going round the territory' (*ryōmawari*) or the 'big going round' (*ōmawari*) or 'boundary renewal' (*kyōkai aratame*). Thus the village is both a geographical space and a social space. In other words, in Japanese rice-growing society, the possession and management of mountain forests, farmland and water are viewed as inseparable from the village. The idea here is that farmland is private property but at the same time, it is also the village's communally owned property. As Ishikawa Rikinozuke observed, the idea of buying back 'land sold out' (*tabarai tochi*) – land that was originally owned by the village, but has become the possession of another village – is based on this dual concept of ownership (Soda 1980). However, as the rural population continues to decline there is a shortage of agricultural labor, and it is increasingly common for farm work to be contracted out to individuals from outside the village. First preference is always given to relatives in neighboring villages, but since they are likewise affected by rural depopulation, land is increasingly sold to any interested party. At the same time it is becoming more common for several villages to join together to run communally operated farming projects. Hence, while the village remains a basic social unit, there are increasingly numerous factors that transcend the borders of any single village.

The importance of family and home

Agricultural production in Japan and around the world is overwhelmingly dominated by family management. Of course, there are also corporate forms of agricultural management, such as the large cattle farms in Western areas of North America, sheep farms in Australia, coffee plantations in Brazil and rubber plantations in Malaysia (Noguchi 1987: 1752). However, the typical example of corporate management in the US is in fact nothing more than a small business with a few employed laborers when compared to industrial production, and therefore agriculture in America is basically family management (Chō 1997). Historically speaking, the following advantages of family management have been identified compared to collective farms or corporate farms: (1) greater freedom, independence and diligence; (2) work security as a product of owning and residing on the land; and (3) a framework that guarantees free and fair competition.

At the same time, since the 1980s, the collective farms and state-owned farms in the former Soviet Union and communes in China have been broken up, and there is movement towards family management.

We can therefore say that, in general, family management is dominant in agriculture, although there are differences in terms of scale and types of commodity production. Unlike dry field farming such as wheat, corn and soy beans that are predominant in the West and which depend mainly on rainwater, in Japan, where rice crops are of primary importance, overcoming severe natural conditions and managing water supplies requires community cooperation – hence strong village unity and family management are central characteristics of agriculture in Japan.

Recognizing the centrality of family management in agriculture has other important implications for the meaning of the household unit in the village, too. The family is basically founded on the husband-wife relationship – which is, in turn, based on erotic love between a man and a woman – as well as the parent-child and sibling relationships that result from this erotic love relationship. In other words, the family is based on two factors: marital relations and blood relations. But research into farming village society in Japan only recognized the importance of considering the nature and dynamics of families after researchers recognized the crucial importance of the household rights to farmland and commonly owned forests, and the accumulation of social connections that unified the households within the village as a single unit. These relationships still characterize village society today, despite the fact that they have been weakened by farmers taking other jobs, the diversification of household economies, individualization, the 'nuclearization' of families, and the provision of a stable supply of water over broader (regional) areas.

The farm household economy

Family farm management has a different logic of behavior than business. Oguni Hiroshi (1978) argues that 'the farmer's ultimate aim is the family's happiness and wellbeing. Income is merely an interim target. When the income is extremely low, happiness and wellbeing cannot be achieved. But income is not pursued at the cost of the family's happiness and wellbeing. Farm households do not necessarily pursue the largest net profit or the maximum income'. If they can achieve a steady income, farmers look after one another's wellbeing, their leisure time increases, quality of life improves, and happiness can be pursued. Of course, no family is entirely free of disharmony and conflict, but members are generally happy to work for the sake of other family members. Hence, farm households have an important investment in a safe natural environment in which they engage in production and life with the family's happiness and wellbeing as their primary purpose. In a family, regardless of whether one is young or old, male or female, each works according to the condition of his (or her) health and capability, each receives according to need, and family members share the ups and downs of life together (Oguni 1978). With income derived from a combination of non-farming and farming activities, the income level that a family aims for differs according to the size of the family, its age structure, capability and motivation, but we can still say that the ultimate purpose is the maximization of family welfare.

The village community, as mentioned, is a local society in which farming households gather together. It is an organization based on blood and geographical ties that aims, through cooperation, to achieve the productive, ecological, and living environments that enable each family to achieve its aims. In short, the purpose of the village is to maximize the total wellbeing of the villagers.

Village life: the comprehensive nature of farm work

In agriculture, as in the market society, management pursues net profits to secure and increase income through increasing productivity as an economic management principle. In this sense, human activities in agricultural production fall under the category of economic labor. Farming labor may be small management, but it is still defined economically in a market society. However, while it is therefore not incorrect to call agriculture an industry, the social and cultural meaning of individuals actively involved in farming labor has begun to be a focus of research attention. That is, in addition to its economic dimensions it is increasingly recognized that agricultural production has other, non-economic dimensions. From this perspective, the everyday meaning of farm work for farmers, and humans more generally, has begun to be considered. This dimension, which I call 'comprehensive nature', can be summarized in four points: (1) cyclical, (2) diversity, (3) mutuality, and (4) self-creativity.

Cyclical

Farm work is characterized by a cyclical rotation that reproduces the circular movements of the cosmos, where one year forms a single unit, typically with four changing seasons. Unlike industrial (manufacturing) production that uses inorganic raw materials, agriculture uses living organisms: plants and animals. To maintain those lives, complex environmental conditions are required. Agriculture is thus dependent upon the life cycle and the circular movement of the cosmos.

Farm work also has a generational rotation. Most crops are grown and harvested within a single year. However, many fruit and nut trees have a productive life of at least 20 to 30 years from planting, while trees grown for timber typically require 30 to 50 years before harvesting. The latter point is reflected in an old Japanese proverb that 'one grows trees for one's grandchildren'. At the time of planting either fruit or timber trees, it is virtually impossible to predict the future price, and thus as an economic investment it involves great risks (see Chapter 3). Any potential income from trees remains merely a possibility, not a certainty. One receives an inheritance from one's parents or grandparents, and leaves an inheritance to children and relatives.

For annual crops, the cultivated land that is required only becomes available through the cultivation of fallow land and mountains, by soil improvement and irrigation, and by long-term preservation of soil fertility. In other words, 'cultivated land is not merely nature – it is created through the blood and sweat of today's generations combined with that of past generations' (Kawaguchi 1995: 14). In this way, farm work with its heavenly rotation, life cycle, and generational rotation is cyclical, which is why Kawaguchi Akira calls agricultural society a 'lifecycle world'.

Through their embeddedness in this cycle, people become aware of the precariousness of life in nature, learn the rhythms and leisure of life, experience feelings of awe in front of a pitiless nature, and learn humility. They also learn to feel gratitude toward their ancestors, and wish to leave an inheritance to their children and grandchildren. This can be seen as a process in which spiritual and material treasures accumulate together.

Diversity

Farm work, especially in rice production, consists of periodical and seasonal processes – 'bringing land into production, cultivation, growing seedlings, transplanting, observing weather conditions, managing water and fertilizers, weeding, harvesting, handling and/or processing, selling' – in which a diverse number of different tasks are successive, change, and continue.

A large number of different types of crops have been cultivated, each requiring different processes and cycles. At the same time, the human capacity to predict weather conditions is limited. Farm work is thus highly complicated: differences in seasonal rainfall and temperature determine whether crops will be abundant or scarce. The range of possible choices in work, crops and animals suitable to different ages and climates are virtually limitless. Choice is a product of the great diversity that can be seen in types of crops, animals, and work: crops can include grains, fruit, vegetables, flowers, and industrial crops, while livestock can include cows, horses, sheep, rabbits, chickens, and pigs, etc. Farming provides opportunities suitable for young and old, men and women. Moreover, the scale of activity can vary according to the contents and form of one's desires – there are full-time and part-time farmers, hobby farmers, and small vegetable gardeners. It is possible to position farming as a 'lifetime industry' or 'life-style industry'. When differences in personal inclinations are effectively combined with different kinds of farm products and livestock, it is possible to achieve adequate (economic or spiritual) agricultural management.

In the case of non-agricultural industry, too, there are innumerable types of products and ways to produce them. However, such industries are not dependent on seasons, and are released from the constraints of sustaining life. Products emerge from processes in which components are manufactured and then assembled, with both processes mechanized and simplified through human enterprise. As discussed in Chapter 4, this is a world of simultaneous parallel production. Even where a diversified range of work exists, it remains a world of simple work in which there is a complete division of labor and specialization, and in which a single operation is continued day after day.

In the case of farm work, because a diverse range of tasks are cyclically required, workers' lives are subject to changing rhythms. Moreover, participating in complicated cultivation processes and with the natural environment provides proof of a lively human existence in a modern society with its simplifications and conformity. The daily lifeworld provides stability and permanence with diversity. In the same fashion, diversifying the contents of life guarantees humans a stable everyday life for maintaining healthy minds and bodies.

Mutuality

The targets of farm work are plants and animals, or crops and livestock. The purpose of farm work is best achieved when the environmental conditions, such as earth, water, air and sunshine, interact harmoniously with the life force that shapes crops and livestock and with people's ideas and work. In short, agriculture is co-operation with nature. As discussed, in other industries, where inorganic matter is processed, the vagaries of nature are largely irrelevant, or can be overcome through technology. Thus, farm work is a special case, characterized by its mutuality with living nature. As discussed in Chapter 2, one dimension of this mutuality is emotional exchanges between people and livestock. Although there are differences in degree, plants and animals exist within a realm of possible communications, as in Imanishi Kinji's 'analogical inference' and Bollnow's 'understanding' (see Chapter 10). Farmers such as Hinoue Heiichirō believe that the best rice crop is achieved by standing in the furrows every day and 'talking to the rice' (Kawaguchi 1995: 17). Farmers who work in solitude in the mountains and fields say it is only possible for them to do so because, while working, they speak to the mountains, rivers, grass, trees, birds and insects. In this mutuality, people are taught by nature, learning about cooperation and mutual harmony, and come to appreciate the pleasure and emotions of life. They cultivate it with prayers and thanks.

Self-creativity

The main purpose of farm work is to produce farm goods. Rice plants are grown to harvest rice; chickens are bred to produce eggs; cows raised to yield milk or beef, etc. Farm work includes creating the instruments and facilities that are required in these processes. Miki Kiyoshi states that 'through creating things one creates one's self', and 'all of our acts have the meaning of creating things. That is to say, it is creative' (Miki 1966–86). All technology is human creation, in the process of which people are forming themselves. It is impossible to separate technology from the laws of nature. However, creating technology also pulls nature closer to humans, altering it and using it: technology is an act of artificial creativity. We can therefore say that farm work is self-creative and self-formative.

Many both in Japan and overseas argue that farm work has a significant influence on raising children, and that it is essential to people's lives.

The problematic relationship between work and education, or labor and education, has been absorbed into the broader context of technical education, vocational education and lifetime learning. However, there is a long history of claims that experiencing nature through farm work has the significance of educating the whole person, in contrast to the truncated or compartmentalized vocational education that has become the norm in highly industrialized society. Recently this perspective has been revived as 'the educational power of agricultural work'.

It was not that long ago that schools in farming communities took holidays during the farmers' busiest seasons, closing their doors so that children could help with the farm work. Until the 1960s, it was taken for granted that children in farming villages would help with the farm work after school throughout the year as well. However, today with mechanization, chemicalization, and the development of other labor-saving devices, such help is almost completely unnecessary. Moreover, children today are too busy studying for further education. Their games have also changed: they now have a greater interest in watching television and playing computer games than in playing in the fields. Whether they live in towns or farming villages, children today have radically fewer opportunities to engage with nature and agriculture than in the past. At the same time, statistical data show that children's average height has increased and their physical build has improved, even as they have become overweight and have decreased body strength. There are also increased incidences of atopic dermatitis, lifestyle diseases, social 'withdrawal',[1] truancy, anorexia, refusing to dine with others, bullying and a breakdown of classroom discipline. Not all of these problems can be directly related to the lack of experience of agriculture and nature, but there is no doubt that smaller family sizes and changing lifestyles are at the root of these problems. Thus the importance of educational farming experience is beginning to be recognized.

Western education theorists and practitioners who have emphasized the value of farm work include Rousseau, Pestalozzi, Goethe, Fichte, Owen, Bairy, Dewey, and Steiner. In Japan, the educational values of work have been promoted by Andō Shōeki, Takamine Hideo, Izawa Shūji, Sawayanagi Masatarō, Obara Kuniyoshi and Kobayashi Sumie (1969: 1971), and more recently Sakamoto Keiichi (in Katō 1986) and Shichinohe Chōsei (1990).

Rousseau (1762) maintained that the more people flock together the more they corrupt. The city is the abyss of the destruction of humankind; the countryside is the place for revival. Nevertheless, agriculture is still the first occupation of man; it is the most honest, most useful, and consequently the most noble one he can practice. The great secret of education is to work like a peasant and think like a philosopher (Kobayashi 1969).

Pestalozzi (1801) thought that education could redeem people from poverty and other misfortunes, and is the basis for social reform. Farmers, he says, have 'the school of nature' – so even if there are no schools, one can become what one should become through the education of nature and work. His intent was to thoroughly form people through sound homes and versatile work. Pestalozzi devoted his life to the education of children, opening orphanages for homeless children in which farming was the means of livelihood. Goethe (1794–96) designed an 'educational home' (*Pädagogishe Provinz*) that would teach farming and hand-manufacturing techniques. Similarly, Fichte's (1846) idea of a 'state of education' (*Erziehungsstaat*) idealized education in the countryside. Owen (1813–14) argued that an agricultural-industrial cooperative which idealized devotion to agriculture and industry was not only fitting for education, but for human society itself. He designed a new society, 'New Harmony', and traveled to the US to realize it.

About the same time as Rousseau, Andō Shōeki (1982–87) in Japan argued that the ideal society was an agricultural society that has no feudal or stratified relationships. He said that work, especially farm work, formed human character. Although Japanese educational theory has tended to center on ideas introduced from Europe, more recently, authors such as Shichinohe Chōsei (1990) have attempted to combine European and Eastern thought. Along these lines, various practices such as schoolchildren's farms, citizen's farms, field trips to mountain villages, and experiential studies have been initiated, but their real development has only just begun. In short, the educational and human significance of farm work has been continuously acknowledged in educational thought both within and outside of Japan.

As mentioned, from an everyday life perspective, farming and farming villages have much deeper significance than we may have otherwise thought. Farm work promotes holistic self-formation and is overflowing with 'comprehensive human nature'.

Life and cultural activities in farming villages

Rice growing society and traditional events

The folklorist Yanagita Kunio framed his research on the belief that rice crop agriculture lay at the foundation of Japanese society. This perspective has been highly influential, but various authors have argued against it. Isobe Toshihiko (1993) notes that the social character of dry field farming in Kagoshima prefecture is different from rice crop society, and argues that it was the application of dry field farming methods to rice paddy farming that liberated Japan from Oriental stagnation and laid the path towards constant technological innovation. More recent research has begun to acknowledge the cultural differences of people in Japan who make their livelihoods in the mountains or from the sea, and argues against seeing Japanese society in homogeneous terms (e.g., Amino 1994: 44–5; Takemura 1981).

While recognizing this diversity, though, the fact remains that the majority of Japanese farming villages depend upon rice paddy farming. Furthermore, although many are becoming specialized niche or part-time farmers, moving into other sectors such as flowers, vegetables, fruit, or livestock, farming households and the Japanese people in general remain fixated on the importance of rice farming. Also, the majority of villages are organized around the working cycle of rice cultivation, which traditionally determines the cycle of traditional events. In other words, the community calendar is centered on various rice farming ceremonies, which thus determine the cyclical rhythm of rural life.

However, the traditional events that once grounded the culture of the village have gradually declined through the high rate of rural-to-urban migration during the period of high economic growth, the industrialization of farming, the increasing proportion of part-time farmers and their higher (off-farm) incomes, the free and equal flow of information to both farming villages and cities, free trade and rising imports of agricultural products, reduced labor hours and more holidays. The accumulation of these factors has led some to argue that the farming village cooperative society has collapsed. Yet, the disappearance of traditional cultural activities in farming villages does not necessarily mean the end of cooperative society; for many of these activities have acquired new forms and many new cultural activities have been created or adopted in the villages.

New cultural activities in the farming village

In the 1980s, the Japanese people rapidly began to pursue spiritual relaxation rather than material goods. Table 6.2 summarizes the result of a 'survey of traditional events and group cultural activities' that I conducted in 69 village societies across Japan (Soda 1993: 212). Traditional activities include New Year's celebrations, ceremonies

Table 6.2: Summary of surveyed communities

Item	Average of all 69 village communities	Average of the seven most active village communities
Number of general households	145	113
Number of farming households	66	63
Percentage of farming households	45.5%	55.8%
Distance to the center of the (closest) small-to-medium sized town or city	27.6 min by car	25 min by car
Locals	15 of 69 villages are suburban	Two of the seven villages are suburban
Activity rate in community activities	Normal or somewhat active	All are extremely active
The number of different cultural activities	17.3 (100%)	29.0 (100%)
Including group cultural activities	6.8 (39%)	11.0 (38%)
Traditional events activities	5.0 (29%)	6.7 (23%)
Other activities	5.5 (32%)	11.3 (39%)
All sport related activities	36%	13%

to mark the beginning of the farming year, temple flower festivals (*hanamatsuri*), burning New Year's decorations (*tondomatsuri*), doll festival (*hinamatsuri*), the drive away birds (*torioi*) procession, the torch procession to drive away crop-eating insects (*mushiokuri*), the star (*tanabata*) festival, the (Buddhist) *bon* festival and dancing, and the autumn (harvest) festival. Group cultural and sports activities include: calligraphy, reciting Chinese poems, playing *koto*, braiding, knitting, playing i*go* and *shōgi* (Japanese chess), travel circles, *haiku* (Japanese poetry), history research circles, Japanese painting, ceramics, oil painting, baseball, golf, softball, volleyball, hiking, and gate-ball (a type of croquet popular among senior citizens), and so on. In addition to young wives' clubs, children's clubs, and elderly people's clubs, which have various forms of activities, adult men participate in activities linked to agricultural production such as research clubs for rice crops, flowers, melons, tomatoes, and fruit trees. In sum, there are a very large variety of activities taking place. The survey recorded an average of 17.3 types of such activities across the 69 communities, and 29 types among the seven most active communities. Traditional events accounts for 29 percent of the total across all communities, but only 23 percent in the most active communities. In addition to this kind of community activity, there are also social and cultural activities for the entire municipal (city and town) community.

Clearly, older, traditional activities are withering away and even disappearing from farming villages. But the survey reveals a large number of new group activities that have emerged as substitutes. Of the activities that are commonly engaged in today, 80 percent were introduced after WWII, and about half of these in the 1980s and '90s. As mentioned, many have interpreted the decline of traditional activities in Japanese villages to mean that the 'good old ways', the cooperative nature of the village, is being lost. As the results of this survey indicate, though, we must recognize that numerous new types of activities are being adopted in the villages. Clearly this new trend requires further research. The more detailed discussion of the survey findings below should provide some direction for future research in this field.

Supplementary note

The results of the cultural activities survey are presented in Tables 6.2 and 6.3.

(1) Looking at the 'different types' of activity (Table 6.2) – the average number of: group cultural activities are 6.8 (39 percent); traditional events

are 5.0 (29 percent); and other activities are 5.5 (32 percent). The total is 17.3 (100 percent). The ratio of sport related activities is 36 percent. In the most active communities, the overall number of activities (29) is higher. The number of traditional events in both groups is about the same, but the proportion of group cultural activities is higher in the active communities, and there are many voluntary, free and diverse modes of communication in place. There are also many activities across the municipality; so we can say that the activity rate for the whole region is high.

(2) Regarding the 'reach of the activity' (Table 6.3) – 30.1 percent of the activities pertain to groups within the community, 37.8 percent involve the whole community, 17.4 percent involve groups formed from several communities, and 16.2 percent involve the wider area. In the seven most active communities, groups within the community account for 44.6 percent of all activities, and municipal activities are 23.5 percent – considerably higher than in average communities. It is also apparent that activities both within the community and the wider area are thriving. Group activities are gradually moving from the local community to the town-village area. For example, village-based doll making and gate-ball have developed into town level exhibitions and games.

(3) 'The number of participants' in 4.9 percent of activities is less than five people, in 14.7 percent is five to ten people, in 26.9 percent is ten to 20 people and in 50.1 percent is over 20 people. In the seven most active communities, there is a higher proportion of activities with groups of less than ten people, which indicates that small group activities are added freely. In the past, whole community festivals and *bon* dancing were typical, but this has gradually changed to a versatile range of smaller group activities. Furthermore, individual hobbies are becoming especially versatile.

(4) The main participants in 21.2 percent of the groups are elderly people, in 30.9 percent adults, in 10.3 percent youth, in 19.1 percent women only, in 7.3 percent children, and in 9.4 percent everyone. In the seven most active communities, the ratio of adult participants is higher (36.8 percent) than the average, which indicates that the adults who are at the center of production are contributing more to the community's higher activity rate, including cultural activities. Group activities are diversifying in all age groups, and the ratio of traditional activities among all participants has fallen to slightly less than 10 percent.

(5) Looking at 'the type of leadership': 35.9 percent of the groups do everything equally, while in 39.0 percent of the groups a leader plays a central role, 10.8 percent of the groups have a specialist leader, and 11.4 percent of the groups are under the government's or farming co-op's leadership. In the seven most active communities, the number of activities

Table 6.3: Cultural activities of all 69 communities and the 7 most active communities

	All 69 communities		Seven most active communities
	Number of activities	Percentage	(Percentage)
Reach of the activity			
A Groups within the community	358	30.1	44.6
B The whole community	425	37.8	24.8
C Groups of several communities	207	17.4	6.6
D Wider area's groups	192	16.2	23.5
E Unclear	6	0.1	0.5
Total	1188	100.0	100.0
Number of average participants of one group			
A less than 5 people	57	4.9	9.4
B 5 to 10 people	170	14.7	16.4
C 11 to 20 people	312	26.9	23.9
D More than 20 people	581	50.1	45.5
E Unclear	39	3.4	4.8
Total	1159	100.0	100.0
Main participants			
A Elderly people	313	21.2	19.6
B Adults (not incl elderly)	455	30.9	36.8
C Youth	152	10.3	8.4
D Women only	282	19.1	18.8
E Children	107	7.3	3.6
F Everyone	138	9.4	12.8
G Unclear	26	1.8	-
Total	1473	100.0	100.0
Type of leadership			
A Everyone equal	432	35.9	30.7
B Group leader	469	39.0	50.2
C Specialist leader	130	10.8	10.6
D Government, farmers co-op	137	11.4	8.5
E Unclear	35	2.9	-
Total	1203	100.0	100.0
Frequency of the activity			
A Weekly	233	19.4	19.3
B Monthly	282	23.5	24.1
C Several times per year	294	24.5	29.2
D Annually	382	31.8	27.4
E Unclear	9	0.8	-
Total	1200	100.0	100.0

	All 69 communities		Seven most active communities
	Number of activities	Percentage	(Percentage)
The activity began			
A Recently	222	18.5	14.6
B About 10 years ago	350	29.1	31.4
C About 20 years ago	264	22.0	18.6
D About 30 years ago	119	9.9	11.9
E Before WWII	233	19.4	21.7
F Unclear	13	1.1	1.8
Total	1201	100.0	100.0
Activity level			
A Extremely active	526	44.7	43.8
B Quite active, usual	589	50.0	55.2
C In stagnation	57	4.9	1.0
D Unclear	5	0.4	-
Total	1177	100.0	100.0

centered on a group leader is increasing. We can see that activities that are carried out by people who have independently gathered together, rises to 80.9 percent. If we assume that hobbies and cultural activities in the cities are often centered on courses and lectures in cultural centers, then cultural activities in the farming villages are probably more independent and creative.

(6) Regarding the 'frequency of activities': the percentage of activities held weekly is 19.4 percent, monthly is 23.5 percent, several times per year is 24.5 percent and annually is 31.8 percent. In the seven most active communities, the figures are about the same, but there is a slight tendency towards a higher frequency.

(7) Regarding the time when each activity was established: the percentage of activities introduced recently is 18.5 percent, a decade ago 29.1 percent, 20 years ago 22.0 percent, 30 years ago 9.9 percent, and 19.4 percent of the activities date from the prewar era. In the seven most active communities, activities appear to have been adopted relatively earlier than the average. In recent years, the number of participants and the vigor of traditional events like New Year, the *ujigami* (village deity) festival, and the *obon* festival that were central to prewar farming village communities have decreased – they are generally in decline or stagnating. However, the density of these traditional events (if we see prewar activities as such) is not more than about 20 percent. About 80 percent of the communities' cultural activities were adopted in the postwar era, especially since the 1980s (about half of

them). This means, compared with relatively closed societies that cultural activities and traditional events in farming villages are being supplanted by new ones that are spreading rapidly – creating friends, associating groups, recreating farming village life, and making it enjoyable.

(8) All groups were asked about the present activity levels in the community: 44.7 percent reported that they are extremely active, 50.0 percent that they are somewhat active or usual, and 4.9 percent are stagnating. In the seven most active communities, of course, all activities were continuing. Lately, community centers and small squares have been built in most of the communities. In one active community (in Shimane prefecture), the community center rooms were fully booked every day; people were coming together and forming groups, interacting with one another and enjoying themselves.

The lifeworld: creativity and reorganization

Since the 1950s there has been a drastic depopulation of farming villages in Japan, with some effectively ceasing to function as communities. Nevertheless, people still live there in pursuit of the good life. As we have seen, we must recognize the creativity of farming villagers. Farming villages do not conform to the towns-peoples' view of the 'good old countryside', or to the common researchers' perspective of a 'stagnating or collapsing farming village'. The villagers themselves speak of the changes to their lifeworld as a 'creative reorganization of the farming village'. That is, the residents of farming villages are forming a new image of the farming village as an enjoyable place to live, both today and into the future. This means not only focusing on economic value, but on total human happiness, which is reflected in the villagers' new view of the farming village.

Of course, any conception of life in a farming village must be broad, including 'individual and family life, the entirety of social life, and the conditions of the facilities and services that supports it' (see Chapter 7). This extends to leisure activities, shopping, marriage, education (including lifelong learning) and so forth (refer to Table 7.1). Village life is undergoing large-scale transformation, being restructured and revived, becoming more diverse and active. Believing that it is possible for people to live happily, and that they can choose where they live, the ideal village pursues material wealth (economic-value), a beautiful healthy village environment (ecological-value) and enjoyment (life-value). As R. Schmidt (1912) notes, people are 'working, living, and playing' 24 hours a day, and have diverse desires. The farming village has the potential to fulfill these diverse desires (see Chapter 8).

Towards an open society based on regional bonds

When measured against the comparatively 'closed society' of the prewar farming village, the postwar era of high economic growth and social changes has made farming village life more of an 'open society', with very little difference from city-life. However, the present state of the village society's production, natural environment and living standards reserves the basic nature of farming village society, too, despite differences in degree.

From now on, farming village society, as discussed, will have three internal value ideals: economic-, ecological- and life- values. The question of how to harmoniously and creatively realize these values will be a central issue. At the same time, there are two questions that must be addressed: (1) As advanced information society rapidly extends its reach, what will become of farming villages? and (2) How will this affect the relationship with cities and towns? The second question will be addressed in Chapter 8. Here I briefly explore the first.

Information society attributes important values to information, and many social functions are served by various information media. Information becomes increasingly important as people expand their social relationships. With rapid advances in electronic information dissemination technology, whether one lives in a city or a farming village, one can access the same broad range and quality of information.

Broadly speaking, there are two ways that a society may respond to this flood of information: either individuals can independently choose and process a large volume of information, creatively forming their life, and pursue personal affluence; or the nation and other organizations control everything by manipulation of information, forming a controlled society in which individuals are drowned in a sea of information. The information society will evolve from the tension between these two directions.

As previously discussed, village residents live in a society with certain characteristics, but many of them also have jobs outside of the farming professions. Full-time farmers are now the minority in Japan. Village society is becoming more professionally diverse, but also more diverse in terms of human relationships and values. Therefore, the communities' interconnectedness is gradually loosening. Due to their other commitments, many are no longer available to participate in the collective work that is performed several times a year, and money is increasingly being used as a solution.

The diversity of occupations means a dispersal of interests. For example, young people who specialize in management may feel a need for a place to create groups and organizations to collect and share information about farming management, farming weather, the distribution of farming products, the market, farming technology, and farming policy, and to discuss the ups and downs of farming life with colleagues. Where there are shared interests, equal and independent individuals can connect and organize through the Internet without ever physically meeting.

Such organizations, of course, are not exclusive to farming, but are rapidly forming wherever people share purposes and problems. It is not uncommon for a single person to be a member of a number of different networks. This type of purpose-oriented network, knotted in pluralistic and multilayered nodes create social exchange formations that are evolving into wide area networks, and are typical of the 'information network society'. It enhances the possibility of exchanges between city and village dwellers, facilitating the participation of city dwellers who wish to begin farming, as well as the so-called 'returning retired people' (Imai 1984 and others).

In the future, village society will continue to be forged by blood and geographical ties, and thus be a geographically linked, autonomous and centripetal community. But at the same time, it will be continually reorganized as the information network society's centrifugal force spreads. Here the principles that at first glance may seem to be contradictory, such as the 'society based on regional bonds' and the 'open community', are unified. Hence, the farming village, like the city, is a place where traditions and new creations constantly interact, and where society is reorganized and reformed. In my view, the village community is founded on the principle of a regional (geographical) bond, where multi-layered face-to-face relationships exist. To this, the new richness of the network society has been added. The structure and living patterns of farming village inhabitants includes many unknown aspects. Hence new research should include agricultural economics, agricultural sociology and agricultural management studies.

7 Forming Sustainable Farming Villages: Pursuit of Comprehensive Values

Although we have established the importance of attempting to reconstruct agriculture to be sustainable, this means nothing without the farmers, the farming villages and the formation of agricultural management structures that are responsible for this attempt. In order to realize a sustainable agriculture, it is essential to form sustainable farming village regions within which people are willing to live, work and play. While there are powerful voices calling for a revitalization of agriculture, this chapter addresses the question of who is to be responsible for it and where it might be possible. First, I examine the general concept of the region, and outline the significance of research into farming village regions. Second, I explain my own more specific concept of the region (*ba*), defined as a place in which there is an integration of production (economy), the ecological environment, and everyday life. Third, I outline a direction for the self-formation of such a region – the self-realization of farming village residents. As explained in Chapter 8, the sustainable farming village can only be fully achieved through integration with small- and medium-sized cities. Although presented as separate chapters, Chapters 7 and 8 should be understood together as a single piece.

The general concept of the local region

According to Kiuchi Shinzō, the formal characteristics of the region can be summarized as follows: (1) it is part of a land surface, (2) it has a unique geographical location (3) it has spatial extension, (4) it can be distinguished from surrounding space, and (5) it is part of a larger region (Kiuchi 1968: 83). However, this formal conception gives no indication of the region's contents.

In Germany, academic studies of the region developed alongside the formation of geographical studies. However, mainstream regional research in Germany viewed the Earth as an organic whole, and considered each region to be an organic entity as well. Although there are differences in approach, with Richter arguing from a theological and teleological perspective, Humboldt from a physical geography perspective, and Ratzel from an anthropological and political geography perspective, all broadly define the region as a single organic life-form (Kiuchi 1968: 85–7).

Ratzel (1887) especially made this clear in his organic theory of region (*Raum-organismus*). From Ratzel's perspective, just as each living organism is formed from organs that have their own functions and are connected by nerves or blood vessels, so too each state, individual region, and city is seen as an entity that functions with differentiated elements – such as organizations or institutions – connected by and growing through means such as transportation, communication and commerce.

In France, La Blache (1992) built on Ratzel's theory, adopting an environmental possibilism theory, which argues that nature is passive – in the sense that it can be subjectively changed by human action. He argued that the objective of geographical science should be to explain each particular region's individuality and unifying principles. In contrast, Hartshorne (1939, 1957) criticized the organic theory, stating that a region's boundary is unclear and its substance undetermined, so that a region does not develop as a living being does, and it is therefore inappropriate to consider it as an organic being. The mainstream regional research in the US adopts the functional theory of region. This theory differentiates regions in a way that suits various research purposes, such as allocating and comparing complicated ground surfaces into groups according to shared qualities. The differentiation is based on, for instance, climate, topography, ethnic distribution, or land use. Here, the mysteriousness of organism theory is stripped away and the grouping of regions is completely subordinated to each particular research purpose. This method is undoubtedly useful for the research purposes it has in mind, but is less useful for a general understanding of regions.

Recently, however, with the appearance of environmental problems, biological and ecological views of nature and society have once again become influential. These include theories such as Lovelock's 'Gaia Hypothesis' (1984), which view the Earth as a single living being that has a self-stabilizing function. This can be understood as an extension of the theory of the region as organic. Although the Gaia Hypothesis has been vigorously criticized, such theories remain popular as responses to the predominant modern view that human beings can conquer and control nature through science and technology – a view that has proven itself patently inadequate for resolving our current environmental problems. It has become abundantly clear that for human beings to assure their continuing existence we must accept the macro-perspective that despite the human capacity to be a nature-changing technological being, we are at the same time part of nature and must coexist with it or perish.

Hence, although it is not strictly speaking a biologically functioning entity, it is appropriate to consider the region 'as a unity that formed spontaneously from the commonality of place, and that emerged from the direct or indirect interaction between human and human, human and nature, and nature and nature' (Kiuchi 1968: 88). The effectiveness of the concept of region depends on accepting the significance of the limitations on research imposed by the element reductive perspective based on the foundations outlined above.

In these terms, my own position presented in this work can be understood basically as an extension of the theory of the region as space organism, with the difference that the region is not a closed living being, but an autonomous and open entity in the modern sense of the term. My general definition of the region is that it is a place of human productive and living activities, an autonomous and individual geographic space that has economic, social, and natural consistency as a unit. As this does not fully explain the subject of our present research, let us examine it in more detail.

The farming village region as a '*ba* of total life'

In an earlier work, I observed that modern society is moving towards the realization of new wealth or, as I called it there, towards the realization of comprehensive values. In short, this means a comprehensive and harmonious realization of economic values, ecological values, and life (socio-cultural) values. From this perspective it is possible to explain the demands made by the Japanese people on agriculture and farming villages after decades of rapid economic and social development in the postwar period. Agriculture is a 'multi-value industry' that can satisfy the diverse demands of the Japanese people in the early 21st century, while the farming village is a 'multifunctional space'. From this perspective, I demonstrated the formation of new perspectives of various Japanese strata towards agriculture and the farming village, or in other words the development of a national consensus about agriculture and the farming village. I argue that without this formation or development, it is not possible to unveil any future prospectus of Japanese agriculture and farming villages, the agricultural administration that supports these, and Japanese society itself.

However, when viewing the region from the perspective of farmers and the residents of farming villages who shoulder the burden of agricultural production, their efforts to realize comprehensive values are considered to be an appropriate goal in the transformation of the rural region. Only such places can become sustainable regions. As discussed in Chapter 3, Keīchi Sakamoto's position on 'human life' is to a large extent incorporated in what I call the realization of comprehensive values, which identifies the region as the lifeworld best able to achieve this.

In the following sections I examine the region as a *ba* of life in more detail, focusing in turn on the three aspects of production (economy), natural environment, and life.

The village region as a place of production

Formerly, farming village regions operated with agriculture as their core industry, but at present agricultural production accounts for a mere two or three percent of Gross National Product (GNP), and even in those regions known as agricultural prefectures this is at most about 10 percent. Because agricultural production requires a wide surface area, the dominant view of the countryside and farmland is of wide and broad expanses of land, but the productivity of these expanses has significantly declined. Therefore, farming village regions have also begun to incorporate various non-agricultural modes of commerce and industry.

I have previously analyzed the economy of the Tango region of Kyoto prefecture, defining it as a 'region with a diverse production composition' incorporating an organic linkage between the agriculture, forestry and fisheries industries, and non-agricultural commerce and industry. The diversification began with the development of a textile industry, soon followed by tourism and other industries. At the same time, it was defined as a 'regional industrial complex' drawing on Keiichi's notion of a 'regional complex' centered on agriculture but also consisting of other social and cultural aspects (Sakamoto, Takayama and Soda 1986).[1] The regional labor market enjoys strong relations with other regions and especially large remote cities. Moreover, Tango's textile and tourist industries are integrated with national markets, especially nearby Kyoto, Osaka and Kobe. This type of regional economy thus consists of an individual and autonomous farming region, Tango, that is interconnected with other regions.

However, as Watanabe observed, the 'city and factory are typically understood as the exact opposite of anything to do with the farming village, so the idea of the region has developed as an anti-city one' (Watanabe 1978: 121). The modernization of Japan from the Meiji period can also be understood as a separation of, and friction between, city and country. Again, during the postwar era of high economic growth, industry and cities sapped the energy of agriculture and villages. Moreover, residents of the regions frequently rejected local industrial development because it threatened the cooperative society of the farming village and was seen to be a cause of environmental pollution. At the same time, as Robinson (1973) and Masahiko Aoki (1975) (among others) have noted, economics has applied homogenous concepts of time and space from classical physics, ignoring regional particularities. This worsened the problems of urban areas and gave rise to negative attitudes to the introduction of industries to rural areas.

Farming villages, on the other hand, with the high rate of economic growth, could no longer rely economically on agriculture alone; thus many residents were faced with the choice of either remaining in the region and becoming part-time farmers, or relocating to cities to find work. As Watanabe Hisashi notes, a simple anti-city and anti-industrial position that 'aims at a self-sufficient and…autonomous space may recreate the closed nature of the factory space, but in a simpler form'. Watanabe argues that the Japanese people should pursue 'the possibility of industrial space' in the local regions as well (cited in Kamisato, Kiyonari and Masamura 1978).

Thus it is important to understand the farming village region as a place of production (economy) that is open to the outside, but is at the same time a multi-industrial complex with an autonomous and individual constitution.

The village region as a place of an ecological unit

The farming village is also an ecological world, the place of an ecological unit. Forests, rivers, terrain, weather, and living things all form their own natural ecosystems within each particular region. These environments offer much to human beings, some of which is useful, some of which is not. Moreover, in each respective region, the ecological environment is an original and individual climate. Today, however, the regional environment is no longer 'natural' in the sense that it existed prior to human intervention. Not only are 'people created by the social world', but an artificial environment has been created by human interventions in nature (albeit with significant differences in degree) (Gibson 1985: 137–41).

Japanese farming villages in the past were surrounded by *okuyama* (deep mountains) and *satoyama* (forests owned in common), from which fuel (firewood and charcoal) was secured. The mountains also stored much water. This provided the fields with plentiful water and reduced flooding. Even with modernization, villagers cooperated to a considerable degree in managing the regions. 'In the regional community of the past, when agriculture played the main role, the local nature was carefully managed in accordance with non-market laws and customs, as a result of which a continuous balance with nature was realized' (Kamisato, Kiyonari and Masamura 1978). The primary basis of the resources that are necessary for human society's continuing existence is first and foremost the regions' natural ecosystem. The appropriate management of this ecosystem must reproduce coexistence and circulation.

But the industrialization of agriculture brought to an end the long history of coexistence and circulation. Population growth, mass urbanization, and the trend towards mass consumption of food stuffs have contributed to the industrialization of agriculture, comprising, among other things, expanding the scale of agricultural production, specialization, mechanization, monoculture cropping, chemicalization, and facilitation.

Mechanization and facilitation have created a dependency on oil as an energy source. These trends were accompanied by frequent use of (petro-)chemical fertilizers, the specialization of farm products, and hence a concentration on single-crop production. The repeated cultivation of a single crop on the same ground, in turn, led to technical problems such as sick soil – a decline in fertility. These practices triggered outbreaks of harmful insect infestations, for which a diverse range of agricultural chemicals were created and heavily used. Meanwhile, an enormous range of water management facilities have been constructed (Ikegami 1991), including dams, irrigation systems, and concreted waterways. One side effect of this has been a major disruption of water circulation which separates water into that which can be used and that which has been used. Combined, these changes have disturbed the natural environment, causing river and air pollution, killing plants and animals, and negatively impacting food safety and the human living environment.

Furthermore, globalization, with its demands for trade liberalization and further entrenchment of an international division of labor, has led to the disruption of the natural environment on a global scale. In those advanced nations that export agricultural products, there has been a steady push towards the development of excessively large-scale industrialized agriculture. For example, the US and others have increasingly adopted high input and non-sustainable agricultural practices to sustain large-scale agricultural exports. Dethier notes that current American agriculture consumes 'chemical fuel calories that are equal to five to six times the food calories harvested'. In England the comparable figure is three times. Hence, modern agriculture is 'an energy drain' (Dethier 1979: 309). In developed countries such as Japan, which depend on imported agricultural goods because of their disadvantageous environmental conditions, the price of domestic farm products remains relatively high, resulting in further imports and a marked decline in agricultural production and farming village life. The principle of comparative advantage that underpins the theory of the international division of labor does not take into account differences in natural geographical conditions such as topography, climate or water supplies. Therefore in regions with less favorable conditions, agriculture is in decline. While this means that the negative externalities that agriculture generates will disappear, the positive externalities will also be lost. Hence, the international trade of agricultural produce based on maximizing economic efficiency appears to be undermining farming village regions and the natural environment in both exporting and importing countries.

This economy-centric view of agriculture, as well as the vicious cycle of mass-production–mass-consumption–mass-waste that characterizes modern societies, is becoming a common evil in developed countries. In cities and farming villages, human coexistence and circulation with the natural environment is being lost. Its recovery is a significant issue.

Academic views that advocate the formation of a socio-economic system with the region as a unit that recognizes the importance of the ecological environment are therefore essential. Tamanoi Yoshirō argues that 'a common norm that unifies the community is based upon the living environment of the unique ecological region, founded on nature and the ecosystem' (1978: 57), and notes that 'there are calls to recognize the need to consider the region – and more widely, the state – as an ecological unit' (1978: 33). On this basis Tamanoi developed a regionalism movement that aimed at bottom-up regional independence and self-reliance.

It has become almost common knowledge that the ecosystem is incredibly complex (Shimizu 1992: 129) and diverse (Iwatsuki 1993: 167), and that its stability is dependent upon this diverse complexity. Furthermore, recognizing and respecting this in the mutual relationship between people and nature is a basic condition for forming a sustainable regional society. Nakamura Hisashi argues that 'circulation, diversity and relationships are abstract concepts, but also the fundamental concepts that support 'regionality' in the widest sense' (Nakamura 1993). Kamisato Isao and Tamanoi Yoshirō were the first in Japan to introduce such an understanding to the field of economics (Kamisato, Kiyonari and Masamura 1978; Tamanoi 1977).

It is important to recognize, however, that over-emphasizing the rehabilitation and preservation of nature can lead to forgetting, or even discounting, human existence altogether. Such positions are formulated in opposition to the idea of human mastery of nature through modern science and technology, but go too far in posing a choice between humankind and nature. The region as a place of the ecological unit is 'nature as a *ba* of human life and self-realization' (Noe 1987: 283); not a mere 'ecotopia' (Takagi 1989: 299), but a place for humankind and nature to coexist.

Thus it is clear that the 'modest management of nature' (Dethier 1979) is a significant issue at present.

The village region as a place of daily life

The farming village region is a place where people live. Here, 'to live' means those human social and cultural activities that are relatively independent from economic activities. Thus 'to live' refers to the general areas of personal, household and social life, such as shopping, a family enjoying one another's company, dressing up or playing, leisure activities such as engaging in hobbies, communication of all kinds, health and medical care, study and education.

Bergson referred to 'the time which is concretely lived by people', opposing it to mathematical time. With Bollnow, I want to emphasize the concept of lived place (*der erlebte Raum*), that is to say 'place, as it opens to people in their concrete lives' (Bollnow 1978: 17). This is to emphasize the importance of the 'spatiality of human life', as opposed to homogenized mathematical space. Herein lies the significance of the region as a human 'place of living'.

In everyday life, people relax, set down roots and enjoy peace of mind by settling down in a particular house or village. First and foremost, human beings are 'inhabiting creatures' (Bollnow 1978: 120), or 'an individual is place' (Relph 1991: 77). The German word *wohnen* (to inhabit) has connotations of feeling comfort, satisfaction, a place to come to a stop, to stay, and to be in a particular place. People situate themselves through unifying relationships with particular houses, village squares, shrines or temples, mountains, forests, rivers, and the roads that connect these, and through memories of childhood and significant relations that are connected to these places.

In the modern world, however, for myriad reasons many people have been uprooted from their places of origin, and thus have no peaceful homes. Bollnow argues that if 'denied this type of peace, it will be difficult to avoid an inner collapse of the person' (1978: 136).

According to Relph, modern society is characterized by its 'placelessness'; it replaces diversity with homogeneity and experienced order with conceptual order (Relph 1991: 143). According to modern economic theory, a 'sense of place and attachment to place are not merely unimportant, but their very absence is an economic virtue and placelessness is to be sought after to make possible the attainment of greater levels of spatial efficiency' (Relph 1991: 117).

From my perspective, to reside in places where we enjoy peace of mind, such as our homes or hometowns, is the 'essential nature of man's existence' (Bollnow 1978: 122). We have 'a deep human need...for associations with significant places' (Relph 1991: 147). This is as true for city-people, for whom the backyard may offer peace of mind, as it is for the people in farming regions. Thus regardless of short term adaptability, in the long term, 'to live in an environment which has to be endured or ignored rather than enjoyed is to be diminished as a human being' (Gauldie cited in Relph 1991: 147).

In this sense, we ought to form regions that can provide a greater total satisfaction and happiness than mere economic happiness. As Bollnow observes, 'experienced space is not a neutral sphere to values' (1978: 18). We need to autonomously pursue values that are unique and peculiar to our own region, and plan to realize these values. I examined this process of planning and moving towards forming such regions in *Jizokuteki nōson no keisei* (The formation of the sustainable village).

The formation of new *ba* of total life

So far I have considered each living space not simply as a place of production or the economy, but also as a natural environment and site of human life. However, the dominant economic reductionist view of society has resulted in an accumulation of problems of the natural environment and has distorted life. I believe that modern society has entered an age in which there is a harmonious pursuit of economic values, ecological values, and life (social and cultural) values; or, in other words, an age in which the pursuit of comprehensive values is viewed as ideal. I have noted the importance of agriculture in both state and local region as a multi-values industry, and the *raison d'être* of the farming village as a multifunctional space that includes agriculture (Soda and Ōhara 1994).

Like the inhabitants themselves, each region or *ba* of life is in a continuous process of creation and reorganization, aiming at a realization of comprehensive values which include regional individuality. The *ba* of life that is the region does not exist separated or segmented into the aspects of 'production (economy), ecology, and life'. Rather, the place of true total human 'life' will be formed in its attempt to find its own unique way to overcome the mutual contradictions of each aspect (see Chapter 3). The perspective that argues from each standpoint for a unification of ecology and economy, or an integration of life and economy, or a union of people and nature, is in effect arguing for a unification of two of the three aspects discussed above. Eventually, a harmonious unity of the third aspect will also be pursued.

It is necessary and inevitable that the comprehensiveness and creative transfiguration of the *ba* of life be discussed from the perspective of the totality, formation, and history of human existence.

Schmidt's plan for an 'industrial–residential garden city' (*Industrie-Wohn- und Gartenstadt*) (1912; cf. Soda 1997) was highly influential in the history of German regional planning. I will examine this in more detail in Chapter 8. For the moment suffice to say that Schmidt begins by rejecting the perspective that is almost universally held in economics today, that views humans as creatures that merely labor eight to ten hours a day, rather than as beings that live 24 hours a day. Schmidt recognizes people as total beings 'with fundamental desires to work, live, and play', and was an early pioneer in the movement towards creating a modern living space that can fulfill such desires.

Elements of Schmidt's argument can be traced to many different social philosophers. For example, Fourier claimed that human beings have twelve basic emotions, and criticized modern society because it prevented people from experiencing all twelve fully, instead forcing them to limit themselves to socially acceptable emotions. He developed the idea of the cooperative community, or 'Phalange' of total emotional satisfaction. The ideal of the Phalange is that people should spend their everyday lives in diverse and constantly changing ways, including both industrial and agricultural labor, hunting and fishing, enjoying pleasant meals and conversation, artistic and religious activities, etc (Gotō and Sakamoto 1975).

According to Kropotkin (1968, 1970), modern society managed to separate agriculture and industry. This division of labor led to economic profits, but at the cost of many other things that were lost. Implicit in this claim is an emphasis on the social benefits of unity. He claims that the social ideal is a society in which each person takes part in both physical and mental labor and that there must be a unification of agriculture and industry, rural labor and regionalized industry. Each member of society should be a rural farmer and a factory worker as well, and each combines within himself scientific knowledge and industrial art knowledge. This is exactly what is desired today by people of civilized nations. Here, as in Fourier, the total human life formation is emphasized. These ideas were later incorporated in Howard's 'garden cities' and by Schmidt.

More recently, Leiss (1976) argues that capitalism and socialism both consume an enormous amount of energy to expand industrial production and material satisfaction, and both have proceeded down the path of pursuing happiness through endless self-destruction. Material satisfaction is merely economic happiness, a partial satisfaction, a passing happiness. The wealth we created is false. Nowadays being oppressed to power, different possibility of the satisfaction, possibly taken from production activity of various forms or relations of people and nature in various forms, can contribute to our needs even more effectively than any new combination of goods. In short, seeking a broader, total satisfaction is an issue for future societies.

Polanyi (1975) offers a different perspective on the same problem. Market societies are economically determined societies in which every possible aspect of human life is judged by means of a single standard, the 'economy'. Even labor and land have been commodified, treated as if they exist solely in order to be sold.

The 'essential value of human life' is oppressed, and in its place the illusion that economic man is the true man is created. Polanyi argues for the need to begin work to rebury the economy within a more broadly defined society, a society that embraces and enables widely diverse modes of being.

All of these positions stress the importance of a comprehensive human life. They all support a position that considers the creation of regional society as a multi-faceted living space and a total everyday lifeworld to be important. As discussed, modern scientific technology and economic activities have curtailed and even obstructed such developments by ignoring the complexity of the natural environment, forcing complex human beings to conform to a one-dimensional economic model, and thus dislocating people from their 'homes' ('placelessness'). Philosophers of science such as Takashi Maruyama explain that what is described as a 'return to the lifeworld' is an attempt to rescue the simple world of life-knowledge, or tacit knowledge, from a narrow world of scientific knowledge that views itself as *doxa*. In other words, it is an attempt to reassert these other knowledges as an 'ultimate foundation' for creating ways of life that can overcome the problems created by modern scientific knowledge (Maruyama, Ogawa and Noe 1987). This lifeworld, in fact, is what I have referred to as the region.

Some have referred to the lifeworld that develops in the local region as a human 'life-size society' (Schumacher, Yoshirō Tamanoi and others), as 'relationships with a human face', a 'face-to-face society', or 'a world that shares climate and culture'. It is a place where people can enjoy direct communication, a world in which it is easy to engage in conversations and to act together. The need to overcome the maladies of market society can be recognized as concrete and urgent, check the excesses of industrialization and urbanization that are determined by hyper economic rationalism, while protecting the ecological environment and valuing the logic of everyday life. In short, we are discussing the formation of a comprehensive living space that suits the full spectrum of human nature, that is rich in terms of diversity, stability and sustainability, and will thus create a new type of wealth.

The above presents my theoretical understanding of the region. It might be useful now to outline what this would mean in concrete terms for those who live in farming villages, using the concepts of production (economy), ecology and life.

As we have seen, these three aspects are all interrelated. Without rising economic standards, the quality of life will not improve. But when the economy is the sole priority, the environment is damaged and life's fullness is lost. Also, attempts to improve the environment demand changes in life styles, economic regulations and farming methods. How to do this is inseparable from the question of whether the concept of the ecological environment should be restricted to 'nature as it originally existed', or includes 'human-made nature' as well. All three aspects intertwine, and it is difficult to clearly distinguish them. Hence, while I have employed these three concepts to describe the region as a place for life, by doing so, I have risked reproducing a conceptual schema that lies at the heart of the various problems facing us today. At the same time, though, I think that using this schema has served to highlight the intricacies of these problems.

Table 7.1 presents a concrete description of the three aspects of economy, ecological environment, and life. The economy is 'the distributive quantity of economic resources and the level of production activities of the various industries that make use of these resources, as well as the condition of the various facilities and services that support it'. The ecological environment is 'the condition of the natural ecosystem – including geographical features, climate, and plants and animals – of the region as a place of human productive activity and life actions'. Life is 'the entire area of the family and social life of the people of the region, together with the various facilities and services that support this'. To improve these conditions is to facilitate the activities of the people who live within the region, and therefore to improve economic value, ecological value, and life value as much as possible, thus providing greater satisfaction. At the same time, to even partly overcome the contradictory relationship between these three values and to move towards a harmonious relationship where each value supports the others is to form a sustainable farming village region, to engage in creating a new region. The important point here is, as discussed in the next chapter, the premise that sustainable farming villages need to be combined with and merge into small- and medium-sized cities for the sustainable farming village region to be complete as a region.

Of course, we cannot expect the creation of such local regions to be sufficient to rectify the lifeworld, nor can we assume that a harmonious coexistence with nature is inevitable. If it were, the mad leap away from these total values through the twin pursuits of technology and economy could not have occurred. In relative terms, however, in order to identify the problems of today as 'problems', and to reconstruct our own reality in such a way that we can address these problems, the region as I have defined it in this chapter promises a way forward. The resolution of the various problems of the modern age requires that the residents of each region become aware of the problems, create new plans and concrete practices, and join together to recreate the region.[2]

Table 7.1: Specific contents of 3 aspects (economy, ecological environment, life) of the local region.

Economy

The economic resource potential in the local region and production activity levels of various industries using such resources, and the conditions of various facilities and services supporting such activities.

Actual conditions of agriculture, forestry, fishing industries, commerce and manufacturing industries, service industry, etc; production and sales; income (agricultural income, non-agricultural income, etc.); efficiency; total acreage under cultivation; average farm size; workforce; presence or absence of successors; agricultural methods; means to obtain information (organization information and information systems, training, inspection); working environment; price level; production system; distribution system; leaders; producer-consumer cooperative organization; irrigation ponds; water supply; transportation (general road, farm road, forestry road, railroad, bus); distance from town; etc.

Ecological environment

Conditions of the local region's natural ecosystem including geography, climate, flora and fauna that constitutes the place of people's production and living activities.

Geographical features; soil; altitude; climate; forests; rivers; ponds; flora and fauna (trees, grass, animals, insects, fish, microorganisms, etc.); aquatic environment (water holding, water storage, springs, water quality, water quantity, etc.); atmospheric environment; seasonal landscapes; etc.

Life

Residents' personal, family and social lives as a whole and the conditions of various facilities and services supporting such lives.

Personal and family lives
Details of food, clothing and shelter; size of residential land; room structures; shopping; health care; childcare; education; culture and training; amusements; recreational activities (hobbies, cultural group activities); security, quiet sleep and rest in life; happy circles of family and relatives.

Social life
Communication (within the local region, with town, with foreign local regions, between generations, between different industries, etc.); traditional events; religious associations; mutual aid; various social activities; religious activities; group work (road repair, river dredging, etc.), meetings and management of neighborhood associations and settlements; crime/fire prevention organization; secure access to various information.

Facilities, etc.
Educational facilities (childcare facilities, kindergartens, schools of various levels) etc.

---

8 The Combination of City and Country

In this chapter, I investigate the relationship between city and country, in relation to the formation of sustainable rural areas or 'total places of human life' (*ba* that have comprehensive values) as introduced in the preceding chapter. This relationship is the precondition to this formation. First, I provide an overview of conventional theoretical examinations of the relationship between city and country (urban-rural theories) and establish this chapter's position: the marriage of the city and country. Second, I will examine relevant European thought and practices about combining small-and-medium cities with country areas, with a focus on Germany, which has relatively good relationship of city and country. In this examination, I pay careful attention to four stages of development of urban-rural theories: (1) the stage that identifies the need to incorporate pastoral elements into highly congested industrial cities; (2) the stage in which such theories become concerned with the combination of cities and their location; (3) the stage in which cities and the country are viewed on an equal footing, and a desire to combine the two is vocalized; and (4) the stage in which policies emerge that aim to correct the disparities between rural villages. Third, from this vantage point, I investigate the current situation and long-term prospects of the urban-rural relationship in Japan.[1]

The genealogy of urban-rural theories

The problem of the relationship between the city and country is both old and new. In sociology, Sorokin and Zimmerman (1929), among others, proposed a dichotomous urban-rural theory that typologically separates urban areas from rural areas based on variables such as population size and density, residence patterns and the heterogeneity of community. Others have theorized an urban-rural continuum, rejecting the dichotomous framework in favor of a continuous difference between rural and urban 'poles', and placing all community systems on different points along this continuum (Wirth 1938). After the publication of Wirth's 'urbanism theory', the urban-rural continuum theory became the mainstream position of urban-rural theories. But another position was formulated by Galpin and others who saw rural and urban areas as a single 'rurban community' (Galpin 1915; 1918; 1924). In my understanding, in intellectual history, rural areas have often been understood merely as a periphery of the city, providing land, labor and capital to the latter. Conversely, seen from the countryside, cities are parasites that feed off the products of the farming village, and are sites of human and cultural corruption. Some theories argue that the city and country have a confrontational relationship in which the two are in constant conflict, while others note the functional reasons for their mutual existence, and argue for their complementary combination – hence, urban-rural combination theories.

While distinguished by different social, economic and natural conditions and characteristics, rural and urban areas coexist in a relationship of mutual influence vis-à-vis one another. Furthermore, the relationship is simultaneously confrontational, discordant, complementary and integrated. In other words, none of the urban-rural theories are wholly adequate; the various theories about the urban-rural relationship have had more-or-less influence depending on their time, circumstances, and ideological positions. I maintain that one can only correctly understand the urban-rural relationship if one pays due attention to division and combination, confrontation and cooperation, and comprehensively interprets history from both quantitative and qualitative perspectives.

A comprehensive understanding is all the more indispensable today when, due to advances in information technology, people's life spheres have expanded and the news of the world spreads as images, letters and words to every corner of the world instantaneously. Furthermore, it has become quite clear that the problems of industrialization and urbanization that confront societies today, while showing some variation, transcend the politico-economic structural issues of capitalism and socialism, and should be understood as problems of modern civilization *in toto*.

This is the position adopted in this chapter, my aim being to discuss the combination of the city and country from a comprehensive perspective. By focusing especially on European cases, I aim to clarify the historical stages of urban-rural relationship step-by-step (cf. Soda 1997).

European thought on urban-rural combination

The garden city

The garden city theories of E. Howard (England), Fritsch (Germany) and others emerged as measures to address the various problems caused by advancing industrialization and urbanization after the Industrial Revolution. These included the disorderly expansion of cities (e.g., London), noise, smoke-laden air and the formation of slums. In the first stage of the garden city movement, resolving urban problems was the priority. While rural space and scenery was deemed to be necessary for this purpose, agricultural areas themselves were treated as only secondary concerns; until Schmidt pointed a way to a new stage.

Howard's 'garden city' in England

Lewis Mumford argues that two great inventions were bestowed upon humankind at the beginning of the twentieth century: the airplane and the garden city. In short, the airplane gave humans the freedom to fly the skies, and once on the ground again, the garden city promised them the most humane living environment (cited in Howard 1965: 29).

Howard's garden city theory (1965) has indisputably had a significant influence on the inhabitants of congested modern cities and remains influential today as a model for resolving urban problems. In Germany, stimulated by Howard's *Garden Cities of To-Morrow* (1898), the garden city movement emerged as early as 1902.

Let me explain the background and main points of garden city theory.

The Industrial Revolution that began in late 18th century England transformed society in fundamental ways. For Adam Smith, this meant freeing the individual from the shackles of feudal society and establishing a system of natural liberty that would guarantee people the right to engage in self-interested economic activity and enjoy urban freedom.

However, this was at the time of the emergence of a capitalist society that obeyed the cold logic of capital; a society that soon revealed its darker side by gradually separating capitalists and workers, as well as cities and rural villages, and pitting these against one another. The population began to be concentrated in the cities, starting with London, where factories were built in large numbers, and which were therefore shrouded in soot and smoke. Workers had to contend with harsh living conditions and low wages. Eventually, sunlight and fresh air disappeared from the cities and most working-class neighborhoods became slums. Infectious diseases such as cholera and typhus were rampant in these areas, and cyclical economic depressions of varying intensities regularly played havoc with people's lives. The intellectual classes gradually started to take note of this aspect of the economic system and its affects on the living conditions of city dwellers, and began to demand reform. The garden city theorists paid close attention to these urban living conditions and environments. Central to their position was the idea that the city could be resurrected through integration with nature and agriculture. Examples of this approach can be seen in: R. Owen's Community, which adopted the cooperative association system; Fourier's Phalanges, communes that prioritize agriculture and subordinate industry; Buckingham's Model Independent Industrial City, Victoria; and Cadbury's 'Garden City'. Penberton's Happy Colony, Richardson's Healthy City, Le Corbusier's Honeycomb-like Garden City and Lever's Port Sunlight are further examples of this idea.

Howard's garden city theory aimed to resolve the urban-slum conditions of his day based on the ideas of the day.

The Garden Cities and Town Planning Association and Howard defined a 'garden city' as 'a town designed for healthy living and industry; of a size that makes possible a full measure of social life, but not larger; surrounded by a rural belt; the whole of the land being in public ownership or held in trust for the community' (1965: 26). Therefore, in such a city one can lead a life where 'the advantages of the most energetic and active town life, with all the beauty and delight of the country, may be recurred in perfect combination' (1965:45).

Rather than having to choose between either urban or rural living, a third option has been made available. This position challenges the almost universally held view that it is now, and must forever remain, quite impossible for working people to live in the country and yet be engaged in pursuits other than agriculture; as though crowded, unhealthy cities were the last word of economic science; and as if our present form of industry, in which sharp lines divide agricultural from industrial pursuits, were necessarily an enduring one (Howard 1965: 45). In short, Howard proposes a marriage of agriculture and industry, and of rural villages and urban cities. Figure 8.1 graphically summarizes his arguments. In brief, the goal here is to realize a form of human social life that has erased the shortcomings and combined the advantages of cities and rural areas by creating a 'third alternative', i.e. the garden city.

Figure 8.1. Howard's town-country magnet

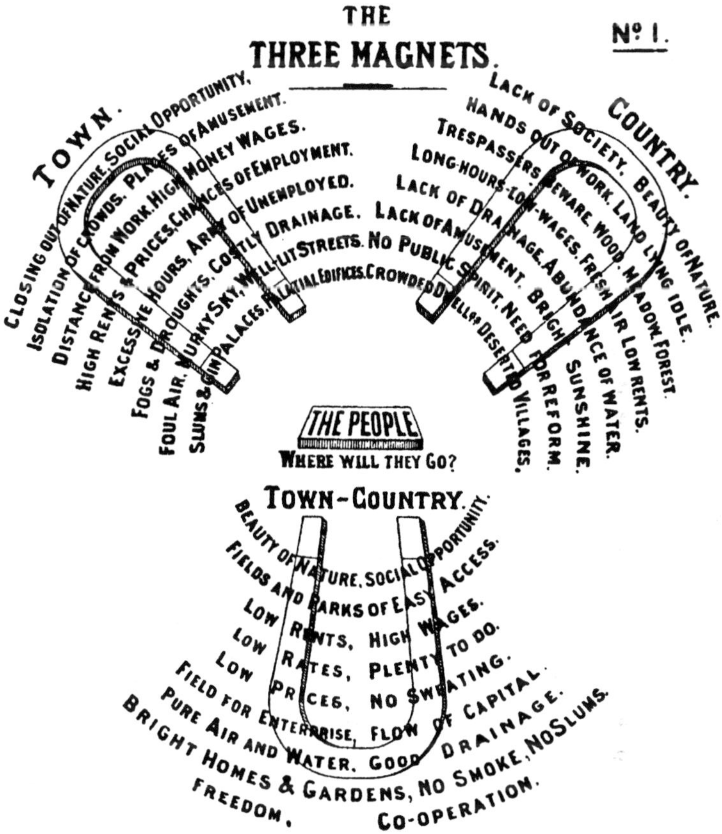

Source: Howard 2000: 46.

Howard's garden city consists of an area of 5,000 acres, including a circular urban district covering 1,000 acres. Its areas are connected by radial and circular roads and railways, with 2,000 people living in the rural district and 30,000 in the urban neighborhoods, a total population of 32,000 residents. Public gardens, a public hall, a theatre, a library, a museum and a gallery, a hospital, recreational facilities and shopping centers are located in the central district, surrounded by a residential area consisting of 5,500 house plots of approximately 238 m^2 each, plus schools, churches, and parks. The roads throughout are tree-lined. Furthermore, this district is ringed by a variety of factories and a railroad. Farmland surrounds the city area, scattered

Figure 8.2: Howard's garden city and rural belt

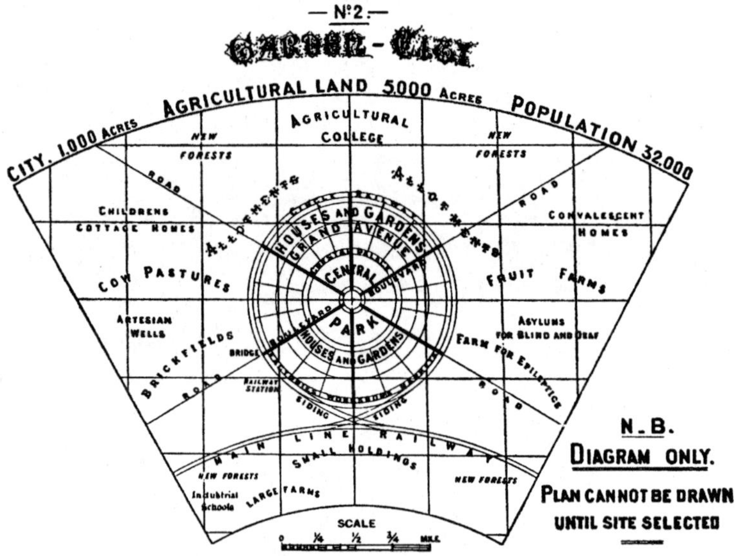

GARDEN CITY AND RURAL BELT

Source: as per Figure 8.1.

with vocational schools, brick factories, day-care and welfare facilities (see Figure 8.2) (Howard 1965: 51–3).

Two garden cities, Letchworth and Welwyn, were built in London's suburbs based on this design, with the intention of alleviating the overcrowding of the metropolis. Howard's *Garden Cities of To-Morrow* had a wide international impact, being translated into myriad languages and influencing, for example, the creation of the garden city at Chōfu, Japan.

Fritsch's 'city of the future' in Germany

While there is no doubt that Howard's garden city theory influenced Germany's garden city movement, a similar proposal had already been put forward by Fritsch, a German citizen, in his *Stadt der zukunft* (City of the Future) in 1896, two years before the publication of *Garden Cities of To-Morrow.*

Like Howard, Fritsch focuses closely on the problems brought about by the progress of industrialization and urbanization: the concentration of population in cities and the accompanying urban problems. He views the culture of large cities as unhealthy. According to Fritsch, large city culture attracts and encourages various evils; it is a 'pigsty of a culture' that makes people imbeciles. Rural areas, in contrast, are viewed as the source of national vitality and health (1896: 5–6).

However, Fritsch does not reject cities outright. They are centers of production and distribution that supply products necessary for a cultured life. They also function as centers of administration, higher education, and art. Only cities can adequately satisfy these demands. In sum, while recognizing the importance of rural areas, Fritsch cannot envision modern life without cities (1896: 6). For Fritsch, the problem is unplanned and aimless city construction; the objective must be to build humane and beautiful cities where people can lead a healthy life.

Fritsch's design for the future city is depicted in Figure 8.3. In his model, the semi-circular city is covered by parks, surrounded by

Figure 8.3: Fritsch's future city

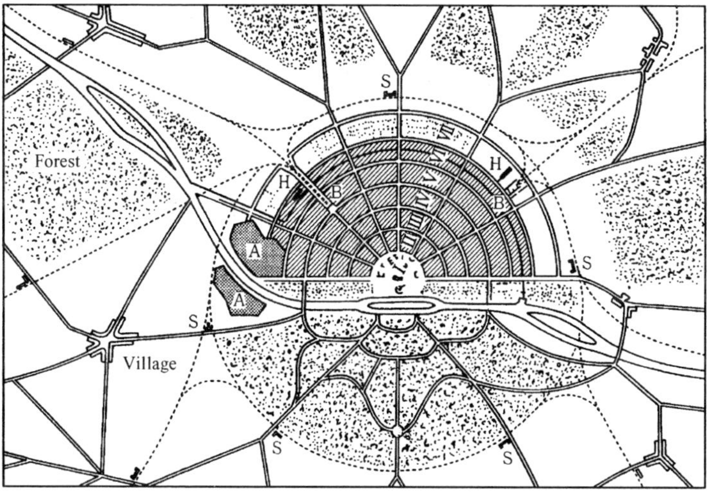

I public facilities
II memorial buildings
III upper class residential areas
IV houses and shopping centers
V working-class houses and small factories
VI factories and warehouses
VII gardening and residents' rented gardens (Klein-garten)

A preserved urban area
B railway station
H waterway port
S loop railway station

Source: Fritsch 1896: 25.

farming villages and forest areas. The city is connected to the outside world by railways, roads and canals, and is internally linked via circular and radial roads, with a railroad running around its circumference. A public facility area is located in its center to symbolize the city. Moving away from the center of the semi-circle towards its borders, the following districts can be found: a commemorative building area, an upper-class residential area, a residential and commercial area, a working-class residential and small industry area, an industrial and warehouse area, an area of gardens and citizens' farms, and an agricultural area (1896: 11, 25). This was used as the model for planning the southern-German town of Karlsruhe (cf. Bergman 1970: 149).

However, Fritsch's model city of the future was not widely adopted in practice by the German garden city movement or in city planning for a number of reasons: first, because Fritsch was anti-Semitic; but perhaps more importantly because Howard's theories were considered more comprehensive and socially valuable; and because England – where Howard's theory originated and was first adopted – was generally regarded as the most advanced European nation.

R. Schmidt's industrial–residential garden city

Amidst the processes of modernization, industrialization and mass urbanization, Schmidt observed that cities nevertheless develop autonomously. He describes the new cities as *Industrie-Wohnstadt* (industrial-residential cities) (1912: 6).

Schmidt conducted a field trip to the first garden city, Letchworth, five years after Howard's successor, Unwin, had begun working on its construction. Thereafter, with Essen as his starting point, Schmidt set to work on city construction in the Ruhr region. Schmidt's urban theory, in essence, had contents that justified calling it a garden city theory, as noted by Phannschmidt, who claims that Schmidt's '*Industrie-Wohnstadt*' (industrial-residential city) should be called an '*Industrie-, Wohn- und Gartenstadt*' (industrial–residential garden city) (1970: 62). Schmidt's urban theory discusses the city within the context of planning wide areas and regions to integrate city and country, and is thus fundamentally different in character than the isolated city constructed at Letchworth.

Schmidt's urban theory takes the principle 'humans instead of factories' as its point of departure. Schmidt states that humans have a fundamental need to work, live and play within every 24 hour period (Schteinhauer 1967: 12). The 'violent expansion' of factories, destroyed the general living environment of the cities. The air was polluted, the sun stolen, and green areas were retreating. Factories must be situated not solely for economic convenience, but with consideration to their relation to residential areas. Schmidt stresses that 'pure' residential areas with favorable environmental conditions must be built separately from areas of industrial production, one-quarter of the city area should be reserved for greenery and recreational use, and that as well as municipal roads, roads in residential areas must be expanded.

Schmidt recognized a tendency for propertied people to remain content with the status quo and not consider future needs. Hence, he argued, to provide for the expansion of roads and public facilities such as stations, green areas and graveyards, it is necessary to set-aside considerable amounts of public land. Limitations on private landownership are therefore unavoidable. Schmidt was not a dreamer but a practical planner (Schteinhauer 1967: 14–15).

Schmidt also took note of the German tradition of vegetable gardening (Schteinhauer 1967: 25). The laborers who had flocked into the Ruhr region from agricultural areas had been raised on farms and had spent a lot of time doing farm work. For them, according to Schmidt, it was important to be able to work in vegetable gardens even *after* their arrival in the city. This understanding was later put into practical action by the Ruhr Coal Mining Area Development Union and applied in many locations.

Schmidt's urban theory thus addresses many of the problems of the 'industrial city', arguing for the creation of industrial-residential garden cities that provide a comprehensive human living environment.

Figure 8.4 depicts Schmidt's model industrial–residential garden city. From the perspective of urban planning theory today, some inadequacies are evident. Nevertheless, Schmidt's ideas are an important contribution for having established a clear position that encompasses the three aspects of industry, living and nature, and for stressing the importance of planning cities in which people could 'work, live, and play' 24 hours a day.

Schmidt's planning theory (1) progressively and comprehensively synthesized all of the (then) current streams of thought about solving urban problems, (2) introduced the idea of wide regions that integrate cities with one another, (3) established the concepts of industry, living and nature, and sought their integration, and (4) was applied in practice in Ruhr's regional plan, thereafter becoming the model for Germany's national land improvement policy (*Raumordnungspolitik*) and regional policy (Frotriep 1970: 53).

Figure 8.4: Robert Schmidt's 'industrial–residential garden city'

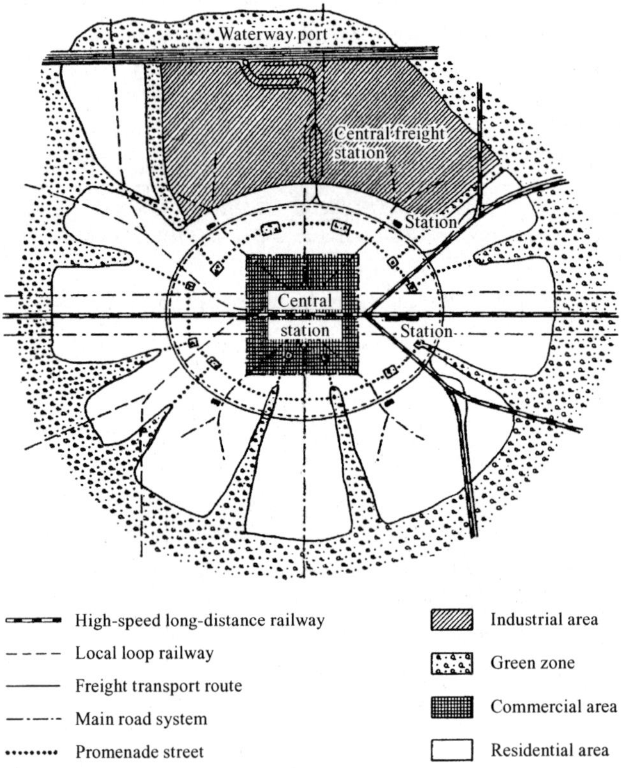

Source: Schmidt 1912: 84.

The dispersed location of cities theory

The first moves towards intercity relations can be found in Schmidt's work, proceeding from the garden city theory that proposed a new, humane living space through the integration of rural and urban areas, and by relieving congestion in large cities. Christaller theoretically clarified the key issues of this second phase of urban-rural theories. Christaller's hexagonal or net-like city (central place) allocation theory provided the theoretical background for subsequent German land improvement policies.

Christaller's central place theory

The core of the theory behind Germany's land improvement policy was proposed by Isbary and others, who applied Christaller's central place theory to that policy. The central place theory emerged from a theoretical investigation of the 'already-developed' spatial structure of Germany. As a theory for areas that 'must be consciously developed', this became the core theoretical background for land improvement policy in postwar West Germany (Figure 8.5).

Figure 8.5: An illustration of Christaller's central place theory

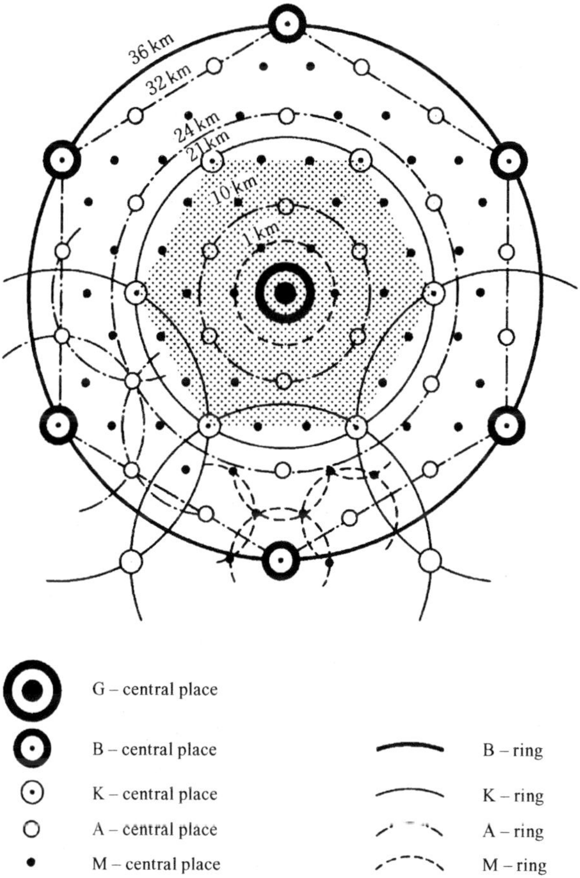

Source: Kroner 1964: 421.

The central place theory had been developed by Müller, Thünen, Kohl, Schlüter and Gratmann, and then completed by Christaller. Although his study, *Die zentralen Orte in Süddeutschland* (The Central Place of Southern Germany) (1933), was not initially well received in Germany, it was well-regarded in the USA and was re-evaluated in postwar Germany when its land improvement policy was being established. It received support from Kroner and Isbary, especially in the latter's thesis of the 'Central Place and Service Areas' that made central place theory the centerpiece of land improvement policies as well as plans for the dispersal and allocation of small-and-medium cities (Isbary 1965).

Röpke's theory of 'true decentralization'

Röpke led postwar West Germany's economic policy, but he also had a theory of national land policy which introduced the terms 'spurious decentralization' and 'true decentralization'. Röpke calls 'the ribbon development (suburbanization)' of large cities false decentralization. This is better than congestion in the city center but consists only of a simple dispersal of sleeping quarters. This means commuting to work, shopping and entertainment, and generates a waste of time and money. Traffic problems increase, upper-class and working-class residential areas are segregated, and other specialized kinds of land use emerge. As early as 1944, Röpke predicted the generalized 'traffic hell' that characterizes mega-cities today (Röpke 1996: 162–3).

In a nutshell, Röpke's 'true decentralization' refers to 'the creation of fresh small cities in lieu of the big city'. 'Only thus shall we create real communities and the conditions of a natural existence for human beings' (1996: 163). Under these circumstances, the evils of the concentration of power and financial wealth, excessive systematization, specialization and division of labor can be prevented. 'They must be directed in only one direction...to the bringing back of all dimensions and proportions from the colossal to the humanly reasonable'. The economy must be reconstructed according to 'human factors' and be suitable to the 'healthy form of life'. Furthermore, the social point of emphasis should be shifted 'to the restoration of property; to shifting of the social centre of gravity from above downwards; to the organic building-up of society from natural and neighborly communities in a closed gradation stating with the family through parish and country to the nation' (1996: 154). Röpke's aims, then, include curbing concentration and growth, dispersing into small-and-medium cities, and bottom-up management and reintegration.

By integrating the theories of the social market economy and regionalist local planning, Röpke wielded a great ideological influence over the formulation of postwar local planning by Bülow and Dittrich. His theory of local planning rejects large cities in favor of a scheme of regionally dispersed small-to-medium cities integrated with the surrounding agricultural areas. Röpke argues that only under such conditions can the acquisition of property, i.e. land, homes and gardens be realized and the humans who own them enjoy a humane life in accordance with natural rhythms.

Röpke's *Civitas humana* (The moral foundation of civil society, 1944) not only lead postwar Germany towards the social market economy model, but its great achievements include directing the country into high economic growth under efforts to create a dispersed form of national land use. Japan's high economic growth, on the other hand, was accompanied by drastic rural depopulation and urban overcrowding that distorted its national land use structures.

Dittrich's multiple-nuclei decentralized space

According to Gildemeister, the guiding principles behind West Germany's current land improvement policies were created and formalized by Dittrich almost single-handedly. Dittrich's thesis is based on Röpke's and Bülow's regionalist land improvement policy principles. In 1961, he became a central member of the federal government's expert committee to establish regulative strategies for regional policies (SARO, established in 1953). Dittrich compiled the famous SARO report (1961) and in 1965 contributed to the establishment of West Germany's Land Improvement Law (*Raumordnungsgesetz*) (Gildemeister 1973: 40–1). Let us briefly examine the relevant additions he made to urban-rural theories.

In contrast with early liberalism, Dittrich argued that the distinctive characteristic of modern economic theories was the high density of 'social streams of thought' (*soziale Strömungen*) contained therein. Focusing on the 'social' momentum evident in concepts of the 'social market economy', 'social constitutional state', and 'social land improvement', Dittrich's land improvement theory is rich in terms of social policy.

In the SARO report, Dietrich provides a chapter titled 'Principles for Land Improvement as Part of the Guiding Ideals of Social Policy' and examines three points: (1) the principle of freedom, (2) the principle of social balance, and (3) the principle of security.

The principle of freedom refers to, in concrete terms, guaranteeing the freedoms to decide one's 'location, residence, consumption, occupation, reproduction, and workplace', among others (Dittrich 1961: 52–3). The principle of social balance points to correcting the disparities between different regions, taking as its goal the 'sufficient equalization' and development of West Germany's entire citizenry. This does not mean that life in all regions would be uniform, but rather 'the identical value of living environments' (*Gleichwertigkeit der Lebensverhaltnisse*). Acknowledging the various contrasts and differences between 'areas of high population densities and agricultural areas', this principle attempts to correct the qualitative gaps between regions (Dittrich 1961: 32).

The principle of security is concerned with security in the workplace and security of health. Security of the workplace refers to minimizing the various employment-related risks that arise from a regionally unbalanced industry structure. Security of health concerns the protection of the citizens' health from various environmental pollution problems such as soot and smoke, exhaust gases, noise and water pollution that often occur as a result of unrestricted production practices and locations. If the economy is viewed as essentially a tool to secure the existence of humans, this order of priorities is natural.

It is clear that the two principles of social balance and security entail limitations on the principle of freedom. Dittrich calls them 'regulators of freedom' (1944/1961: 42). It is understood that land improvement policy, while taking these principles as its basis, takes the regulation of freedom as its principal aim.

In a nutshell, we can call this land design pattern a 'multiple-nuclei decentralized space'. By contrast, France, England and Japan exhibit land design patterns that can be understood as 'single-nucleus centralized space'. During my stay at West Germany's Regional Planning Academy, this point was repeatedly emphasized by the academy's head, Haupner, who drew the illustration in Figure 8.6. Underlying these differences are natural, historical and policy-related factors.

Attempts at regional policies in England and France that aim to alleviate the congestion in London and Paris have been based on a reactive and passive 'principle of decentralization'. By contrast, as well as the 'principle of decentralization' that rejects large cities, I argue that a more advanced and active 'principle of combination' is manifest in

Figure 8.6: Types of territorial space

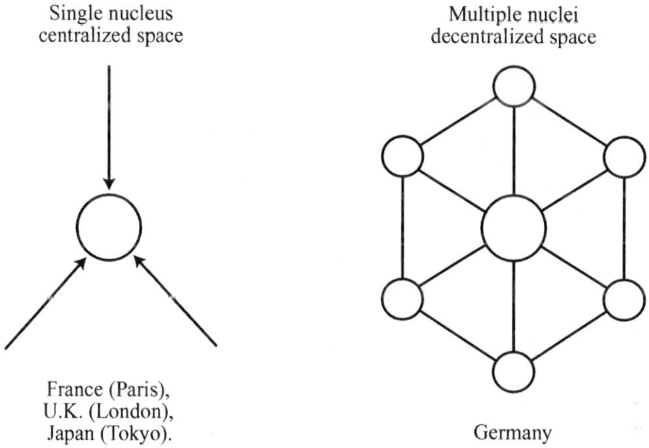

West Germany's regional policy. This principle addresses issues such as how to develop scattered small-and-medium cities and combine them in an organic way. Furthermore, as I explain in detail below, it is concerned with combining these small-and-medium cities with farming villages, and agriculture with industry, thereby progressing to a 'third stage' in land planning. I believe that West Germany's regionalist land improvement policy must be seen as a formulation that harmonizes these principles of decentralization and combination and takes advantage of the tensions between them.

The urban-rural combination theory

The regional planning policies of the 1960s were markedly different from previous planning theories that, while paying some attention to rural areas, still took cities as their central focus. These new policies clearly placed cities and rural areas on an equal footing, or prioritized the latter, and adopted national regional planning based on equality, thereby clarifying the ideas of the so-called 'third stage' of urban-rural theories, encompassing the combination of city and country.

Integrating agricultural and regional restructuring policy

In an effort to restructure agriculture, the importance of regional economic structuring policy within land improvement policy and broader regional structuring policy cannot be neglected. The West German Land Improvement White Paper, titled 'Restructuring Regional Economies' (*Gemeinschaftsaufgabe 'Verbesserung der regionalen Wirtschaftsstruktur'*), adopted in 1968 as a common policy of both the federation and the states, can be considered to be the most fundamental policy in land improvement.

Similar to Japan, during the period in which Germany went from reconstruction to high economic growth (1951–65), problems emanating from the widening gaps in productivity and income between agriculture and industry gradually began to appear. An agricultural law aimed at rectifying this problem was passed in 1955. Its objective was to strike a 'balance' between agriculture and industry by way of agricultural restructuring policy, and to prepare the economy for the overwhelming trend towards trade liberation. Agricultural restructuring in this case meant the redeployment of arable land, expanding the scale of enterprises, mechanization, the provision of cheap agricultural materials, the financial measures required for this, and revitalizing agricultural education, among other measures (Kunigō 1960: 319). Similar policies were promoted in other EU countries as well. However, with West Germany's traditional orientation towards decentralization, corrective land improvement policy was successfully enacted as an early response to the concentration of population and industry in large cities. The agricultural plan of 1968 acknowledged that: the (then) current principle of maintaining constant numbers of agricultural laborers was contradictory; it was necessary to expand the scale and improve the efficiency of agriculture; this should not lead to an outflow of rural residents; and the provision of steady non-agricultural work for farmers with insufficient livelihoods was the key to successful structural policy (Zenkoku Nōgyo Shinbun 1961).

Herein lies a major point of difference. In Japan, it was accepted as axiomatic that declining numbers of agricultural laborers cause a corresponding decline in the number of rural residents. Under pressure from large cities, small-to-medium cities experienced unstable employment conditions. This resulted in massive flows of young people from the country into the large cities that continued their ferocious expansion, free from any outside interference. However, in West Germany, it was decided that even though declining numbers of agricultural laborers was inevitable, falling numbers of rural residents was not desirable, and that in order to maintain the numbers of rural residents, special policies were required. It is self-evident that if one is to expand the scope of individual agricultural businesses and simultaneously sustain the number of rural residents, stable non-agricultural employment opportunities in nearby small-to-medium cities must be provided before any reductions in the size of the agricultural workforce can be affected. This idea can be called 'stationary industrialization'; that is, a way of thinking that proposes creating stable working conditions to enable people to stay in their villages. This philosophy conceives of cities and agricultural areas as integrated, rather than separated.

Barner's urban-rural theory: commutation possibilities

By Barner's time (1975), an urban-rural combination theory formulated from the perspective of rural advancement provided the foundations for all land improvement policies in West Germany. Barner and Wehling theorized and schematized an 'urban land-use model' (Figure 8.7), which examines urban-rural problems from the perspective of bringing the issue of commuting to the fore.

Barner argues that 'this urban concept is clearly created with regard to the possibility of commuting, taking as its basis a roundtrip

Figure 8.7: Schematic diagram of urban area

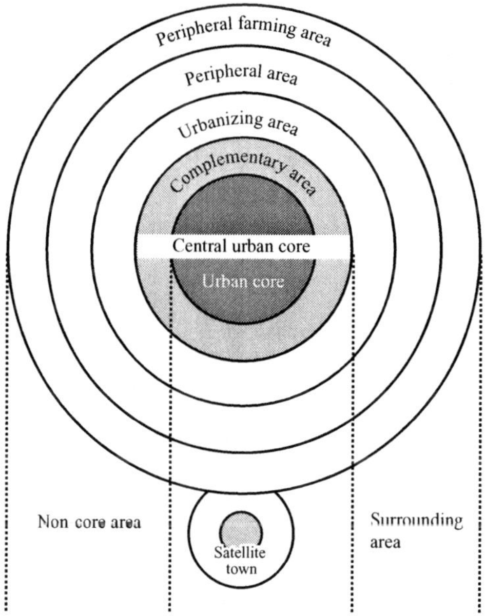

Source: *Geographische Zeitfragen*, 1982, No.7, p.6.

commuting time of no more than two hours (one hour each way)' (1975: 89–90).[2] This conceptual development of the possibility of commuting perceives of urban and rural areas as a unified whole. It allows for some agriculture within the city and assumes that farming families will be at least partially employed in non-agricultural businesses as well as part-time farming; hence, 'rural city' would be an appropriate term here.

The 'regional economic restructuring' policy launched in West Germany in 1968 aimed to secure the preconditions for this conception. The key points of this policy, to repeat, are to: designate the small-to-medium cities scattered throughout economically weak rural areas as 'priority districts'; grant federal and state subsidies for infrastructure investment by cities, towns and villages; subsidize companies that expand to or initiate new activities in targeted districts by 15 to 20 percent of the amount invested (with some variation between states) through grants or reduced taxes; and, create employment in non-agricultural industries such as commerce and manufacturing.

The prototype for this model had been established much earlier; in 1959, sixteen locations had been designated as 'priority districts'. In 1968 this increased to 81 locations, following the adoptions of the regional restructuring policy. In 1975, 178 existing labor market spheres were subdivided in consideration of commuting access, to create 312 'priority districts'. Although these new 'priority districts' did not necessarily coincide with the existing 'central cities', it was mainly the medium and large urban centers that were targeted by this policy. While the ideal was to make it possible to reach a medium center (or small-to-medium city) within 35 minutes, the concrete task of the regional economic restructuring policy was focused on reducing the number of areas that did not conform to this aim.

Rectifying disparities between different rural areas

The production conditions in different agricultural areas differ in geographical conditions – for example, some are flatlands and others are in valleys. The proposition that some financial compensation is necessary to strike a balance in such cases is gaining momentum across the EU. This can be seen as evidence of a new, fourth-stage in regional planning theories that acknowledges the disparities in conditions between various agricultural areas and attempts to eliminate them. Here, the severity of environmental problems and the multiple functions of agriculture are given due attention.

The locations characterized by disadvantageous conditions (less-favored areas) are, in economic terms, marginal and peripheral areas that are subject to a high risk of agricultural withdrawal. To justify the maintenance and development of such areas through any sort of compensation scheme requires a new theoretical orientation. Such a re-orientation is provided by the multi-value or multifunctional theory of agriculture; or, in economic terms, the theory of external economies. In short, where a particular area is considered to be marginal in strict economic terms, these theories propose that it be maintained and developed for its positive externalities, such as its role in national land conservation, disaster prevention, storing and cleaning water, scenic benefits and other functions provided by agricultural production activities. In other words, this is an area in which the corrective principles targeted at rectifying market failures must be applied. Although I am unaware of any cases where such a payment scheme has been adopted, having stipulated that agriculture is a multi-value industry and that agricultural areas are multifunctional spaces, I support corresponding direct income compensation policies (Ministry of Agriculture, Forestry and Fisheries 1999).

In this way, implementation of direct payment policies for less-favored areas marks the birth of a completely new approach to the application of corrective principles in agricultural policies. From now on, measures to minimize negative externalities, such as the provision of aid for the transition towards environmentally friendly agriculture that supplies safe and environmentally harmless food products are likely to become one of the most important policies. In the EU, these kinds of measures have already started to be implemented.

Establishing the EU's land improvement policy

So far I have examined the following four stages of relevant theory: (1) the garden city theory, (2) the city combination and allocation theories, (3) the theory to facilitate urban-rural combination, and (4) the correction of disparate conditions between agricultural areas.

In the 1990s, EU member countries launched a common land improvement policy. This was effectively an extended application of Germany's regional restructuring policy, adopting the ideas of 'multiple-nuclei decentralized space' and the combination of small-to-medium cities and rural areas based on German-style regionalism (Bundesministerium für Raumordnung, Bauwesen und Städtebau,

1995: 10; Federal Ministry for Regional Planning, Building and Urban Development; 1994: 2).

As we have seen, in a little more than a century, urban-rural theories have gradually been refined and systematized. Born in England and brought to maturity in postwar West Germany, they have now been applied across the EU.

Implications for Japanese rural area policies

From this perspective, several issues concerning Japan's rural area policy come to the fore (Soda 1999; 1997; Soda et al. 1996).

First, while Japan does have an agricultural structural policy, it has never had a regional structural policy. The first and second national land policies both assumed that concentrated investment in the Tōkaidō megalopolis (the urban belt connecting Tokyo and Osaka) was required to facilitate Japan's economic recovery and rapid growth. The consequence was an extreme polarization: a heavy population concentration in large cities and a corresponding rural exodus. Agricultural expansion was constrained by the sudden disappearance of rural labor. The outcomes were therefore drastically different than in Germany, which provided employment opportunities in small-to-medium cities through its regional structural policy, building well-balanced spaces on the principle of multiple-nuclei decentralization. In contrast, the highly urbanized situation in Japan today constitutes a fundamental planning problem.

Second, the absence of a regional policy in Japan means that rural areas were not developed into humane living environments and were therefore significantly depopulated.

Third, if a multiple-nuclei decentralization-type national land policy is to be adopted in Japan, it would need variations of the policies employed in Germany. However, no concrete design for such a policy has as yet been formulated in Japan.

Fourth, such a plan cannot be realized by the Ministry of Agriculture, Forestry and Fisheries alone, but requires the development of conditions that transcend the vertically divided functions of Japan's individual ministries.

Urban-rural relations in Japan: now and the future

As argued in Chapter 7, despite efforts to build a total and sustainable rural community as a *ba* of production, ecological environment and everyday life, various restrictive conditions exist that transcend individual regions and hence cannot be countered on the regional level. As we have seen in our survey of European, and especially the German, history of urban-rural relations theory, a national land policy that appropriately integrates cities and the country as well as a regional restructuring policy are needed for the realization of sustainable rural areas.

The formation and problems of single-nuclei centralized spaces

The problems of large city concentration

As indicated by Figure 8.8, during the high economic growth period – from the mid-1950s to the mid-1970s – the Japanese population flowed into three large cities – Tokyo, Osaka and Nagoya – at a rate that was nothing short of extraordinary. During this period, the population of Nagoya increased by approximately 50,000 people per year, Osaka by about 100,000 per year, and the greater Tokyo metropolitan area increased by roughly 300,000 people per year, or almost 1,000 individuals per day. It was as if a moderately sized city was being added to Tokyo every year. The enormous influx of investment into the area created a situation described by

Figure 8.8: Changes in migration rates to and from Japan's three major urban areas

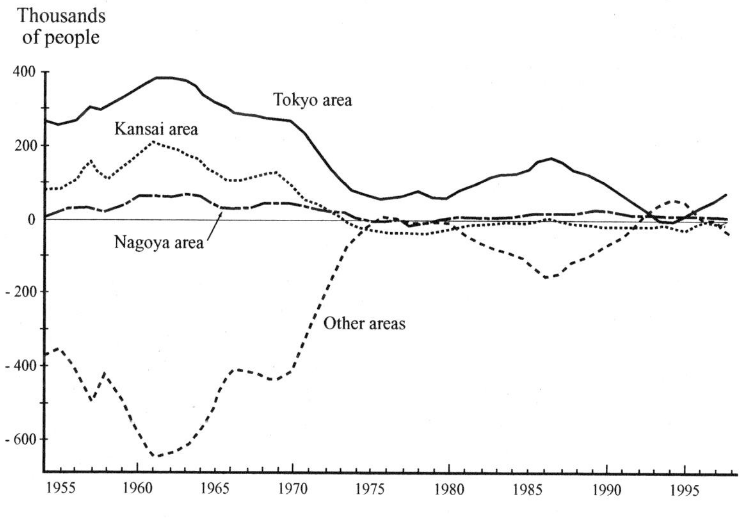

Source: National Land Agency Data, 1999.

the Economic White Paper of 1960 as 'investment attracting more investment' (Economic Planning Agency 1961).

The situation cooled down somewhat after 1975, but in Tokyo, the population inflow began accelerating once again around 1985. At the same time, Osaka's population decreased while Nagoya's remained unchanged, resulting in an even greater concentration of the nation's population in Tokyo. Many corporations moved their headquarters from Osaka to Tokyo and reduced their operations in the former. Accordingly, there was talk of a so-called 'sinking' of the Kansai region (Osaka is the central city of Kansai). While the pace of population movement into Tokyo slowed at the beginning of the 1990s, partially due to the economic bubble bursting, any predictions of future trends are unwarranted.

In contrast to the large cities, rural communities experienced incredible outflows of population during this period, producing a proliferation of depopulated regions. As Figure 8.9 indicates, social decline during the period of high economic growth occurred at a fierce rate. Statistics indicate that this happened in 60 to 70 percent of all cities, towns and villages. From the 1980s onwards, the number of cities, towns and villages experiencing a natural decline in population (deaths exceeding births) increased rapidly. The period up to the

Figure 8.9: Depopulation in 3 phases

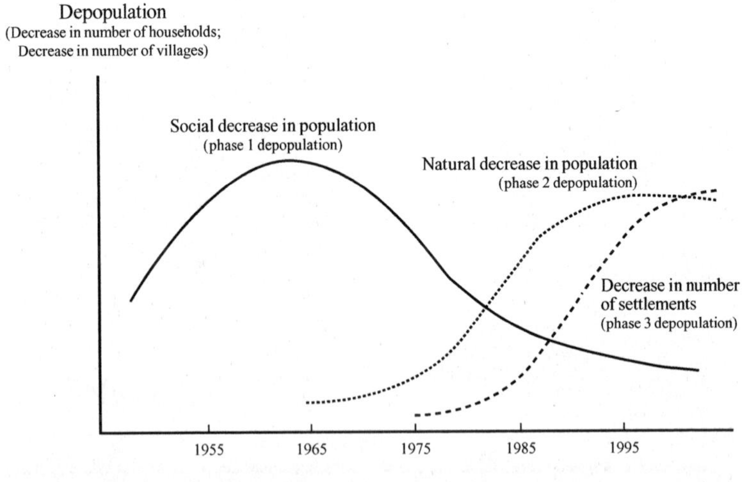

Note: The figure schematically shows, based on results of population statistics, settlement (village) statistics, etc., that decreases in population and in the number of households and settlements in rural areas, especially hilly and mountainous areas, occur in 3 phases.

1980s, typified mainly by 'social' population decline, can be called 'the first depopulation phenomenon', and after the 1980s – with mostly 'natural' population decline – 'the second depopulation phenomenon'. Subsequently, besides a gradual and continuing decline in population, the outright disappearance of entire villages marks 'the third depopulation phenomenon'. Comparing areas classified as 'depopulated' in 1970 and 1989, we find that within this period, the proportion of towns and villages increased from 32.3 to 37 percent, the land area they covered grew from 41.5 to 47.7 percent, and the number of inhabitants fell from 8.6 to 6.5 percent of the total national population (National Land Agency 1972, 1993). This trend does not merely imply a population decrease, but encompasses the aging of the remaining population, abandoning cultivation and declining agricultural activity, and the gradual cessation of the regional functions.

At first glance it might appear that these problems only affect rural areas. But, as we have seen, the concentration of population and industry in urban areas generates myriad problems for city-dwellers. On the other hand, many people in cities hope to live in small-to-medium cities or rural villages. I will now examine this situation more closely.

Preference for life in the countryside: nature and peace of mind

Large cities attracted burgeoning populations by offering plentiful attractions, especially employment opportunities in a vast array of professions. With the exception of plots favorably located in the suburbs of cities where farmers can get non-agricultural income as a part-time farming household, it has become difficult to sustain a family by farming, when the average cultivated plot is only about 1.4 hectares. This is the primary reason behind the exodus of rural inhabitants.

However, while enjoying the benefits of large cities, the people who migrated to these cities are now expressing a preference for living in the countryside as the myriad problems of urban living emerge. These problems include high land prices, high rents, small dwellings, difficulties in gaining house-ownership, long commutes, traffic congestion, lack of sunlight, water and air pollution, long working hours, cruel competition for success, the absence of community, shallow neighborly relations, and loneliness. For these and other reasons, more and more people report a preference, if given the opportunity, to escape to a regional capital to live and work. At the same time, the number of salaried workers, or 'salarymen', who report a desire to start working anew in agriculture, is rising (Soda 1992:

8). All in all, urban residents report a significant preference for living in the countryside, and it is clear that this preference is gradually gaining strength.

Supplementary note

According to a survey conducted by the Mainichi Shinbun (Mainichi Newspaper) in 1991 about population over-concentration in Tokyo, 82 percent of the respondents (chosen to represent the whole of Japan) answered negatively when asked if they would prefer to live in the greater Tokyo metropolitan area. No more than 17 percent of respondents clearly gave a positive answer. The latter group reported reasons such as 'close proximity to things and information, plentiful work opportunities, flourishing cultural activities, greater availability of schools and cultural facilities, and uncomplicated human relations'. Respondents who did not want to live in the greater Tokyo metropolitan area, reported reasons such as the 'shortage of natural areas, high living costs, difficulties in owning a house, commuting, heavy traffic congestion, air pollution, and superficial human relations'. Furthermore, with regard to population concentration, 71 percent of the respondents expressed the view that it was something to be corrected, while 23 percent maintained that it was not a bad thing. Nearly identical results emerged from the National Land Agency's 'Survey on UJI-turn Facilities of Local Public Organizations' conducted in 1992. In a survey by the Prime Minister's Public Information Office in 1994, 25 percent of the respondents chose the three largest metropolitan areas as the ideal place to live, while 70 percent opted for the countryside and five percent for other locations. In the 'Large Urban Region Survey' conducted in 1987, 43 percent of the residents of the greater Tokyo metropolitan area indicated a preference to live in the countryside, indicating that this trend has already existed for a long time. In the National Land Agency's 1994 survey, 'Questionnaire Survey on the Effects of the Location of Businesses and Trends in the Employment System on the Regional Migration of Workers', 56 percent of the businesses that responded predicted their workers' preference for life in the countryside would strengthen. In response to such trends, many companies are considering the introduction and expansion of a localities-only employee system and a limited-time personnel relocation scheme. It also appears that, especially among the younger generation and under a declining birth rate, there is a preference to live near one's parents. In the 1990s and especially after the burst of the 'bubble economy', the preference to live and work in the countryside upon retirement manifested as a large-scale returning to the countryside, while the desire among the urban youth to work in the country has increased.

The small-to-medium city orientation of rural residents

As previously explained, during the period of high economic growth, the young age cohort had no choice but to leave their rural villages where it was no longer possible to earn an average income because there was not enough suitable work. Following the explosion in investment flowing into the large cities, as well as income, the convenience and possibilities for cultural activities, information, and excitement attracted young cohorts.

However, the examination of the 'countryside orientation' above indicates that a rather large group of people would prefer to stay in their original hometown if there was a small- or middle-sized city within a commutable distance. Many surveys have been conducted on life in farming villages and relevant conditions, and all have found that securing of non-agricultural work is the number one concern. During the postwar period of high economic growth, similar trends appeared in Germany, but the difference was that Germany had long been pursuing 'stationary industrialization' (*Industrialisierung im Sitzung*); or, in other words, had adopted a regional policy that made it possible to find work in commerce and industry while continuing to live in one's hometown. Also, while Japan's basic agricultural policy tended initially to deny part-time farming, in Germany a philosophy and policy that accepted side-work was formulated and incorporated into its regional structural policy (Soda 1997; Priebe 1970).

All of this suggests that, if it were possible, rural residents would prefer to look for work, stores in which to purchase everyday items, entertainment, and cultural and educational opportunities of a certain standard, in a small- or middle-sized city that is easily accessible from where they live.

The concretization of a new urban-rural relationship

The pros and cons of rural and urban areas and their relationship

It can be assumed that not only economic but various other factors underlie the urban residents' preference for communication with the country and the rural residents' desire for communication with small- and medium cities.

I have previously advocated that from now on it will be necessary to focus on developing a 'humane' living environment. As a precondition to this, we will need to realize economic, ecological, and life values in a harmonious way. Table 8.1 displays the pros and cons of rural and urban areas from the point of view of these three aspects. This table is based on a number of literary sources on rural villages and urban areas, interviews of five people from each region, and letter-based questionnaires among others, to which I have added some of my own opinions. It is important to recognize, however, that completely different evaluations result from different sets of individual values. It may be the case, for example, that the 'cooperativity' found in rural villages is either highly valued or alternatively experienced as troublesome. Or, more generally, there is a sense in which the rural villages' pros are the cities' cons and vice versa. However, I have only presented those answers that demonstrate strongly felt shortcomings. In either case, this figure attempts to categorize and organize the general tendencies of cities and the country.

As can be seen from the table, both the city and country can be characterized by a range of pros and cons. If, as Howard noted with reference to the 'three magnets', people think, make decisions and act along the presented four dimensions (see Table 8.1), urban residents desire the pros of the country while rural residents long for the pros of the city. It can therefore be argued that there is a demand for an appropriate integration of these two areas. This of course does not mean completely homogenizing cities and rural villages. Rather, the positive qualities of each must be maintained while aiming for the mutual integration of the two regions.

We thus detect expectations for mutual exchange and integration in both cities and rural villages. The integration of cities and countries can therefore be seen as a precondition for the formation of living spaces that guarantee the realization of 'comprehensive value' by linking agriculture to industry, humans to nature, and people to other people.

Regional policy in Germany and Japan

As discussed, the history of urban-rural integration is the history of Germany's regional policy, constantly reverting back to the principles first outlined by Schmidt.

Christaller designed a 'hexagonal multiple-nuclei decentralized space'. He called the cities that function as nodes 'centers' and divided them into several classes. Adhering to this schema, postwar Germany's regional policy classifies cities into the four categories of large, medium and small cities, and 'small centers', and aims to disperse these across the national landscape. The functions of the centers in varying categories differ, and the policy aims to allocate various public and cultural facilities as well as everyday conveniences, when possible, according to city size and its combination. The policy also takes into consideration, with reference to everyday needs, the time it takes for a resident to commute to cities of different categories via public transport or private car from their place of residence. The ideal is to make it possible for any resident to get to a large city (more than 100,000 thousand residents) within 50 minutes, to a medium-sized city (20–100 thousand residents) within 30 minutes, and to a small city (10–20 thousand residents) within 20 minutes. Through such planning, it is possible for anyone living in the countryside to satisfy certain minimal economic and cultural needs. In short, while some regional differences and other special characteristics remain, Germany's regional policy has been developed according to the principle of making it possible to provide each resident with a 'total and humane life' and 'total value' (Soda 1997).

The German philosophy of regional policy contains the concept of a 'healthy region' (*das gesunde Gebiet*) (Akademie für Raumsforschung und Landesplanung 1970: 2460; Soda 1985: 20). The word 'healthy' here refers to the conditions that make it possible for people to live their daily lives with 'total satisfaction'. In concrete terms, a 'healthy' region is required to (1) have a diverse and organic economic structure springing from the integration of agriculture and industry, (2) provide residents with various everyday conveniences through the integration of cities and the country, and (3) integrate the human environment with the natural environment. In a region with such a structure, people have opportunities for work (including side-jobs as in 'stationary industrialization'), to enjoy social and cultural conveniences, and to live in harmony with the natural environment. This German 'regional restructuring policy' is thus the prototype for the regional development philosophy I am advocating.

When implementing an agricultural restructuring policy – i.e., transferring agricultural labor to other industries, while gradually expanding the scale and improving the efficiency of agriculture – it goes without saying that new jobs must be provided for the former agricultural workers who are released in the process. Whether these jobs are available in a nearby small or medium city, or only in large cities, is an issue of decisive importance for the residents of rural areas. In Germany, effort was made to constrain the emergence of megacities by linking rural regions to dispersed small-and-medium cities. In short, in Germany, agricultural restructuring policy was seen to go

Table 8.1: The pros and cons of the city and the country

The pros of the country	The pros of the city
Economic	
Better opportunities for running a family enterprise as well as having a side-job Cheap land, spacious houses, and detached living quarters Low cost of living Ample storage space Availability of fresh food and locally made food products Space for a fruit/vegetable garden (family's work and hobby) Short travel times, few traffic lights and traffic jams etc	Greater business opportunities and chances for success Better access to information on the markets, business circles, politics and consumption conveniences Opportunities for higher and vocational education, abundant human resources Greater job opportunities and variety Higher wages and income Cutting-edge consumer life Economic advantages arising from population concentration, convenient traffic and transport
Ecological	
Fresh water and air Pure water from springs, wells and streams Abundant nature and beautiful scenery Four seasons and a real sense of a growing nature Plenty of sunshine Good environment for gardens The presence of, and contact with, diverse plants and animals The self-supply of fresh foodstuffs that contain no agricultural chemicals The sight and song of birds Hot springs Greater possibility to recycle things The possibility to utilize local energy etc	Completely artificial urban environment with the natural ecosystem eliminated (stylish high-rise buildings, roadside trees and flower boxes, city parks) etc
Everyday life-related (social and cultural)	
Cooperation and solidarity in social duties and obligations Friendly and warm human relationships (internal openness) Relaxation and peace of mind (agricultural and natural rhythm of living) High public safety Good environment for raising children (experiences of nature, abundant space to play outdoors) Sense of gratitude Maintaining a balance between mind and body Diversity, stability, permanence Farming work available for the elderly The revival of 'total human-ness' Independence and freedom arising from self-employment Freedom during the agricultural off-season Participation in traditional events and group work The opportunity to work on fine days and rest on rainy ones (as a lifestyle) The richness of agricultural art, hobbies, research activities and raw materials The joy of tending a garden or a vegetable plot, growing trees etc	The availability of various everyday life services and facilities (water and sewage, gas, roads, information) Convenient shopping Cutting-edge, glamorous living Keeping up with trends easily High-quality medical care Abundant sports and entertainment facilities (soccer, baseball, amusement parks, restaurants, movies, theaters, dancing, amusement centers) Contact with diverse people, plenty of social acquaintances Wider opportunities for romance and marriage Abundant cultural facilities (museums, art galleries, libraries, aquariums) Great educational facilities (universities, vocational schools, research centers, various lifelong learning facilities, professional training facilities, preparatory schools for children) Greater availability of information about various kinds of knowledge, art and culture Variety of ways and opportunities to participate in society (voluntary social service, social work, the Red Cross, NGOs, various citizen movements) Freedom from traditional practices, customs, conventions and old morals; and greater opportunities for social mobility

The pros of the country	The pros of the city
	Anonymity and confidentiality A certain sense of irresponsibility Change and fluidity Stimulation and refreshing excitement More tolerance of diverse lifestyles etc

The cons of the country	The cons of the city
Economic	
Absence of job opportunities Little diversity with regard to available types of work Few or no general shops and convenience stores No glamorous consumer life Inconvenient traffic and transport arrangements etc	High land prices High office and housing rents, small premises Long commuting times Garbage and energy problems and other wastefulness etc
Ecological	
Not much gorgeous architecture Vermin damage commonplace (wild boars, monkeys, birds, dear)	Lack of nature Pollution of air and rivers Pollution and depletion of ground water Pollution of drinking water Flood of ready-made food products with additives Shortage of sunlight Possibility of a high death toll in the event of an urban disaster Green house effect resulting from the use of air-conditioning Never-ending removal of agricultural plots and green areas from the city and their conversion into residential and industrial land
Daily life-related (social and cultural)	
Depopulation Absence of social and cultural facilities (libraries, assembly halls, sport facilities, theaters, art galleries, museums etc.) Poor level of general infrastructure (water and sewage services, roads, railroads, buses) Few places for shopping No facilities for basic or vocational education Shortage of entertainment facilities Low awareness of human rights (especially with regard to women's social position) The structure of houses and lack of privacy Closed attitude towards outsiders Close adherence to tradition etc	Uniformity and the loss of individuality Overcrowded living spaces Disparities in quality of schools; school violence Refusal of children to go to school and domestic violence Increase in, and aggravation of, juvenile crime Little sense of community Oversupply of information Excessive competitiveness and a higher death-rate from overwork Loneliness and obstacles to good physical and mental health Vibration, noise, hustle and bustle Dominance of nuclear families and alienation of the elderly

hand-in-hand with regional restructuring policy. However, in Japan, although agricultural restructuring policies were formulated and implemented, regional restructuring policies remained weak, leading to the uncontrolled expansion of megacities with the corresponding depopulation and overpopulation problems discussed. While it would be incorrect to say that there is no depopulation problem in Germany, compared to Japan, it is a relatively minor problem. Herein lies the fundamental difference between Japanese and German thinking about policy, underpinned by the German philosophy of regionalism (*Regionalismus*).

Enhancing small-and-medium cities through regional revitalization

When examining Japan from the above perspective, it is apparent that the country needs to adopt the principles of 'multiple-nuclei decentralized spatial development'. Addressing this point, I have designed and will present a 'comb-like multiple-nuclei decentralized space' as the model for decentralized social development in Japan. This model aims to resolve the problems of both rural residents and the residents of large cities.

In our present civilizational situation, the satisfaction of economic needs alone is not a sufficient condition for human existence. Neither is access to beautiful nature and a healthy environment. Furthermore, even when blessed in terms of both economic and natural conditions, an individual still cannot live and satisfy all of his/her everyday social and cultural needs without maintaining communications with other people. This is why I have defined the total realization of economic, ecological and life value as the central issue facing both modern agricultural villages and modern society more generally. A place that minimally satisfies all of these needs, however, permits people to settle down, young people can form many friendships, and agriculture can be revitalized. At some level it is correct to argue that agricultural revitalization may stimulate regional revitalization. However, I think it is more practical to emphasize that it is through the economic and cultural revitalization of entire regions that agricultural revitalization can best be achieved.

Here, one important issue is the distance between an agricultural region and the nearest small- or medium city center. Negishi Takeo's research (1993) on the levels of activity in villages, for example, has made it clear that, according to all indicators, levels of agricultural activity are highest in areas that lie within 30 minutes' travel-time from the center of a small or medium city.

While it is clear that population movement from agricultural villages to urban areas continues, differences can be observed between cities, towns and villages of varying scope. Between 1975 and 1990, cities with more than 50,000 inhabitants experienced considerable population increases, while those with less than 30,000 people experienced decreases. By contrast, cities with populations between 30,000 and 50,000 experienced neither increase nor decrease. At the same time, towns and villages experienced considerable increases where the original population stood at more than 20,000, while those with less than 10,000 people dwindled, with the turning point being settlements of 10,000 to 20,000. The relative stability of towns and villages with populations of 20,000 to 30,000 inhabitants and a town centre is noteworthy.

To sum up, let us examine statistics from 1990 concerning the travel-time from agricultural settlements of various sizes to the closest DID zone (a densely populated area: defined as an area with a concentration of more than 5,000 inhabitants that is constructed with at least one or two areas with more than 4,000 inhabitants per square kilometer). Looking at percentage distribution, 61.7 percent of agricultural settlements lay within 30 minutes of a DID while 38.3 percent were further (Ministry of Agriculture, Forestry and Fisheries, Japan 1990: 27). Based on our earlier considerations, we can assume that the latter group suffers the most from problems regarding vitality.

This figure closely corresponds to the size of the 'mountainous regions', which cover approximately 42 percent of Japan (according to estimates by the Ministry of Agriculture, Forestry and Fisheries). In these areas, it has become extremely common for sons/daughters to leave for employment in the large cities and for parents to remain in the village. The 'isolated farming and mountain village pattern' on the left side of Figure 8.10 indicates this situation. In such cases, there are no small- or medium cities nearby, but at most a small village center with no employment opportunities. Hence, with no non-agricultural sector work available, and no possibility of supporting oneself and family by farming alone, it is unavoidable that most people will move to the larger cities for work. It is under such circumstances especially that social aging occurs, the local community's functions degrade and various other problems emerge.

As indicated by (1) in the right side of Figure 8.10, the revitalization of a region and its agriculture becomes possible if the region is made into a commutable area. As indicated in (2), problems can be alleviated at least for a while if a region or village is within a reasonable commuting distance for weekends and emergencies. Where commuting remains

Figure 8.10: Present situation of the relationship between the town and the rural/mountain village and desirable directions – combination of small or medium city and country

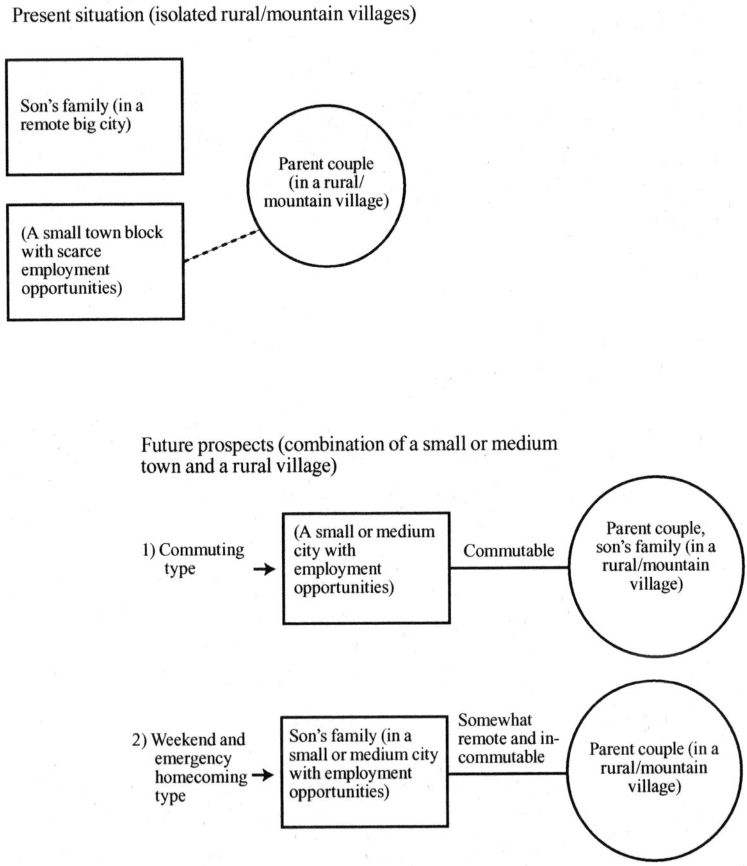

impossible, if there is a small city close by where villagers can secure non-agricultural work, he/she can communicate with his/her parents using videophones or other communication technologies. As long as it is possible for workers to return to their parents' village on short notice in the weekends or in the event of an emergency, it should be possible, albeit with limits, to maintain the local community. These kinds of concrete measures are needed alongside agricultural policies. The revitalization of a region's agriculture is no longer possible without combining with cities and enhancing the entire regional economy and society.

The formation of multiple-nuclei decentralized space in Japan

Leaving aside the immediate future, how should the use of national land be planned in the 21st century? I have been arguing for the development of a 'comb-like multiple-nuclei decentralized space' for some time. As shown in Figure 8.11, this would mean reviving the 'river basin socio-economic area' that declined under the high growth

Figure 8.11: A proposal for multiple-nuclei decentralized space in Japan

A comb-like social economic sphere

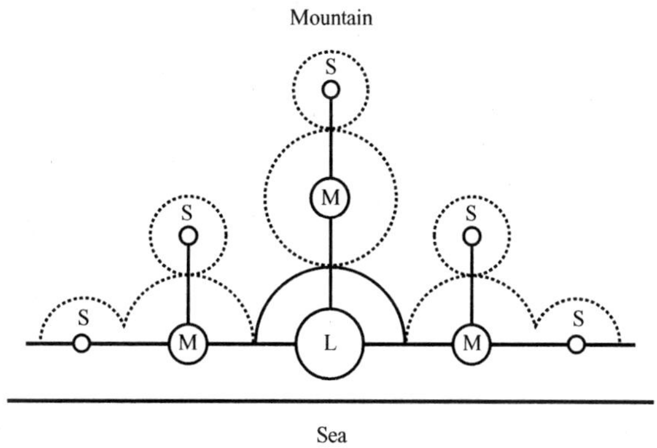

'L', 'M', 'S' stand for 'large', 'medium' and 'small', respectively, each of which represents the relative size of the central place (city, or town block). 'L' does not mean a big city but the central city of the local region.

A comb-like multiple-nuclei decentralized space

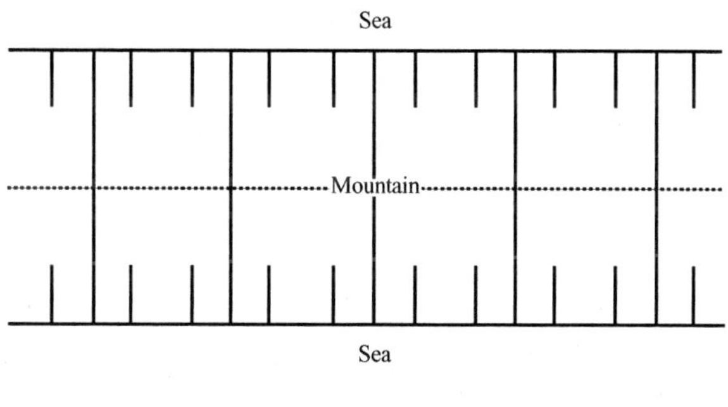

A space consisting of a series of linked comb-form social economic spheres.

period, reforming the countryside and the coast belt socio-economic areas that have accumulated growth, and tying these regions together by forming comb-like well-organized socio-economic areas of say approximately 300,000 residents. As the result of allocating these comb-type socio-economic spheres across Japan's main islands, a 'Japanese-style comb-like multiple-center decentralized-type space' would emerge.

This chapter has combined theoretical and empirical discussions of the relationship between rural and urban areas. Alongside environmental issues, it seems clear that the urban-rural problem needs to be recognized as one of the central issues of the 21st century. Correcting the distortions in Japan's national land policy is likely to take much longer than it took to create them during the period of high economic growth. Combining rural and urban areas is indispensable to the sustainable development of rural villages. And without the proper combination of rural and urban areas, realizing 'comprehensive value' will not be possible even for the urban residents of modern society.

9 Innovations and the Diffusion of Agricultural Technology

Technology is essentially invention and imitation, which to be socially effective must be transmitted and diffused. In this chapter, I discuss the processes of innovation and the diffusion of agricultural technology primarily in terms of the pursuit of economic values. I also briefly discuss the role of agricultural science. The aim is primarily to clarify the process of creating specific farming village areas as places that pursue comprehensive values, developing in response to circumstantial changes, and then striking a new social/economic/technological balance. Furthermore, I aim to elucidate the relationship between different behavioral types of farmers and this process of transformation. First, I explain the general characteristics of farmers and the genealogy of theories of agricultural/farming village development. Second, I will examine the functions and meaning of three behavioral types of farmers (i.e. pioneering behavior, adoption behavior and imitation behavior) within agricultural/farming village development. Third, I comprehensively discuss the meaning of the process of agricultural/farming village development as it evolves in the interplay of the above-mentioned factors. Fourth, I will explore the epistemological implications of this approach, by way of laying the foundations for the next chapter, which details the relationship between the study of agricultural science and agricultural administration.[1]

The three developmental stages of agriculture and agricultural science discussed in this chapter correspond to the three-stage methodology of agricultural science which we discuss in the following chapter. Specifically, the practical recognition that is considered in chapter 9 corresponds to the scientific recognition addressed in chapter 10. In this sense, the two chapters are inseparable and mutually linked.

Spontaneous development and farmers' behavior

To understand the process of agricultural technology formation and its spread – by way of finding a solution to the emergence of new contextual problems or expectations in a particular agricultural production site – it is helpful to use the behavioral types of farmers, including their economic and social meanings. That is the 'spontaneous' process within a production site of the development of a new technology, which is valued and disseminated.

My purpose here is to theoretically elucidate the process and context of a farming village's dynamic formation, motivated by the quest for profits or economic benefit, attributable to technological innovation. However, at the same time, I intend to construct a theory that can explain the process of a society's developmental transformation – which includes farming village problems such as the reform of village administration and restructuring the village, as well as issues related to agricultural science (e.g., ecological values, the pursuit of lifestyle values, and finally the search for a harmonization of the three values; i.e. total values).

Generally speaking, agriculture is family-centered and run as a very small management unit in comparison to other enterprises. Hence, the farmer is an economic subject who acts as both an entrepreneur and laborer. Moreover, the farmer is not only an economic subject but also a social actor within a village. The individual's behavior as a farmer is also the behavior of an enterprise; but unlike an ordinary corporation, the farmer's economic actions are vulnerable to social constraints, since the place of production and the place of everyday life is the same.

For instance, when a corporation considers adopting a new technology, it decides based solely on the economic value of the change. However, the farmer might recognize an economic value in the new technology, but will also be influenced by psychological and social processes, such as concerns about what other villagers will think, worrying that others might talk about his/her strange ideas, or ridicule him/her. Hence, the farmer might sometimes be reluctant to innovate. This situation is quite common in the daily life-world that is the farming village. Dewey observes that 'his activities are encompassed in an environment that is culturally transmitted, so that what man does and how he acts, is determined not by organic structure and physical heredity alone but by the influence of cultural heredity, embedded in traditions, institutions, customs and the purposes and beliefs they both carry and inspire. Even the neuro-muscular structures of individuals are modified through the influence of the cultural environment upon the activities performed' (Dewey 1938: 43). In short, the economic behavior of farmers is strongly influenced by the norms of social behavior.

Furthermore, the three types of farmers that are defined in this chapter – pioneer-adopter-imitator – correspond to the process of scientific recognition that I, following Dewey, define as 'hypothesis formation based on a framework, the establishment of logical validity based on inference, universalization and actualization based on the repetition of the verification'. In other words, the development of empirical knowledge (practical recognition) rests on the processes of scientific recognition.

Many have expressed negative views about the capacity of Japanese farmers to instigate or adopt endogenous innovative actions. Such views are clearly reflected in the philosophy of agricultural administration under the (now superceded) Fundamental Agriculture Law. For example, according to Tōbata Seiichi, 'those who have pushed Japanese agriculture forward' or 'the pioneers' include industrialists, large-scale merchants, some farmers, dynamic landlords, agricultural organizations, and the government. In particular, the government is typically depicted as the most significant driver of agricultural innovation. The government is in fact seen as the entity that 'replaces the "simple farmer" who lacks creativity, and generates the development of agriculture'. 'That is to say, the government takes various initiatives and makes the "simple farmer" follow them' (Tōbata 1936: 30).

In short, Tōbata's theory attributes the pioneering spirit behind the development of Japanese agriculture to the government's economic activities. He sees Japanese farmers as nothing but 'simple farmers' who follow these activities, or 'subordinates bearing the risks'. The creative activities of the farmers themselves are given very little credit. Their economic activities are seen to follow a simple and repetitive reproduction cycle.

From the perspective of rural sociology, Fukutake Tadashi argues that 'the attitude of farmers, in a word, sets high value on traditions and submission to authority' (Fukutake 1964: 246). Similarly, Kamiya Keiji claims that 'the principle of small profits and large security' (security first and small profits second) and 'resignation' are the two main defining characteristics of the behavioral patterns of farmers (Kamiya 1964: 50–1). Clearly Tōbata's, Fukutake's and Kamiya's views of Japanese farmers overlap considerably, even though they approach the question from different perspectives.

Likewise, under the agricultural administration regime of the Fundamental Agriculture Law (established in 1961), the bureaucratic leadership maintained a top-down view of agricultural development, which failed to recognize the creative subjectivity of the farmers who are the real agricultural reformers. Through the medium of subsidies, the central government assumed the lead in developing agricultural and farming villages, creating a policy system that restrained farmers' initiatives and technological innovation, and controlled the dissemination of agricultural developments.

It is perhaps already clear that I attach much greater importance to the pioneering enterprise of the farmers themselves, seeing this as the primary driving force in agricultural and farming village development. From this perspective, I will focus on farmers' behavioral patterns and outline the theoretical origins of rural economic development in terms of farmers' behavior.

Schumpeter was the first scholar to outline the dynamic theory of economic development (1912). He explained the now conventional concepts of 'static' and 'dynamic' economic systems. In fact, the concept of 'stationary state' or 'circular flow' includes a wide area of economic phenomena, referring to a balanced state/condition that remains unchanged as long as external conditions remain unchanged. In effect, the economic situation tends towards a balanced state because the majority of economic actors believe that repeating the business cycle unchanged is the most rational approach. Schumpeter includes adaptations to changes in external conditions within the concept of 'static state (usual repetition)', and defines the economic subject, which is based on 'experience' and has adaptive or repeating behavioral patterns, as 'simple farmers' (*Wirt Schlechtweg*).

Schumpeter explains that 'by development, we shall understand only such changes in economic life as are not forced upon it from without but arise by its own initiative, from within' (1912/1989: 63). Such change does not result from changes in external conditions, but is generated from within the economy itself. It refers to an economic activity that destroys the existing stationary flow and pioneers new technology, new products, new organizations or new markets. In short, such change amounts to 'carrying out new combinations (innovation)'. Schumpeter ascribes this type of economic activity to the 'entrepreneur' (*Unternehmer*), but I prefer to call it 'pioneer', to invoke creative activity beyond the merely economic.

However, Schumpeter concentrates on the differences between a 'static state' and 'dynamic state' and on devising models of these two states, and therefore 'does not pay enough attention to the interaction between the two' (Nakayama 1939: 248). To more accurately theorize economic society, it is necessary to go beyond simply detailing the characteristics of these two economic states, and to position them within the actual developmental processes.

Tōbata Seiichi (1936), skillfully adapted Schumpeter's theory of economic development to explain the developmental processes typical to Japanese agriculture. Elaborating on Schumpeter's theory, Tōbata separates Schumpeter's (stationary state) adaptation of simple farmers, distinguishing two different types which follow innovation: 'most skillful adaptation' and 'simple adaptation'. By separating these two levels or degrees of adaptation, Tōbata positions it within the 'circular flow' of the economy (Tōbata 1936: 42). Whether 'skillful' or not, 'innovation' and 'adaptation' are obviously dissimilar – as they have different functions.

Nevertheless, I would like to resolve this opacity and better explain the process of agricultural development by establishing an intermediate economic subject category, that of the 'adopter', who sees value in a new innovation, adopts it, skillfully implements it, and propagates it widely. That is to say, I intend to explain the process of agricultural development in the light of the behavioral patterns and the functions of three categories of economic actors: pioneers, adopters and imitators.

Supplementary note

Some sociologists and cultural anthropologists have attempted to classify social actors in a similar fashion. For instance, the cultural anthropologist Murdock argues that cultural change undergoes four processes: 'innovation', 'social acceptance' (by a few individuals of groups), 'selective elimination' and 'integration' (reorganization of the whole) (1971: 319–32). Similarly, the rural sociologist Rogers, discusses 'innovation and its adoption', explaining its transformation and development through five categories or stages: 'innovator', 'early adopter', 'early follower', 'late follower' and 'late comer' (Rogers /Fujitake1962: 200). These classifications, however, although suggestive, are limited in their approach. They categorize innovation according to the speed at which it is adopted, that is, according to the rate (time) and breadth (amount) of diffusion. In so doing, they fail to incorporate the function and meaning of innovation. In contrast, the three categories 'pioneer', 'adopter' and 'imitator' identified here to elucidate the process of agriculture and farming village progress include not only elements such as rate and breadth, but also emphasize the economic function and meaning of innovation, and add social and psychological meaning as well. In a way, the taxonomy identified here is socio-economic, describing farmers' behavioral patterns, as well as epistemological.

The types and meanings of farmers' acts

In order to reveal the meaning of the three types of farmers' economic behavior I will discuss them in terms of the 'circular flow' of the economic system. This entails an examination of the circumstances in which they gradually appear, perform their functions, generate processes of technological and economic development or societal change and eventually, after struggle, become complete. This section presents the conclusions of my research on the basis of my interest and theoretical genealogy and my investigation into the actual conditions of farmers.

The meaning of pioneering acts (pioneer)

The economic meaning of the pioneer

I would like to begin by presupposing the 'stationary flow' model when problematizing development and change. This means that economic activity moves at a constant speed, simply reproducing itself in an eternal/unchangeable economic process or similar. This may also be called 'stationary state' or balanced equilibrium state. It refers not only to economic activity, but includes the spheres of social customs and individual practices, which I will mention in more detail later. It has, at its basis, the actors' unconscious preference for repeated actions that continually follow the customary path, taking experience as a guide. In an equilibrium situation of this type, the existing value system is the only criterion for judgment – and behaving repeatedly like most people without deviating is considered to be normal everyday behavior, or the dominant norm. Conforming to the dominant norm supposedly offers stability and a sense of security.

Now, let us assume that a change occurs within this routine – not a gradual change from outside, but a subjective and endogenous innovation from within the economy that changes its original data, and thus its given conditions. Schumpeter calls this 'carrying out a new combination' and perceives it to be a unique entrepreneurial function that appears discontinuously on the customary path.

As mentioned, I will call those economic subjects who engage in pioneering behavior 'pioneers', an ideal type that refers to an endogenous and developmental power of action. The term 'pioneer' used here is the same as Schumpeter's 'entrepreneur', but it excludes Tōbata's 'most skilful adaptation'. The 'pioneer' is an innovator who brings about a spontaneous change in the economy; the subject who through creative economic activities such as inventing new technology, adopting new production methods, producing new goods and services, opening new markets and creating new systems – in other words, through 'carrying out new combinations' – pursues profits. The 'pioneer' emerges on the customary path as a dot; endogenous and yet isolated.

Schumpeter saw the construction of new factories, the foundation of new enterprises, and the entry of newcomers as typical symbols of innovation. He described their economic roles as establishing new production capabilities, increasing marginal productivity and cutting costs. In Schumpeter's words, a special relationship is established, where 'without development there is no profit, without profit no development' (1912/1989: 154).

The pioneer secures profits ahead of other economic subjects, typically by identifying or creating new sources, or 'locations' of profits. Of course, profits may vary according to the type and level of innovation. That is to say, economically speaking, a great innovation has larger profits while a small innovation has little profits. Therefore, the pioneer is also the 'person who determines the scale of profits'. In this way, economic development is led by the quest for profits.

The social meaning of the pioneer

At first the pioneer appears as a deviant. In general, people rely on tradition, are reluctant to encounter new situations and for the most, they consider it normal to follow customary paths. The existing socio-economic system has strongly established pathways and standards, and it is not easy to deviate from or change this system. According to Dewey, already formed customs 'mean only the increasing fixation of certain ways of behavior through repetition, and an attendant weakening of other behavioral activities' (1938: 31). It is hard to change the existing pathways that are grounded in the belief system and established customs. According Miki Kiyoshi, customs are a 'formation of balance', unconscious repetition built upon sustainable adaptation, habitual in one way or another, passive and are guided by the principle of imitation (Miki 1967: 30). Within such a balance, the more creative a pioneer's innovation, the more it is felt within the existing system as an extraneous thing.

Hence, pioneer's behavior is often seen as deviant and meets resistance from other members of the community. Deviant behavior also encounters legal and political barriers, such that the pioneer is sometimes excluded from financing and government subsidies. At other times, he is socially rejected and met with violent obstruction. According to Schumpeter, 'any deviating conduct by a member of a social group is condemned' (1912/1989: 86). Further, he argues that 'surmounting this opposition is always a special kind of task which does not exist in the customary course of life, a task which also requires a special kind of conduct' (1912/1989: 87). As I mentioned in the beginning, this social pressure has a strong impact upon farmers – for whom their enterprises (economic activity) are the same thing as their social behavior – and is therefore an important factor in the process of developing and transforming rural society.

To be more specific, the repetitive type of behavior which strictly follows the customary path is the antithesis of the creative, individual type of behavior. All of the knowledge and customs that a subject has ever acquired is deeply rooted in his/her unconscious. And this means that we are guided by a survival rationality that manifests in our avoidance of economic and social risks, in our attempts to minimize social frictions, and our reluctance to re-evaluate our values. Hence, in almost any customary sense, the pioneer's behavior is by definition deviant, characterized by estrangement, rebellion and rejection of the existing system. Because of the dominance of this view, I would like to point to the fact that a pioneer must have a personality exemplified by individuality, forwardness, self-control, perseverance and executive ability in order to overcome being viewed as deviant. Without these personal qualities, the pioneer is doomed to failure and their innovations will disappear in the face of social resistance. It goes without saying that innovation needs to be truly useful and meaningful, bringing about progressive technology. Otherwise, it will be viewed as a 'simple deviation' and will soon self-evaporate.

The pioneer is the executor of innovation, locating potential profits and deciding profit levels – and its behavior is typically viewed as a deviant attempt to subvert or displace social customs. Every economic development begins with a hard and isolated germinal stage. In the agricultural sector, this initial stage, when one or a few farmers display deviant and innovative behavior, we witness individual subject types, dynamic behavioral style and the driving force behind the development of rural economy and society.

The meaning of adoption acts (adopter)

The economic meaning of the adopter

During the second stage, after the emergence of the pioneer, the adopter emerges, endeavoring to promote and spread the innovation. In regard to technology, Miki Kiyoshi argues, 'once something good is invented, it needs to be spread and popularized within the imitative/mechanical society' (Miki: 1967: 230). Technology fulfils its role through dissemination.

Although the path to development begins with the new innovation, it begins in isolation and is regarded as deviant by the community. The majority of economic subjects are conservative; they cannot fully comprehend the meaning of pioneer's innovatory actions, if at all, and often respond with suspicion and repulsion. To overcome this resistance, it is the adopter's role to recognize merit in the innovation, analyze its value, and further improve it. All this must happen for pioneer's innovatory actions to bring about progress and be meaningful. If we depict pioneer's emergence as a '(sporadic) dot' within the existing system, adopter acts as a go-between to make this dot more acceptable to the system, with the aim of ultimately fusing it with the mainstream.

While Schumpeter sees the issue of development as a two-way process, namely the implementation of innovation and the consequent reaction, he also identifies a third phase of the process: 'the successful appearance of an entrepreneur is followed by the appearance not simply of some others, but of ever greater numbers, though progressively less qualified' (1912/1969: 229). But he ultimately describes this process in terms of the contrast between 'statics' and 'dynamics' on one hand, and 'simple farmer' and 'entrepreneur' on the other. For my purposes, I have characterized this as the difference between a conservative mainstream and the pioneer, and want to stress the importance of recognizing the go-between, the adopter, if we are to explain the development of rural economic society.

Adopter emerges at the second phase of this development process and achieves some early returns within the limits of the profits generated by pioneer. Therefore, the adopter's economic roles are the evaluation of innovation, adoption and improvement, and dealing with the eventual disappearance of profits from the spread (success) of innovation.

The social meaning of the adopter

Adopter is the economic subject who stands between pioneer's generation of innovations and imitator, who sees this innovation as deviant behavior. Adopter analyzes the significance of pioneer's innovation, evaluates it, judges it, determines it, and improves it. In this sense, I define his/her behavior as adoption.

Miki Kiyoshi argues that the quintessence of technology can be found in the two principles 'invention and imitation' (1967: 255). From this perspective, if innovation is the job of pioneer, imitation is the job of imitator. However there is a large gap between invention and imitation. The role of adopter is an intermediary one, between invention and imitation, bridging the gap between pioneer and imitator by initiating and driving the popularization of innovation. By establishing the adoption behavior – the function of intermediary between invention and imitation – as a new category, I believe I can properly explain the process of social and economic change.

Adopter is different from pioneer in that it never carries out innovative and/or deviant behavior by itself. It also differs clearly from 'imitator' who takes non-selective and passive sympathizing actions, as I will explain in the next section. Imitator secures its existence by entrusting itself to the general trend and by 'adapting' as well as 'imitating'. However, although imitator's tendency is to cling to the existing norms and value system, if a new, safe and stable track is created, imitator will readily follow the new one. In short, adopter can be found between pioneer's innovative, deviant, challenging and isolationist behavior and imitator's customary, conformative, passive and majority-adapting behavior.

In response to pioneer's behavior, adopter examines this calmly, objectively and correctly, explores it, evaluates it and adopts it, without being biased to any purpose. Adopter's conception is rooted to some extent in the existing system, but unlike imitator, adopter will never show more attachment to it than is useful. In this sense, it can be argued that adopter does not side with the imitator either.

Therefore, adopter is situated between the pioneer and imitator as an economic subject who plays a major role in achieving a smooth and stable development of the social economy. It might be argued that adopter holds the key to more progressively and sophisticatedly creating and converting societal value systems as well as the economic system. The economic subject who performs these kinds of tasks is viewed, in many cases, as a leader within a village and is venerated by the villagers as a calm, rational and trustworthy person. The trust people put in him/her will eventually lead to the positive valuation of, and faith in, the new technology.

Supplementary note

Allow me to explain why I have introduced the 'adoption' behavior type. I am critically examining Parsons' sociological theory which, based on the concepts of 'deviant behavior' and 'conformative behavior', attempts to analytically elucidate the social system. It can be argued that the 'system' concept, which Parsons employs in his analysis of relationship patterns such as social system, cultural system and personality system, corresponds to what Schumpeter refers to as 'circular (or stationary) flow' – the social process of concept equilibrium – from an economic perspective. Parsons' 'system' is one type of utopian idea. More specifically, it assumes that the social system is intrinsically and comprehensively stable and harmonious. It assumes that within this system people will agree about values and norms. Nonetheless, there might be social actions that at least in part do not conform to this system. Parsons calls this kind of action 'deviant', and juxtaposes it to the concept 'conformity'. He argues 'a tendency to deviance is a process of motivated action, on the part of an actor against one or more normative standards' (Parsons 1951: 230). However, Parsons does not pay much attention to any societal change that stems from such deviant behavior. Rather, he focuses entirely on how deviant behavior comes to be sanctioned by the existing system in order to reinstate conformity, harmony and equilibrium to the system. In short, Parsons' notion of deviance is merely a stationary behavior that emerges along with customs and/or equilibrium. Consequently, his systems theory focuses mainly on the maintenance of structure and equilibrium; and therefore cannot explain dynamic development, through which the system/structure is itself changed. Two important factors need to be incorporated into Parsons' theory if it is to be reconfigured as an adequately dynamic theory. First, deviance must be understood as the starting point of an irreversible process towards social/systemic change. Second, an intermediary element is required, i.e. 'adoption', to explain the transition between the antithetical behaviors of 'deviance' and 'conformity'.

The meaning of imitation acts (imitator)

The economic meaning of the imitator

Developments in the social economy go through three stages: first, an isolated innovation emerges, which is then evaluated and adopted by a small number of economic subjects, and finally imitated by a large number of farmers. These three stages can also be described as: the emergence of endogenous 'dots', followed by a 'linear' expansion of these dots, and finally the formation of a 'complete' whole. Imitator belongs to the third stage of social-economic development.

Once pioneer creates the seed of development, it is evaluated and nurtured by adopter, until it reaches a stage in which the mainstream can view the development as socially acceptable. When this happens, the innovation is no longer seen as pioneering or deviant; it becomes normal, everyday custom, which no longer bears risk. The collective adoption of innovation by the majority of subjects indicates the establishment of a new individual custom and social tradition. This implies that the system has fundamentally changed its form and not merely expanded in size. In this case, economy reestablishes its circular flow, but as a new, more sophisticated circular flow. In this way, a developmental cycle is fully socialized and completed. This new equilibrium is the outcome of development (i.e. different types and bigger social products, new production function, new and/or higher net currency income, different value system and a lower price level) that has taken root across the society and reflects real income increases. The new custom indicates the completion of the development cycle, a return to a high degree of stability, and is a sign of rationality.

Even the imitator pursues profits. However, because s/he prioritizes security, imitator can not obtain significant profits.

The social meaning of the imitator

To sum up, imitator follows adopter's lead in moving away from repeating customary patterns of behavior, undergoes a process of re-synchronization through the imitation of the new technology, and attempts to follow the new customs. More specifically, repetitive and imitative behavior on the one hand and synchronizing behavior on the other, are two different sides of the same behavioral pattern. Imitator's criterion for value assessment is based on convention, on the security of having received a positive evaluation from society; in other words, on a 'common' societal evaluation rather than merely economic considerations. Imitator tries not to deviate from the customary value system, and pursues it with determination; but as soon as s/he realizes that this value system has definitively changed, s/he attempts to adapt the new value system by imitation, seeking to establish a new harmony with it.

Most farmers fit into this third category, attaching great importance to societal values, and striving to not deviate from the customary path. This is why farmers are often viewed as 'simple farmers' – their behavioral patterns are largely based on the principles of acceptance and safety-before-profits. Arguably, this behavioral pattern is characteristic of most people, not just farmers. In this sense, imitator's behavior is necessary to complete the process of agricultural technology innovation, and the corresponding transformation of agricultural society, by reestablishing society's circular flow at a new level. Hence, imitator is the subject who completes the development cycle.

Tarde explored 'imitation' at the level of the individual to explain how societies function and concluded that it forms the basis of socialization. This theory, adopted by Miller and Dollard, is still strong today as the 'theory of social learning'. According to Sakuta Keiichi 'The values acquired through social learning serve to connect/unite the personalities of the actor with others. To be committed to these values is a means for the actor to secure a sense of solidarity with others and harmony with the outside world' (1963: 51). However, Tarde's theory is critical of Durkheim's view that social reality is transcendent and dominant in relation to the individual. According to Tarde 'people have seen only half of this truth, and seen that badly, when they have said that social facts were distinguished by their constrained and coercive character' (1903: Preface 16). It is from this perspective that I have analyzed the theory of imitation as the basis for 'the law of the normal return of fashion to custom, the rooting of innovations in customs and traditions' (Tarde 1903: 22).

Miki Kiyoshi, however, sees imitation as playing a more active role than a simple negative conservatism of habits and repetition. More specifically, he argues that if invention is 'the force behind productive imagination', then imitation is 'the force behind reproductive imagination' which stabilizes everyday life and reproduces itself (Miki 1967: 331–3). On this basis, if we consider social change from a long term perspective, we can characterize the role/meaning of the adopter as an 'intermediary imagination' and 'reformist imagination', which connects invention and imitation and contributes to the spread of innovation.

I differ from Tarde and Miki Kiyoshi in having inserted the intermediary process of 'adoption' between innovation and imitation, or between the invention of technology and imitation. Nevertheless, there are still strong similarities in our basic conceptions. The emphasis on recognizing the spontaneous innovation of pioneering people is a critique of elitist views such as the erroneous Japanese view of agriculturalists described at the outset. I believe that we should rather look for the germs of social change in farmers' endogenous, creative and rational behavioral patterns and stop viewing farmers' imitation and synchronizing behavior, i.e. imitation behavior, as the one-way top-down spread of technology, if we are to develop sustainable agriculture and farming villages that enhance comprehensive values. We must also recognize that it is only with the imitator that a development cycle can be completed and the process of technological transformation achieved.

Types of farmers' acts and rural development

Describing the development process from the perspective of social economy, I have assumed, as a precondition, a uniform, balanced system. By 'development process' I refer to an irreversible process that entails a rupture of the existing circular flow, undergoing a certain transformation, and reestablishing a new higher level of circular flow in equilibrium. I have divided this development process into three stages, and have established three types of economic subjects – pioneer, adopter and imitator – each with unique behavioral patterns that characterize their respective stages. Furthermore, I have endeavored to explain their functions and roles. Each economic subject embodies, at the same time, an economic role and a psychological or social role, according to their particular behavior.

Figure 9.1 is a visual representation of the entire process. The supplementary note below provides a summary of the economic process as depicted in Figure 9.2.

Supplementary note

Figure 9.2 provides a visual representation of the economic process. '0' represents the point where innovation begins in the existing cycle. 'T' indicates the point where the new cycle is established. The innovation conducted at point '0' by pioneers brings them profits equal to an indefinite number of 'E_1 to area p'. These profits are maintained only for a period of time equal to 't'. Shortly, adopter emerges and the profits begin a process

Figure 9.1: A general picture of the process of normalizing innovation

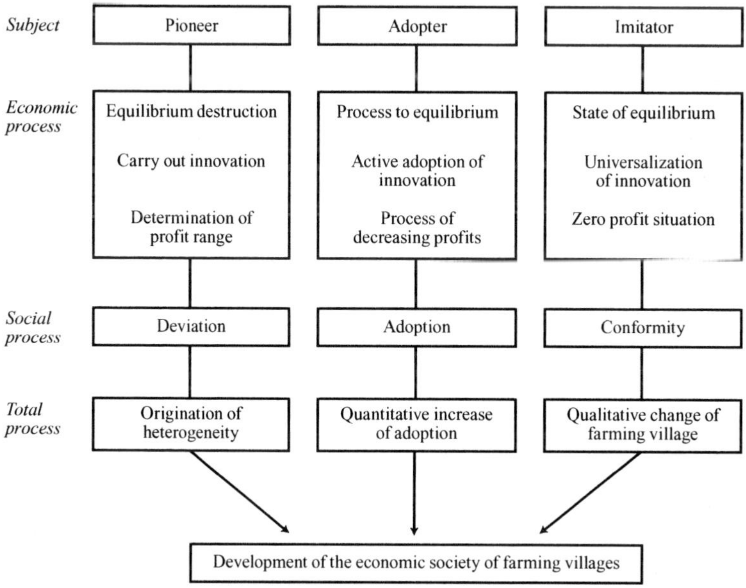

of decline. A small number ('$E_2 - E_1$') of adopters cause profits to fall to almost zero, over a period of time ('T – t'). Furthermore, at point 'T' an indefinite number ('$E_3 - E_2$') of imitators emerge and adapt to the new situation. At that moment, the profits become zero. With this, the economic cycle has been fundamentally transformed. In this way, the development process goes through one complete cycle, having '0' as the starting point and 'T' as the ending point.

The development process described so far, applies to the technological spread from region-to-region as well. That is, I regard the whole country's development as a gradual process, with the country divided into 'pioneering regions', 'adopting regions' and 'imitating regions'.

In pioneering regions, pioneers, adopters and imitators enjoy the profits of the pioneering activity, due to the innovative behavior of the whole region. This is particularly important to note because some adopters and all imitators enter into the process of declining profits. In adopting regions, all economic subjects enter into the process of declining profits, managing to secure only small profits. In imitating regions, with the completion of the whole development cycle, the

Figure 9.2: Innovation and the economic cycle of innovation uptake

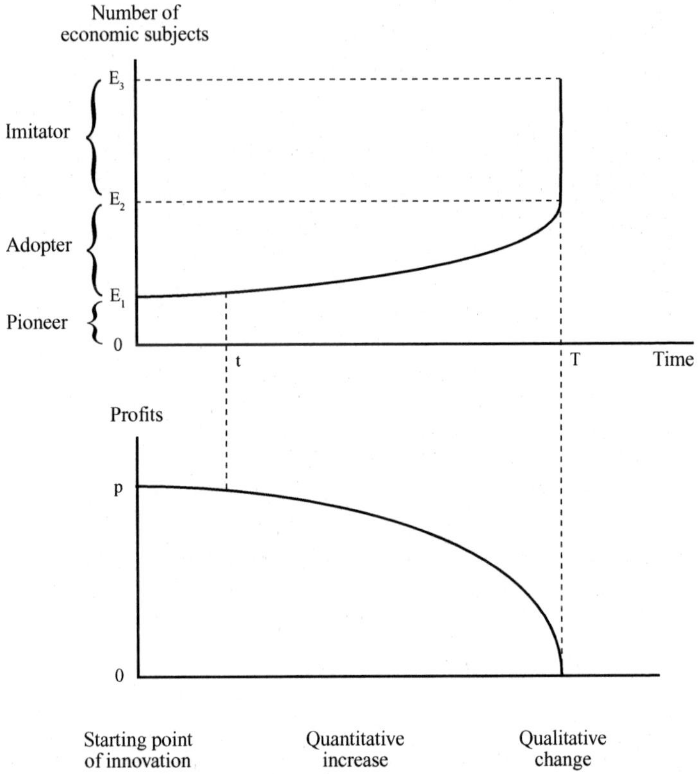

entire region obtains zero profits from the innovation itself, but its normalization brings the overall level of agricultural output to a higher level of equilibrium.

In this chapter, I have mainly focused on the pursuit of economic value (i.e. dynamic profits) as a result of agricultural technology innovation. The transformation process of local community – influenced by the pursuit of three target values (i.e. economic value, ecological value and life value) by modern agriculture, farming villages and agricultural science – can be explained on the same principle. The areas with farming villages are today aiming to develop by achieving comprehensive values in a harmonious way. The functions of regional cooperative organizations are therefore more sophisticated than the ideal-type economic functions I have described thus far. Their range of activities is growing to cover almost all fields, in accordance with local demand. These include: welfare services such as elderly care, children's nature-education and so on, human resource management, restructuring villages, cultural activities, exchanges between cities and villages, international exchange between regions, regional environmental preservation and so forth. There are a growing number of regional cooperative organizations in Japanese farming villages that cover every aspect of life including economics, society, culture and politics. Small-scale farming village societies such as those in Japan and other Asian countries are defined not only by the importance attached to farmer's conformative behavior, but also by agricultural cooperatives, agricultural unions, various agricultural corporations, public corporations, agricultural councils and so forth. These features show that for a small-scale farming village society to develop, it needs to grow into a cooperative society which has various regional organizations (see Chapter 6 as well).

Roles of agricultural science and administration

The integration of empirical and scientific knowledge

The process described above is one of endogenous technological innovation and dissemination. I have explained it structurally, from the perspective of practical behavior. I have also noted ways in which this process roughly corresponds to Miki Kiyoshi's process of 'invention and spread' (1967: 231; cf. Akamatsu 1994), Dewey's process 'from institution of problem to problem solution', as well as the transition 'from an indeterminate situation to determinate situation' (1938: 109–17) and Peirce's 'process describing the transition from suspicion to conviction' which can be viewed as the scientific cognizance process of 'abduction', 'deduction' and 'induction', as I will discuss in the next chapter.

Before moving on, though, I would like to briefly discuss the meaning of the three stages of scientific cognizance process. In a situation when there is a problem that needs to be resolved, the abduction (or hypothesis formation) stage plays the role of transcending the customary path, imaginatively creating a hypothesis that forms a basis for problem-solving. In terms of the typology of farmers' behavior that I have presented, this stage corresponds to the pioneering stage. The second stage, deduction, carefully examines the hypothesis and questions its theoretical validity, thus corresponding to the adopter stage. The third stage, induction, empirically verifies the creative hypothesis and puts it into practice to solve the problem, which corresponds to the imitator stage.

These two processes of scientific knowledge creation and innovation implementation through practical concrete knowledge have strong similarities and suggest a possibility for practical and scientific knowledge to cooperate and integrate in urging necessary reforms within agriculture and farming villages. As Dewey puts it, especially when technology is concerned, knowledge is created as *practical recognition.*

What can agricultural science and agricultural administration, as scientific knowledge, do for empirical knowledge? Needless to say, agricultural science finds its problems in the real world of agricultural life. Its duty is to work independently on technological development. However, it is also necessary for science to contribute to and support farmers' spontaneous innovation. The same holds true for agricultural administration.

If we think of the role of agricultural science and agricultural administration based on the (economic) subject typology I have established in the previous section (i.e. pioneer, adopter, imitator intermingling in regional cooperative organizations), the following can be said:

Roles of agricultural science and agricultural administration
vis-à-vis 'pioneer':

- Start from field evidence, comprehend the problem and set a target to resolve it.
- Identify and respect pioneer's creativity for new technology (value the spontaneous power of development).
- Evaluate the pros and cons of pioneer's technology and management.

Supplementary note

The new agriculture should be one in which problems arising at the production site be brought to the attention of researcher. Research institutes conduct independent research on technological development and people recognize and foster the pioneering attempts of farmers themselves. Pioneering farmers have been the initiators of the most innovative agricultural technology of the past. For instance, *hoon secchu nawashiro* (a rice nursery method for higher altitudes in which the farmer plants the rice seeds and covers the planted area with a sheet of plastic) was an epoch-making development for postwar rice cropping technology in Japan. The method was first tried as an experiment by a farmer in Nagano Prefecture in Japan and the development process was brought to completion as a result of collaboration between the Nagano Prefecture Agriculture Testing Institute and the Ministry of Agriculture, Forestry and Fisheries Agricultural Technology Research Institute. There are many similar examples.[2]

vis-à-vis the 'adopter':

- Demonstrate the scientific basis for recommending the new technology.
- Further improve the new technology, improving its performance and stability.
- Become a good 'promoter of technological dissemination'.

vis-à-vis the 'imitator':

- Promote the scientific perspective and the dissemination of technology.
- Identify attendant problems and help to achieve technological stability.
- Provide assistance by flexibly interpreting agricultural production policies, economic policies and social policies.

vis-à-vis the 'formation of regional organization':

- Research institutionalization and administration, rational management, and processing and distribution issues.
- Enlightenment.
- Aid and support.

By the end of the era of rapid economic development, innovation had come to be seen as normal rather than deviant behavior in Japanese agriculture and farming villages. This has not, however, introduced any essential differences to the structure of community transformation and development. At the same time, the current era can also be legitimately described as individualizing or individualist society, and the information and network society. All of these complicate the composition of any regional populace, with a diversification of farmers and the formation of networking organizations that transcend regions. It has become extremely shortsighted to consider each region separately. Yet I believe that the locally-rooted community remains at the base of society; and the transformations within a networking society can also be explained in terms of the processes discussed in this chapter.

The method of practical science that I propose in the next chapter recognizes the necessity of a practical integration of scientific knowledge with everyday practical reality, based on this examination of the process of forming tacit and *practical knowledge* through the stages of the innovation process as discussed above.[3]

10 Agricultural Science: Characteristic Research Methodology and System

In this chapter, I discuss the characteristics of agricultural science, its research methods, and its system. To do so, I must first clarify its position in relation to modern science as a whole, which includes an outline of the processes by which modern science was established and the relative differences in research methods of different fields of science. More specifically, I am going to detail the defining characteristics of and research methods for three fields of science, namely, the science of nature (natural science), the science of humans and society (human science), and the science of the relationship between humans and nature (practical science). Second, I will locate agricultural science in the field of practical science, as a comprehensive science of life systems, and detail its research methods. Finally, I will present an overview of the system of modern agricultural science in Japan, with consideration of the reorganization of faculties and schools that occurred in the 1990s.[1]

Science of nature

Establishing modern science

Modern science was born of a revolution in thinking during the 16th and 17th centuries. During the Middle Ages, strong forces had prevented science from existing independently from religion, particularly Christianity. 'The medieval mind had explained the unity of the human race through a real bond, just such a bond as a soul unites the parts of an organism' (Dilthey 1988: 302). From this perspective, Copernicus' heliocentric theory, which mathematically explained that the earth travels around the sun, was rejected by the Church. Galileo, a later supporter of the heliocentric theory, was tried by a religious court, before which – under threat of being excommunicated for heresy – he recanted the theory. However, as scholars such as Luther,

Zwingli, Leonardo and Galileo continued to struggle against a religion that had become mere formalities, the conditions for intellectual inquiry gradually altered, 'manifest in the increasing autonomy of religion, science, and art and the growing freedom of the individual against the communal life of man' (Dilthey 1988:290). In the process, science ceases subordinating itself to the medieval schema of religious understanding. Objective examination and positive science came into being (Dilthey 1988: 289). In short, modern science was established by breaking the framework of the medieval social systems.

Thus, the early modern period was marked by 'philosophy separating from theology and natural science from philosophy, each acknowledging the others' field of study and entering into a nonaggression pact promising not to encroach on each other'. Thus science, philosophy and theology, which had been integrated in the medieval Christian world, 'escaped' from this integration, 'which allowed each to follow the route of specialization and independence' (Murakami 1986: 59). Science, furthermore, condemned not only Christianity but also so-called superstitions, mysticism and animism for being contrary to scientific truths (Murakami cited in Uzawa et al. 1990: 52). Various fields of science daringly challenged traditional religious and metaphysical understandings of the world and, by the application of reason through the experimental method, freed people from the 'enchanted world', thus bringing about major social changes.

Bacon, Descartes and Newton were among the scholars who laid the groundwork for the new scientific methodology, thus setting its future course. This system is typically called the Cartesian-Newtonian paradigm.

Modern science is based on the human-centered idea of controlling nature advocated by Bacon. The idea is to establish human mastery of nature, through experimental analysis and exploitation of the ensuing knowledge. To Bacon, nature was an object to be conquered by humans, and modern science was an advanced weapon that would enable humans to achieve this conquest. Descartes' contribution to this 'project of rational-mastery' was to add the 'mechanistic view' of nature, which regards nature as a type of machine. In formulating this view, he eliminated from the world everything qualitative, such as colors, smells and tastes, as well as life and awareness, and reduced everything to materials (Itō and Kaneko 1992: 16). Newton then founded modern physics on the basis of this mechanistic view. In sum, modern science objectified nature, separated it from humans, regarded it as a mechanism, and produced a radical transformation of civilization based mainly on the science of physics.

Clearly, then, modern science was first born as a science of nature, which is now customarily called 'natural science'. In the 17th century, mechanics was born as the science of physical objects (that humans can directly see or touch) and planetary motion. In the 18th and 19th centuries, chemistry was born, focusing on qualitative changes of matter. At the same time, new fields of study appeared in physics that focused on thermodynamics, light, electricity, magnetism, etc. Meanwhile, biology was born of studies of life forms (living matter). In the 19th century, the discovery of the law of energy conservation and the advent of atomic theory produced a systematic perspective on various motions and forms in the material world. At the end of the 19th century, novel forms of motion, such as X-rays and radiation, and new forms of matter, such as electrons, were discovered.

The comforts and conveniences of the material civilization that we now enjoy is a product of modern science and its many discoveries and inventions.

Characteristics of natural science

Natural science primarily aims at discovering the 'laws' that govern the natural world. As Mill observes, nature has a certain regularity or systematic order that produces the same effects under the same circumstances. This is called the 'uniformity of nature' (Tanabe 1918: 231).

Only because of this uniformity of nature, are humans able to understand nature and ensure their own survival. For example, the ancient Egyptians observed celestial bodies and natural phenomena to map their regularity, and used this knowledge to secure a stable food supply through improved agricultural production. If nature had no regularity, people would not be able to survive. People know what season it is from the motion of celestial bodies. With this knowledge, they know when to, for example, prepare for sowing seeds. If they sow seed at the right time, crops mature and produce yields. People understand patterns of wildlife behavior and take informed measures to protect their crops from wildlife damage. They know the characteristics and yields of different parcels and types of land through their experience and observation. They know natural characteristics of various materials and use them in clothes and housing. The more these characteristics of nature and the causal relationships between natural phenomena become consciously known, the more such knowledge accumulates in symbolic acts, traditions and letters, and the more stable and advanced human life has become.

It is modern science, and its methodology of experimentation that has provided humans with the means to obtain, in a dramatically advanced and reliable manner, knowledge about nature and the causal relationships between natural phenomena, and to validate such knowledge as 'law'. Dilthey states that natural science undertakes to 'study reality as it is given in experience by way of the search for causal structure, hence by breaking down composite reality into factors, especially through experiment' (Dilthey 1988: 291), and to discover the laws of nature through such studies. The essence and purpose of natural science is to reduce complex natural phenomena into their constituent elements and to discover and formalize, through experiment, the laws governing such phenomena.

Mathematics has an extremely close relationship with the establishment and development of modern natural science (Tanabe 1918: 183). In order to attain high levels of accuracy in natural science, we must not only meticulously repeat experiments, but also express the laws of nature as mathematical formulae wherever possible. Although claims that mathematical explanations are always possible and valid must be treated with reservation, its effectiveness in particular fields – most notably physics – has been remarkable. Physics deals with various substantial concepts, such as rigid bodies, atoms and elementary (subatomic) particles, describing and explaining these in terms of physical laws and quantitative dimensions, such as length, mass, momentum, force and energy. Physicists tend to maintain that accuracy in the natural sciences can only be assured by developing mathematical models of the findings obtained from experimental observations (Furuta and Nakamura 1968: 69).

In principle, then, natural science proceeds as follows: hypotheses are formulated, which are then tested by experiment, and those which have stood up to such testing are established as laws. These 'laws' must be proven to be universal and objective – that is, not affected by differences in the researchers' cultural or sensory factors. Planck calls this the 'elimination of sensation' (cited in Tanabe 1918: 253–4).

The word 'experiment' is derived from the Greek word *mêchanasthai* via the Latin word *machinare*, which have such meanings as 'to contrive' or 'plot' (Fujisawa in Uzawa 1989: 24). To experiment thus means to make a contrivance against nature and observe how nature reacts to such artificial conditions. Nature reveals and describes itself in the experimental system set by humans. It is not at all easy to understand nature as a whole. Scientists therefore put nature in artificial, manipulated experimental systems and ascertain what it really is in terms of specific elements targeted by humans (Fujisawa in Uzawa 1989: 24). Murakami Yōichirō states that science is basically an effort to 'explain a phenomenon by resolving it into its constituent elements and describe how they behave in space-time, instead of seeing the phenomenon simply as it is' (Murakami 1986: 107). Science has developed through specialization and diversification, and is thus characterized by a certain essential reductivism.

Natural science thus aims to eliminate human sensation from its observations, such that nature might reveal itself through experimentation, producing what has been called the 'self-description of nature'. Hence, the essence of natural scientific recognition lies in the discovery of 'natural laws' – laws that are independent of subjective human value judgments.

Thus, in contrast to the science of humans and society, which is deeply concerned with values, natural science has long seen and represented itself as 'value-neutral'. However, as argued in Chapter 3, although it may be a laudable aim, natural science has never been entirely independent of values. Scientific discoveries often develop lives of their own, independent of 'discoverers'' values and desires, leading for example to certain production technologies that have resulted in environmental destruction, or military technologies such as nuclear bombs. In this meaning some argue that certain values always lie behind science and call them 'undercurrent values' (Fujisawa in Mutai and Kozai 1967: 183).

The significance of experiments in natural science

The intellectual tradition from ancient Greece to medieval Scholasticism relies on argument by syllogism and speculative methods, while the Arabian tradition is based in experimentation. Bacon, who is generally considered to be the father of modern science, argued for unifying the spirit of argument with experimentation, thereby establishing the basic methods of modern science (Noe in Itō 1992: 159).

Argument is a process in which a novel conclusion is drawn from confirmed truths – proven propositions – and experiential data. There are two types of logical argument: deduction and induction. Deduction means to draw a particular conclusion from a general proposition in accordance with rules of logic, without relying on experience. Induction means to draw a general proposition or law from the observation of particular facts that are causally related; or to draw a general truth from specific events. Deduction and induction have their respective strengths and weaknesses, and natural science becomes truly scientific and highly effective in discovering natural laws when it properly utilizes both methods.

The most generally accepted method in natural science today is the hypothetico deductive method formulated by Jevons (Trusted 1974: 61), according to which, natural scientific research must:

> (i) Formulate a hypothesis by inductive means, based on data obtained from previous observation and experiment; (ii) test propositions drawn from this hypothesis by deductive means until a conclusion is reached; and (iii) verify the conclusion by repeated observation and experimentation, after which the original hypothesis is finally accepted (and a law posited), modified or abandoned, as the case may be.

In this process, observation and especially experimentation play an important role. It is safe to say that the demonstrativeness of modern natural science, or what makes modern science science, lies in experimentation. As Kant put it, 'experiment without theory is blind', and 'theory without experiment is empty'. Yamamoto (1972) suggests that a 'fear of emptiness' has driven science to experiment. Observation and experimentation are inseparable in the sense that an experiment must be observed. In another sense, however, they can be differentiated: observation is to examine natural phenomena as it exists in its natural state, whereas experimentation is to examine phenomena under artificially controlled conditions. As such, experimentation assumes the regularity and reproducibility of natural phenomena; whereas observation might search for such regularity, but need not assume it *a priori*. Experiments are conducted in a repetitive test system that has been artificially assuming reproducibility, in which nature unfolds or describes itself in response to questions posed by scientists.

According to Galileo, however, there are two directions in scientific studies: resolutive methods (*methodo resoltivo*) and compositive methods (*methodo compositivo*). Scientists use resolutive methods to resolve a natural phenomenon into simple elements and conduct experiments and observation. The action of a specific element in question is observed and measured with all other factors made equal. They must also, however, use compositive methods to measure and collect data that reveal certain laws concerning the relations between such simple elements, and to consider whether such data accords with results of experiments or not (Nagai 1987: 24). Galileo's experiment to prove the law of falling bodies, an archetypical modern scientific experiment, employed a very simple system. But as science has advanced, experiments have become far more complex, elaborate and ingenious, and have in some cases reached an incredible scale, such as the particle accelerators of today, creating a torturous aspect to the examination of nature.

Results verified by experiment are presented as new propositions and/or laws. Yet, 'a hypothesis established by repeating tests a finite number of times...only assures that the resulting general law holds true with a certain probability' (Itō 1992: 168). That is, when laws have stood up to 'counterevidence', or have been verified by trial-and-error experiments, they may be accepted as truths, but only 'for the time being'.

Science of humans

Mechanical view of nature and the autonomy of human science

We have discussed the characteristics and methods of modern natural science, focusing particularly on physics. Modern natural science might be seen as a rupture in the theological, religious world which has enriched material life by elucidating and utilizing nature in specific ways. At the forefront of science's success, physics had become such a dominant force by the beginning of the 20th century that the word 'science' typically implied natural science, and natural science implied physics.

On this, Dilthey wrote, 'immediately following the introduction of the mechanical interpretation of nature by Galileo and Descartes came the extension of this mode of explanation to man and the state by Hobbes and Spinoza' (1988: 303). It then extended to the world of philosophy, with Brentano claiming that the true methods of philosophy are nothing other than those of natural science. Dilthey states, 'Descartes developed his conception of nature as a huge machine' and, with the contributions of Newton and Galileo, won the battle to eliminate the world of 'psychical relations in the processes of nature' from this mechanism. With the rise of modern science, this 'mechanism took the place of the deity' (Dilthey 1988: 80, 292–5). Fujisawa Norio argues that the characteristics of modern science lie in the 'mechanistic, human-centered view of nature', or the 'Cartesian-Newtonian paradigm' (1989: 27). Dilthey expressed strong objections to the growing tendency to extend this mechanistic view of nature and apply the explanatory methods of natural science to the study of humans, societies, nations, and philosophy (1988: 148).

The mechanistic view of a human being as a machine has had far-reaching consequences. The determined effort to reduce Hobbes' ethics to physics influenced medicine and biology, mainly among materialists. La Mettrie (1947), for example, argued that a human being is an 'extremely complicated machine', where even the soul is driven by a 'material portion of the brain that has the ability to feel', the driving force for activities that works like a spring in a machine.

Dilthey rejected the notion that human and social life could be understood in these mechanical terms. He argued that 'we must meet the challenge to establish human sciences through epistemology, to justify and support their independent formation, and to do away definitively with subordinating their principles and their methods to those of natural sciences'. In other words, we must 'establish the relative independence of human sciences' (1988: 85). Bollnow positions Dilthey here as part of an irrationalist trend in the *Sturm und Drang* (storm and stress), and Romanticism more generally. Human beings have been defined in turns as rational agents and irrational agents throughout the history of philosophy. One task for the philosophical communities of the late 19th century was to establish human sciences from the perspective of a 'philosophy of life' that subsumes irrational factors, in opposition to the dominant tendency toward natural scientific rationalism (Bollnow 1958: 134). What Dilthey called spiritual science shall be referred to as 'human science' here (Maruyama 1985: 24). I now turn to the characteristics of and research methods for the human sciences, focusing on Dilthey's arguments.

The establishment and characteristics of human science

Dilthey (1988) was arguably the first to clearly elucidate an argument for separating the human sciences from the natural sciences, and to outline an independent research method. Dilthey describes his approach as simply to explain human 'nature and understand mental life'. To achieve an understanding of human life, he says, it is necessary to distinguish 'nature from history', the 'minds' from 'bodies', and 'facts of intellectual life' from 'the processes of nature' (1988: 79). Arguing for a perspective that recognizes the 'unity of human life', Dilthey says: 'Human being as a living unity is present to us as a grouping of intellectual facts as far as inner awareness is concerned and conversely as a corporeal whole insofar as we perceive through the senses' (1988: 84). The limitation of the natural scientific approach to understanding humans, he observes, is that 'inner awareness and external perception never occur in the same act, and therefore we never encounter the fact of intellectual life simultaneously with the fact of bodily existence' (1988: 84).

This is not to say that there is no place for the natural sciences in understanding human beings. Dilthey recognizes a natural dimension of human being, in that humans, as physical/embodied beings, must deal with nature in at least two ways: (1) restrictions imposed by the natural course of events, and (2) nature's reaction to human activities. On the other hand, Dilthey declares independence of the mental world and mental science from the natural world and natural science stating that if we take one step into the mental world, 'nature to our mind is exactly the nature we have in our mind' (1988: 83). Dilthey's human science came from a 'philosophy of life' perspective, and he saw psychology as a basis for human science research. But he stressed that this psychology must be descriptive psychology rather than the explanatory psychology of the natural sciences. Explanatory psychology resolves mental phenomena into simple individual elements and attempts to explain causal relationships between these elements from an external perspective. Dilthey rejects this 'psychic atomism' in favor of an interpretive psychology which attempts to understand the depths of mental life and its meanings from the inside, based on the shared experience and empathy that we have with fellow human beings. From this perspective, then, it attempts to analyze the structural relationships between meanings (Bollnow/Takahashi 1986: 163).

Thus, the task for human science is to describe the real facts of human life, and the structure of their meaningful relationships, including culture as an 'expression of life', the people of a nation and their social organization, or laws, languages, religions and art – that is, the 'individual living unity' that binds humans and their social activities (Dilthey 1988: 100). To this end, the methods of the natural sciences – understanding phenomena in mathematical terms while assuming uniformity or regularity of events – are inappropriate (Dilthey 1988: 97). What cannot be adequately understood by natural science will emerge only by understanding and interpretation of the meanings given to individual events.

Bollnow (1958) provides a summary of the anthropology common to the philosophy of life perspective, focusing on Dilthey, but also discussing Fichte, Nietzsche and Bergson. 'Life' here includes, without limitation: survival, living, vitality, and the opposite of death. Specifically, life has been described as bottomless (Nietzsche), constantly occurring, flowing and leaping events (Fichte and Bergson), and a drama occurring in each person (Ortega), among other things. Dilthey's human being is an emotional, active agent for whom being a 'pack of impulses and emotions' comes before being an intellectual agent. It is also an unstable agent that owns, loves, hates, fears, and is changeable. Furthermore, it creates things and itself. A human being can also be seen as a comprehensive combination of rational and irrational forces. Despite such 'unity of life', though, the human being always confronts an external world that often proves a barrier to internal thought. The constant functional interactions between the human being and the external world – namely, other people, the society, nation or the environment – significantly determine the particular mode or style of being human. An individual life is characterized by history and society, and completely dissolves into the processes of history.

Dilthey calls categories that are unique to life and history – that are different from nature – the 'categories of life', and specifies temporality, meaning, intention, values, forces, interaction, development, essence, etc. This includes dimensions that are contrary to, and thus cannot be explained in terms of uniformity, regularity or the measurable quantities of the natural sciences. Dilthey emphasizes social structure as a particularly decisive factor. In this vein, Maruyama Takashi explains that the human (experiential) world – as opposed to the natural world – is characterized by structural laws as opposed to general laws, functional relations as opposed to causal relations, and intentional connections as opposed to mechanical connections (1985: 74). The basic characteristic of human science, then, lies in the fact that it must deal with issues that are not reducible to the terms of natural science.

'Understanding and interpretation' in human science

These peculiar characteristics of the human realm directly affect the methodology suitable for human science. Dilthey (1957) argues 'understanding and interpretation is the method appropriate for human science'. He considered interpretive psychology, rather than explanatory psychology, to provide the most suitable basis for understanding and interpretation. Explanatory psychology has destroyed the classificatory systems of the 'older metaphysical schools' by 'subordinating spiritual facts to the mechanical order of nature... Psychical mechanics sacrifices what we experience in inner perception to a line of reasoning which plays around with analogies from external nature' (Dilthey 1988: 303–4). The explanatory psychology that 'regards mental unity as an independent entity against oneself' results in a type of 'psychic atomism' that is incapable of comprehensively explaining human life and society (1988: 303). Mental life can only be revealed in its true form when we attempt to understand it from the inside rather than the outside. Dilthey thus differentiates between human science and natural science with the claims that 'we *explain* nature and *understand* mental life' (1957). Thus, 'understanding' is the starting point for human science methodology.

Dilthey defines 'understanding' as a 'process through which we perceive something internal' from 'expressions of life' (1957: *Band* 5.318) where the latter phrase refers to any 'expression of something mental' that is perceivable to us. In addition, 'based on, and through constant interaction between, experience and an understanding of the experience, an understanding...of other people's lives and...characters are formed' (1957: *Band* 5.333). Individual experiences are very small fractions of a life, and thus we experience life as a 'fractionalized entirety'. This is a different concept of fractions than that referred to in natural science as a unit by which humans attempt to survey and control nature by reducing it to its constituent parts. In addition, while natural science seeks the causal relationships that constitute the laws governing natural events, human science seeks to understand the meanings of events for the people concerned. Subjective experience and the understanding of it provide the basis for this task (Bollnow 1958: 136–7). Understanding begins from 'concern' about a certain problem and is thus dependent on the breadth and depth of this concern. Hence, the possibility of 'understanding life by life itself' is supported by the 'similarity of human spirit' rather than the uniformity of nature.

As for society, Dilthey claims, 'we know the state of society inwardly; we can reproduce it in ourselves up to a point by observing our own condition, and we intuitively follow the depiction of the historical world with love and hate, with passionate joy, with full involvement of our emotions' (1988: 97).

From this perspective, human science operates as hermeneutics. Dilthey describes interpretation as a 'technical understanding of an expression of life fixed in a sustainable manner' (1957: *Band* 5.319). Here, the 'expression of life fixed in a sustainable manner' refers to literature or related modes of expression. Interpretation, then, means to empathize with, experience vicariously, and reconstitute the particular individuality of a certain person or society based on our own experience and the similarity of human mind, using 'imagination' and its 'modulation' (1957: *Band* 7.213). Modulation is an interpretative technique by which we 'can raise or lower the tone of a behavioral pattern, force, emotion, effort, direction of thought, etc, that is included in the relationships in our own life and, by doing so, can reconstitute any other person's mental attitude to life' (1957: *Band* 7.214).

In this way, 'we can understand the mental world. In other words, we can understand internal relationships in such world as meaningful'. Thus, as previously stated, the causal relationships in the natural realm can only be explained, whereas culture only appears in meaningful relationships that require understanding (Bollnow 1958: 137).

Hermeneutics originated as a theory of interpreting ancient (typically religious) texts (Dilthey 1957: *Band* 5.190), so language is of central importance. In fact, a thorough, understanding of a person's inner world can only be obtained through words. Dilthey expects much from the works and biographies of great poets, discoverers, religious geniuses and genuine philosophers (1988: 95). He seems to be biased in this respect, though; the manifestations of life include products of thought that have linguistic, logical forms, such as concepts and judgments, but also other meaningful human activities.

The three concepts 'experience, expression and understanding' have a Trinitarian relevance for hermeneutic interpretation. Dilthey and Bollnow focus not only on the relevance of each, but the creativity of interpretation and 'dynamism of life' that appear in their mutual interactions. 'Experience produces expressions. Expressions fully describe the experience and bring out something new' (Bollnow 1958: 37). Life is constructed and extended in expressions, while at the same time bringing such expressions to conscious understanding. Expressions of life are often made unconsciously: 'No creator clearly knows what his own creation means'. What lies behind an individual's expression often only becomes known through the understanding derived from others' interpretations. The 'effort of interpretation is another genuinely creative activity in its own right' (Bollnow 1958: 138–9). In historical studies, for example, an interpretation that creatively re-presents the present day through meaningful encounters with the past is considered to have more value or merit than a faithful reproduction of past events. This ability of interpretation to change and extend our understanding further enriches the 'wealth of human life'. This is the purpose of the human sciences, whose methods are deeply rooted in the philosophy of life.

Ideal types and comparison

Granting that humans, society and the historical world can be elucidated by understanding and interpretation, a problem remains: since creativity is necessary for interpretation, how can such interpretation acquire validity and scientific-ness? To begin to overcome this problem, we must recognize that creativity and validity do not constitute mutually exclusive choices.

Interpretation in the human sciences must go beyond the level of mere commonsense and achieve 'scientificity' or 'scientific-ness'. This is achieved by comparing and checking scientific interpretation against social events by observation. Individual meanings are clarified by comparison, and comparison is clarified by classification. These issues were addressed by Max Weber and Alfred Schütz. Schütz argues that all social sciences are studies of type formulation. This is the basis of the social scientific methodology, epitomized by Weber's ideal types and cross-cultural or historical comparison.

Weber stressed that comparative historical studies and sociological studies should aspire to 'value freedom' (*Wertfreiheit*) (in Sumiya 1970: 188).[2] The most important issue in this pursuit was the construction of ideal types (Weber 1904/1922: 190–208). An ideal type is 'the construction of certain elements of reality into a logically precise conception' and 'has nothing to do with evaluations of any sort' (Gerth and Mills 1948: 59). It is constructed not as a depiction of reality, but in such a way as to aid in understanding reality. 'In its conceptual purity, it cannot empirically be found anywhere in reality', nor does it constitute an average image of reality. It is an ideal obtained by theoretically enhancing 'one or a few perspectives'.

By measuring how close or how far reality is to or from the theoretical image, researchers can go beyond merely describing an event to reconstructing it and comparing it to other events. An ideal type is, thus, a conceptual device artificially constructed to help to understand the world, 'not as a goal but as a tool of understanding'. Weber eventually defines an ideal type as 'a concept that has a meaning as a genuinely ideal extreme concept and that serves as a measure against which reality may be measured and as a standard with which reality may be compared, in order to identify, out of empirical knowledge that has been obtained from reality, component parts with specific significance' (Weber 1904/1922: 194). An ideal type must be distinguished from a mere category or classification.

Using ideal types, historical or regional characteristics of human subjects and societies, and shared meanings of individual events are clarified. As globalization continues its present trajectory, classification and comparative methods are increasingly important for understanding human and social life, both locally and globally.

Schütz (1932) attempted to analyze the structure of the social world by drawing Weber's insights closer to the everyday lifeworld. Schütz observes that in the human sciences, the researcher is the scientist/subject, but is also the object of science as a human/social subject with an everyday life. Schütz takes the standpoint that 'the basis of all studies lies in the prescientific lifeworld', Schütz thus represents a methodological development of the hermeneutic science after Dilthey, based on Weber (Maruyama 1985: 136-151).

Philosophy of science and philosophy of life

About the same time as Dilthey, a school called the neo-Kantians, including Windelband and Rickert, argued that the science of humans, society and history should be independent of natural science. In his inaugural address as university president in 1894, Windelband referred to the two sciences as natural science and historical science, and contrasted them as 'nomothetic' and 'ideographic' science. He explained that natural science is characterized as discovering and explaining generalizable laws, whereas historical science is characterized as describing a one-time, special event, or a characteristic relationship (1911: 136). Rickert, who succeeded Windelband, designated the two sciences as natural and cultural, and developed Windelband's work by distinguishing them as 'generalizing methods' and 'differentiating methods'. Introducing the idea of 'nature freed from values and culture associated with values', Rickert (1899) claimed that while natural science is unrelated to values, the mission of cultural science lies in understanding and describing the relationships, meanings and values of cultural events that are worth knowing.

Windelband and Rickert thus appear to agree with Dilthey and Bollnow in many respects, and seem in many respects to argue much the same thing. However, they are quite different from each other in some essential respects.

First of all, according to Maruyama Takashi, there are three possible methodological approaches in philosophical inquiry: 'scientific logic', 'scientific epistemology', and '(epistemology as) human ontology' (1985: 16). The neo-Kantians' argued in terms of scientific epistemology, in which they, so to speak, drew a border on the 'globe of world knowledge' between the natural and human (cultural) sciences, and starkly contrasted the differences between the two. Their arguments primarily addressed the 'formal logic' of scientific knowledge rather than the 'objects' of science. In this they followed Kant's *Critique of Pure Reason* (1781), which Dilthey opposed with what he called the 'critique of historical reason', striving to provide a philosophical basis for a spiritual science. As previously observed, Dilthey's scientific epistemology is based on observations of life itself and questions the basis of the dominant modern epistemological framework. We might then characterize the neo-Kantians' standpoint as a 'philosophy of science', whereas Dilthey's is a 'philosophy of life', from which a theory of science is derived (Kuki 1989: 38).

Miki Kiyoshi states, 'the tasks of philosophy are not limited to criticism of science in the sense conceived by the neo-Kantians, that is, the formality of clarifying the logical basis of science; the ultimate goal of philosophy is to establish a new worldview through a scientific worldview' (1967 [7]: 65).

Bollnow argues that modern epistemology must take into account: (1) the necessity for the 'human viewpoint' – that knowledge should be based on the comprehensive relationships of human life; and (2) the 'hermeneutical viewpoint' because knowledge inevitably starts from the 'previously understood world' (Maruyama 1985: 21). These are precisely Dilthey's two focal points, as well. One problem of the present age has been overly aggressive developments in science without due regard for human life, which calls into question the very *raison d'être* of science from the perspective of the everyday lifeworld. In this sense, Dilthey's work in the philosophy of life and scientific knowledge has a continuing significance for the human sciences.

The second difference between the neo-Kantians and the hermeneuticists is in the concept of 'values'. According to Kuki Kazuto, Rickert's values are 'objective values independent of any subjectivity, supra-historical, formal values, characterized by appropriateness' (Kuki 1989:110–25). For Rickert, values exist independently of human subjects, a staunchly anti-relativist view. In contrast, Dilthey denies the existence of anything that 'rises above time' or that 'may constitute a requirement for the overall relationships existing in the process of life' (Kuki 1989: 110–25). From a historical perspective, Dilthey 'even questions universal validity as a traditional categorical imperative of philosophy, and indeed positively commends relativity as a historical sense' (Bollnow/Takahashi 1986: 176). It is not, however, pure and simple relativism.

If, as in Rickert's argument, values are deemed to be independent of the subject's life, transcendental and timeless, then values will be alienated from the everyday lifeworld. This is not only because the mental world is characterized by 'producing values and achieving goals in accordance with the structure of mental life' (Dilthey 1961: 304), but also because, as described above, the formative and creative meanings of human activities and the interpretation thereof constitute new formations in the goals and values themselves as the 'wealth of human life' – unless, that is, we unconditionally assume that human beings are irrational animals (Bollnow 1958: 47, 141–2). In other words, it can be said that 'abandoning all norms that transcend time is a prerequisite to understanding human beings in terms of their creativity in molding history' (Bollnow/Takahashi 1986: 126).

We must thus take a productive standpoint that departs from Rickert's transcendental values and, at the same time, avoid simple relativism in the context of values, by making formative, creative progress, as Kuki puts it, towards the truer, the better and the more beautiful.

Division and integration of scientific methodology

Polarization of methodologies and the 'transitional point'

I have described the process through which science, particularly natural science, became independent of the medieval religious world, then the process by which human science was separated from and made independent of natural science, and finally the characteristics and methods of the two sciences. With regard to natural science, these issues have been followed through the thought of, among others, Newton, Kant and Jevons. As for human science, we have primarily followed the thought of Dilthey and Bollnow, who attempted to establish a unique realm of human science in opposition to the monistic tendency to apply a single explanatory method to all fields of science. Although my description of each includes my own views, which will inform later descriptions, my aim has been to clarify the essences of the natural and human sciences, and to differentiate them in as generalized a manner as possible.

However, researchers in each of these fields, in their efforts to validate 'their' science and extend its reach and influence, have been eager to assert the alleged uniqueness and superiority of their science and its methods, while emphasizing differences between the two sciences, in order to persuade others of the validity of their own view. Arguably, this has led to the polarization of the two sciences and to the concealment of their common features, resulting in the formation of two theories of science that in many respects appear to be binary opposites. Furthermore, the antagonism between the advocates of natural scientific methods and advocates of the philosophy of life – or historical methodology – has extended into the realm of the human sciences (Maruyama 1985; cf. Adorno 1969). The arguments thus take the form of *either* scientific reason *or* historical reason; *either* explanation *or* understanding, etc. The limitations and contradictions that characterize such either-or arguments have resulted in a loss of diversity and thus diminished the potential capabilities of science.

The problems of this dualistic methodology and its implications for any intermediate field of science have been analyzed by Dilthey. For example, Dilthey characterizes human beings as 'psychophysical living units' that are dependent upon nature in two ways: (1) natural conditions restrict social and historical events; and (2) for the resources necessary for their survival. He then argues that at the 'transitional point' between the natural and mental, 'knowledge of both fields are mixed together'. Dilthey notes that the human sciences must draw upon biology (the science of organisms) to elucidate the psychophysical living unit, as well as inorganic natural science for enhancing the human capacity to exploit nature (Dilthey 1988: 96). Furthermore, he argues, following Fechner, that since human life can be divided into 'corporeal life and psychical life', psychology can be divided into physiological (or explanatory) psychology and interpretative psychology (1988: 93). Yet, although these dimensions are theoretically distinguishable, human beings never live in a world where the mind and body are completely separated from each other. Dilthey is thus clear that both scientific methodologies are useful, and perhaps necessary, for elucidating natural and human events as their mutual interactions modify the relative importance of explanation and understanding.

Dilthey locates economics as one of the sciences in the 'transitional point' between these two poles, recognizing Thünen's theory as a specific result. Though not as precise as natural scientific theories, Thünen's theory states that agricultural production activities change their style of management and degree of intensity depending on their distance from the market. Dilthey adds that concepts such as market demand, economic efficiency, labor and value are 'psychophysical concepts' (1883/1979a: 96, 143). Yet he provides no explanation on how the explanatory principle and the understanding principle work together in this transitional theory.

This conciliatory tone, however, soon disappears from Dilthey's work as he increasingly turns his attention to establishing the independence of the human sciences. This turn may have been inevitable for Dilthey, whose efforts to ensure the independence of human science were continually threatened by the overwhelming forces of the natural sciences, which would, if left unchallenged, submerge the independence of understanding-based science.

Arguments for resolving the polarized scientific methodologies

Dilthey's successors, including Misch and Bollnow, highlight the continuing need to consciously resolve the polarized methodologies. While following Dilthey faithfully, Misch (1947) opposes any dogmatic differentiation of various sciences and contends that arguments about the polarized methodologies should be settled productively, rather than forcing a choice between one or another. Bollnow agrees, adding that 'the two groups of sciences each have the ability to understand and the ability to explain, no matter in what different proportions they may be assigned to different groups' (1949: 95). Nevertheless, Bollnow offers a noteworthy qualification to the effect that understanding is a more 'basic function' than explanation and 'explanation occurs only where the self-evidence of an understanding of life has been lost' (1949: 95). Since then, arguments over explanation and understanding have become increasingly diverse.

Ricoeur attempts to settle this unproductive confrontation between understanding and explanation, arguing that interpretation is an extremely complicated task that goes through consecutive stages of understanding and explanation. Understanding includes explanation. Explanation develops understanding (Ricoeur/Kume 1985). Ricoeur criticizes Dilthey's dualistic perspective, arguing that human action entails two distinct aspects of desire (as a driving force and as a reason for action) and belongs to beings that can be explained as causality (cause) and can be understood as motivation (reason) at the same time.

Ricoeur then attempts to replace the either-or relationship between explanation and understanding with a dialectic relationship, noting a need to find continuity as well as discontinuity, and methodological similarity as well as distinctive characters, between natural science and human science (Ricoeur/Kume 1985).

From a slightly different perspective, Habermas argues that human science must be constantly self-reflexive and apply this reflexivity to formulating critical social theories. According to Habermas, while humans have come to control nature through advanced technology, technological rationality has also extended into the depths of human life, society and even science itself, restricting and affecting humans in new ways. Under these circumstances, classical ideologies have been abandoned; but Habermas argues that this abandonment is itself a modern ideological phenomenon. He presents an ethical demand for human science to contribute to human enlightenment and liberation, in part through a sustained critique of ideologies (1981; cf. Maruyama 1985). Habermas then argues that fulfilling these demands requires the unification of explanation and understanding and, hence, of explanatory science and interpretative science.

Many others have advanced arguments in this debate. Thompson compares the ideas of Ricoeur and Habermas towards formulating a critical hermeneutics which integrates these ideas. Feyerabend argues that both the natural and human sciences are based on hermeneutical principles and that there is no distinction between them. Rorty argues that the distinction between the two sciences exists as a matter of relative validity, preference, interest or morality. In addition, Lovelock has proposed a hypothesis called Gaia theory, which regards the Earth as one life (Takagi in Uzawa et al. 1989: 290).

On the question of hypothesis formulation, both Dilthey and Bollnow argue that the mental world 'does not need to be constructed by laws assumed using hypotheses' (Bullnow 1958/1975: 137). For them, nature can be understood by explanation while mind and culture – i.e., human life – can be understood directly through examining one's own or another's life, which makes the intervention by hypotheses unnecessary. Yet, as discussed, with Misch they rejected the polarization of the sciences and accept interaction between them.

Dilthey further argues that human events are interpreted by 'technical understanding', and that such an understanding is only reached via a creative process. Although interpretation is a process for the 'understanding of life by life itself', it is impossible that there will ever be only one interpretation of an event. Each interpretation will differ slightly depending on various factors such as values and perspective. Thus a diversity of interpretations is inevitable in efforts to understand the extent and depth of meaningful relationships among historical, cultural and social events. But then it would appear that it is the diversity of interpretations that distinguishes the human sciences from the natural sciences, which aim at discovering singular universal laws. Without allowing such diversity, understanding and interpretation would fall into dogmatism. The explanatory value of an interpretation is not only assumed to be contingent on a particular perspective at a particular point in time but is also assumed to undergo historical transformations. This is where the essence and significance of interpretation lies. As Dewey frequently notes, the formulation of interpretation may be understood as a type of hypothesis (Meguri 1963: 130).

In my opinion, although explanation and understanding should not be understood as mutually exclusive, neither can the relative uniqueness of the two sciences be denied. But rather than the polarizing antagonism between them, we should strive for a productive, albeit tense, relationship between them, capitalizing on both their commonalities and unique research methods.

Science of the relationship between nature and humans

The realm and characteristics of practical science

Integration of natural and human sciences

I have thus far described the characteristics of the natural and human sciences, focusing on their contrasts and the history of their respective formation. I have also suggested an attempt to bridge the gap between the two sciences. To the extent that one aims at elucidating nature and the other at elucidating human life, and that each is eager to assert its own independence and individuality, it is perhaps inevitable that their respective methodologies are discussed in terms of their differences. The two sciences may only unite in those realms where humans and nature interact with each other. A specific *ba* where explanation and understanding intermingle can emerge from observing the points of interaction between humans and nature.

The terms 'relationship between' and 'points of interaction' between humans and nature as used here mainly mean production activities such as agriculture and industry, but also development activities to enhance the quality of life and recreational activities. Humans must engage in creative adaptation activities in their struggles to overcome the restrictions imposed by nature. Creative adaptation here does not mean adaptation over a long period, but quite ordinary human activities such as making clothing for protection from the elements, building a hut for shelter, or cultivating crops to ensure food supplies. Of course there may be significant differences in the level of creativity in different times and places. Where humans have simply collected natural products and lived within the restrictions imposed by nature, behaving in a manner not substantially different from apes or horses, they are simply part of nature itself. However, once humans acquire the creativity by which they use advanced tools and machines to cultivate crops and raise livestock, they have departed from the animal world and become human beings. According to Miki Kiyoshi, this is when humans enter the world of 'imagination', or the world of technology. Imagination is the ability of humans to create and construct things in their mind.

The world of human creation has expanded at an accelerating pace, resulting in the highly industrialized, urbanized societies of the early 21st century. After millennia of seeking ways to coexist with nature while bending it to our needs, since the industrial revolution we have increasingly attempted to assert our 'mastery over nature' (Descartes), leading to what can only be seen as a battle against nature. The initial success of science and technology in advancing our mastery over nature has created an excessive dependency on technology, as well as many of the most pressing problems we face today.

These problems compel us to reflect on the present state of technology, such as excessive development, environmental problems, the technological determinism of the economy, and the ensuing restrictions on human life. Yet calls to stop technological progress seem futile. We probably have no other choice than to strive to rehabilitate and change the basic direction of technology on the basis of our reflection if, as Dilthey claims, humans can no longer live without science (1988: 157).

Humans face many problems today, many of which are 'unintended consequences' of intentional activities in the field of interaction between humans and nature that aim to realize certain values. As discussed, the field of interaction between humans and nature is the place where the natural and human sciences meet, and thus where explanation and understanding must intermingle. But this should not be simple intermingling; explanation and understanding must be integrated in a new methodological space of 'conception', as Miki proposes.

In this space, the natural and human sciences do not merely meet and intermingle – they create unique fields by their integration. These include practical, technical fields such as engineering, agricultural science, medicine, pharmaceutics, pedagogy, and economic policy. These fields have generally been referred to as applied science or technology and, as such, regarded as subordinate to 'pure' science. As discussed below, however, each of these fields entails highly complex relationships that involve far more than the mere application of laws. They operate in the world in which human intentions or purposes encounter, confront, contend with, and must compromise with, nature – and perhaps more importantly, with other human actions and intentions. They cannot therefore be considered as merely an intermediate realm between the human and natural sciences. It may be appropriate to regard the sciences of the relationship between nature and humans as occupying an independent realm of science, rather than reducing it to a sub-field, such as applied science. The overall structure of science, then, would be supported by three centers, each of which has its own fields and methods, with various fields of science around the centers, intermingling and covering the entire landscape. The methods of explanation, understanding and conception are employed as appropriate to the respective fields of inquiry, creating a complementary methodology.

In the context of Dewey's practical, dynamic, or so-called pragmatic methodology, to which this work owes much, I call this realm characterized by the pursuit of values and the search for improved technologies, 'practical science', as distinct from the natural and human sciences.

Agricultural science as practical science

Agricultural science has long been regarded as a practical science that is concerned primarily with organisms. Let us briefly survey various authors' descriptions of agricultural science.

Yokoi Tokiyoshi regards agricultural science as a 'composite science' that is based on various fields of science (1925: 399); Nitobe Inazō regards it as a unique 'type of science' that invents 'intermediate principles' (1905: 64); Boehn refers to it as a 'practical science' (cited in Yokoi 1925: 394); Krzymowski, following Goethe, contends that 'all studies are of equal value as science' (1919: 102); and Heinrich Weber argues that the scientific fields of agricultural and forestry sciences, with natural and human sciences, constitute a field of 'normative practical science' or 'intentional science' that is guided by human intention (1919: 19). In Japan today, agricultural science is typically regarded as 'applied science'. Kashiwa Sukekata, however, positioned agricultural science as a 'third science' rather than 'merely' an applied science, based on neo-Kantian views such as Rickert's and Windelband's, developed through an amplification of H. Weber's philosophy of forestry science (Kashiwa 1962: 334–423; 1980: 53–118). It is safe to say that the history of the theory of agricultural science is at the same time the history of attempts to establish agricultural science as a unique field of science that goes beyond mere technology. Tsuno Yukindo (1975) regards agricultural science as a study of science while Sakamoto Keiichi (1994) regards it as an applied science, yet both assert the methodological uniqueness of agricultural science.

Krzymowski's *Philosophie der Landwirtschaftslehre* (Philosophy of agricultural science) and H. Weber's *Grundlinien einer neuen Forstwirtschaftsphilosophie* (Introduction to philosophy of forestry science) were both published in Germany in 1919, the year that the First World War ended. Both works were written in the midst of confusion in the war, when the fundamentals of the economic society and science were being questioned both in Germany and throughout the developed world. The philosophy of science was to play an important role during this transitional period in reflection, advancing self-consciousness and the reorganization of science.

At the time, Wagner characterized the intentional sciences dealing with practical issues – such as agricultural, forestry, engineering and commercial sciences – as business sciences and, hence, unscientific, because they pursue private or national economic interest. Rümker declared that the task of agricultural science was to consistently serve practical purposes, arguing that any science of no practical use is worthless. Krzymowski rejected both of these views, asserting that although agricultural science obviously deals with practical issues, it also includes a field of agricultural studies, cultural history and geography, which are of no practical use, which makes agricultural science a field of pure science. For example, an agricultural scientific study of termites may originate with an intent to find a way to exterminate them in order to protect property. Yet the research takes the form of elucidating the entire ecology of termites, including their food, nest building, and the development of their larvae. This goes beyond the practical to purely intellectual interests. Krzymowski considers that in this case, it is not appropriate to argue that the results of this study – the ecology of termites and the way to exterminate them – are unscientific simply on the grounds that it is an intentional study that began from a practical interest (1919/1932: 99–100).

Agricultural science can be characterized by the same defining features that H. Weber identified in his studies of forestry science: it is (1) natural science; (2) normative practical science, or intentional science; and (3) aesthetics (1919: 19). As a normative and intentional science, forestry must be based on human purposes and ethical principles (1919: 20). It is a science that integrates 'nature and the mental world of human activities' and 'represents a comprehensive knowledge that has been laid down as norms for specific purposes and elucidates systematic, methodological relationships between events' (1919: 19–24, 98). H. Weber's recognition of an essential aesthetic dimension is unique in acknowledging that the beauty of forests is valued. In short, he discusses forestry in terms of the pursuit of truth, virtue and beauty. H. Weber further declares that this normative (intentional) science is based on a method that integrates 'existence' and 'oughtness'. Moreover, he argues that forestry needs a leading idea (*Leitidee*) that integrates the ideal with the real to ground its prospective, teleological method for drawing a hypothetical blueprint for the future (1919: 106–8).

Here we see strong parallels with Miki's theory of 'imagination'. Drawing on and further developing H. Weber's methodology for forestry science, Kashiwa Sukekata positions the field of agricultural science as a 'third science' that ranks equally with the natural and cultural sciences. Kashiwa describes the third science as 'a science that formulates models as rules for human activities…and has an aspect similar to a law-making function'. He asserts that it is a field of the pursuit of truth that embraces the system of logical, rational knowledge that is a requirement of science (1980: 75–82, 92–4). Based on these theories of agricultural science, particularly Kashiwa's theory of the 'third science', the following discussion outlines my view of agricultural science as practical science, while taking into account the views of Dewey, Miki, and others.

Agricultural science is an intentional science for practical use. As a practical science, it stands by practice and proposes activity models. It includes both the natural and human sciences as basic principles.

Practical science and pursuit of values

Earlier I explained that Miki's theory of conceptual powers is a theory of practical science as defined above. I would like now to note

Peirce's and Dewey's methods for the study of practical science, as an opening to the methodological integration of various sciences. While pragmatism addresses the general methodology of science, it stresses that the practical application of science is of the utmost importance, thus focusing on the formulation of a theory that systematically explains the study of practical science. Not unlike Miki's study of technology, it constitutes a theory that 'by arguing for the essential links between thought and action, attempts to provide a viewpoint that the two realms – the knowledge of natural events and the intentional expressions of self – are not mutually exclusive but complementary to each other' (Itō 1985: 4; cf. Dewey 1957, Meguri 1969: 78). Despite some differences, both Miki and Dewey have shown that, in the practical sciences, the methods of the natural and human sciences are integrated. It is through such integration that the uniqueness of the respective scientific fields emerges.

As practical science, engineering aims at creating industrial technologies for processing natural materials and producing anything and everything, including various daily necessities, electronic information appliances, planes, trains and automobiles, and all of the machines required to manufacture these articles. Medicine and pharmacology aim at explaining the human body in terms of its anatomy, functions, physiology, pathology, etc. (basic medicine), and discovering the best means for maintaining mental and physical health and treating disease (clinical or applied medicine) (Omodaka 1960). Agricultural science aims to elucidate natural ecosystems and the physiology and ecology of organisms (basic agricultural science), and to develop various technologies beneficial for developing agriculture, forestry and fisheries (applied agricultural science).

More broadly, such fields of study as pedagogy or economic science can be included in practical science. Pedagogy seeks to develop educational techniques that help to promote sound mental and physical development of children. Economics, particularly business administration, aims to develop effective and efficient methods of business administration, while economic policy studies focus on refining theories and policies that increase the population's economic welfare.

Each of these fields of practical science sets its own value objectives and seeks specific goals (see Sakamoto 1994: 81). As argued, no scientific field, including those in the natural sciences, can be practiced free of all practical values. Today it seems highly unlikely that any credible natural scientist would attempt to argue that his/her study is entirely value-free. Nor is it plausible for a natural scientist today to deny any liability for the ways in which his/her discovery may be put to use by society. While Rümker's (1905) view that 'science of no practical use is practically worthless' may be one-sided, Wagner's (1892–93) contrasting argument that science pursues knowledge only for the sake of knowledge and that anything directed to practical purposes is not science is equally as problematic. Wagner's view discredits production (*poiesis*), leaves science in the world of mere contemplation (*theoria*), detaches science from the everyday world, and regards experiential knowledge as mere opinion (*doxa*) unrelated to the 'meanings and values of life'. These are precisely the attitudes that have caused the various problems facing society today. Science can no longer be allowed to remain aloof from the rest of the world, living happily in the transcendental realms of *theoria* and *philosophie.*

Many, if not most, of the evils of modern science arise from the divorce and confrontation between nature and humans. The way to overcome this situation is to acknowledge the intrinsic value of nature and the need to coexist with nature. Accordingly, the science of nature must explicitly define its values rather than continuing to pretend to be 'value-free'. This will make 'natural science practically useful science with specific values, like the human sciences, and allow the possibility of establishing a 'new science' (Toulmin) that will replace the modern science which has affected the divorce between technical knowledge and everyday knowledge'. This new science will have to overcome ideas such as the separation of 'humans and nature', 'science and values', 'knowledge and practice', and 'natural versus human science' (Murata in Maruyama et al., 1987: 263). Practical science directly pursues values at the point of contact between nature and humans. But because nature and humans today seem, in many respects, to be separated from and in confrontation with each other, the proper approach of practical science is an important issue.

It should be clear from the discussion so far that whether a discipline or finding is scientific or not cannot be determined by whether it is value-oriented or practically orientated or not. Thus the delusion of 'value-freedom' must be done away with.

The different fields of practical science share bases in both the natural and human sciences and attempt to resolve technical challenges to realize specific goals. This is why these fields have been called 'applied science' or 'practical science'. In particular, agricultural science faculties and schools in Japan pursue both technological and social scientific fields of study, which have been integrated to ensure that the various technologies they produce offer the most sociably desirable production methods. In other words, modern agriculture and agricultural science are no longer primarily focused on the pursuit of economic values; they have begun to pursue the sort of comprehensive values described in Chapter 3.

Biological natural law: statics and dynamics

Miki Kiyoshi argues that nature, particularly the biological world, has imaginative, technological aspects that can be seen in longitudinal studies (1967 [7]: 209, 223). The laws governing the biological world can be regarded as the consequences of adaptive formation of a technical sort, as seen, for example, in the various shapes, functions and colors of feet and bills or beaks, and in the daily survival behavior or habits of different birds living in their respective environments.

Studies of apes have now confirmed that monkeys use sticks and stones as tools, wash potatoes with water before eating them, use certain plants as medicine, and even form communities, in which they create, maintain and hand down skills and culture. Among smaller organisms it is now beyond dispute that natural ecosystems can transform dramatically on relatively short cycles, as seen in the emergence of new drug-resistant pathogenic bacteria, and chemical-resistant agricultural pests in response to the advances in pharmacology and agricultural chemicals. These examples indicate that technologies in the biological world play a role in making the laws. In other words, the laws governing the biological world are not fixed or determined for all time, but are constantly being created. As discussed earlier, while there may be a uniformity of nature when we are concerned with the objects of physics and chemistry, in biology any appearance of such uniformity is but a static representation of some aspect of nature at a particular point in time; however, the dynamic aspect of nature entails a technical aspect of creative nature that is constantly renewing or redefining its form. In this vein, Shimizu Hiroshi points out that the essence of understanding life lies in coming to terms with this constant creation in complex environments (1996: 149; cf. Imanishi 1994: 113; Shimizu 1992).

My argument in this chapter so far has been that the natural, human and practical sciences overlap each other, constituting a continuum of both values and laws; and practical science should be recognized as a 'proper' science rather than a sub-category of the 'mere' application of technology. Each of these three fields of science constitute activities in pursuit of truth and human knowledge.

In natural science, researchers formulate a hypothesis about natural events that interest them, draw a conclusion through inference, validate the conclusion by experiment, and hope to discover and establish some new law. In human science, researchers observe and interpret, to understand deeply from the inside, human and social events or issues that interest them, and elucidate the meanings of such events or the determinants of the issues. In contrast, researchers in the practical sciences, who aim at solving certain problems or realizing certain value objectives by working on nature, conceive a logical, practical model, validate the model by experiment or field testing, and establish a normative, feasible model (an exemplar) (Bollnow 1958; Miki 1967; Kashiwa 1962: 325).

All of the historical products of the three sciences – universal laws, meaningful interpretations of human and social events, and practical models for realizing specific value objectives – emerge as truths that have been elucidated, created or reached for the moment. Models can in a sense be regarded as new laws established by trial-and-error application of the 'natural laws' of the natural sciences to empirical nature and the world of human intentions, and constitute an essential element of our system of understanding and knowledge. Although this classification is of necessity only relative, if science focuses on my three categories of events – nature, humans and the interactions between them – in conducting its inquiries, it will be better able to elucidate dynamic as well as static truths, bringing itself closer to realizing its own ideals and purposes

Supplementary note

Yoshida Tamito has proposed, as an alternative to the conventional formula 'engineering science = the physical science of man-made objects', the concept of 'design science' as 'scientific expressions, at human levels, of the self-designing function that is unique to the biological human world' (1998: 11). If, however, such expressions develop a more experiential, localized or individual tendency, they should not be treated as 'mere technology' but scientifically studied and elucidated as complexity. If science intends to comprehensively and systematically understand more complex, various realities, it must, as argued below, tend toward 'element-integrating' in as comprehensive a manner as possible. Science in the 21st century must deepen if it is to procure a more comprehensive understanding of the diversity, complexity and creativity of the biological world.

The birth of practical science

Miki Kiyoshi argues that, when mediated by imagination, technological studies, i.e., practical science, is where humans and nature, mind and matter, subject and object, knowledge and practice, logos and pathos, natural science and human science, causation and teleology, and explanation and understanding can meet and integrate. In my view, the field of practical science can be established, and comprehensive science achieved, through this kind of integration.

Practical science has become too preoccupied with goal-achieving and is in the process of reverting to a focus on the discovery and formulation of natural laws. In this respect, it resembles natural science. Yet, applied science is not a field that produces knowledge for knowledge's sake, nor is it simply based on the pursuit of 'the truth'. It is rather a practice for discovering and formulating knowledge that makes it possible to establish models for the purpose of realizing particular value objectives to improve human lives. Practical science can therefore be described as 'futuristic and visionary, purpose-defined, systematic and imaginative' (Kashiwa 1962: 370–75, 383). In other words, it is a school of scientific inquiry that is, from the beginning directed at achieving specific value objectives. In this sense, while resembling natural science in certain respects, practical science transcends natural sciences' self-defined limitations (Sakamoto 1975: 80).

Furthermore, as the values to be pursued through the use of practical science are determined by humans and tend to vary across historical eras, geographical regions and cultures, practical science appears to be quite humane and individualistic in character. In this sense, practical science closely resembles human science. Yet, efforts to realize value objectives are constrained by natural laws, influenced by (local and contemporary) standards of imagination, and arrived at through concrete practice. Practical science does not stop at understanding and interpreting the past or present realities, or at defining problems: it is a science of intentional and purposive human practice directed at solving problems by overcoming various constraints. In this respect, practical science is akin to human science, but rises above it.

Hence, while founded upon the natural and human sciences, practical science enables us to solve problems that emerge in action-based practical fields such as industry, agriculture, medicine and education, through research that takes an original approach to problems and methods. Dilthey's conception of three spheres that constitute the human sciences are suggestive here. For Dilthey the first sphere expresses reality as given in observation; it contains the historical component of knowledge. The second sphere develops the uniform behavior of parts of this reality which have been isolated by abstraction: it contains the theoretical component of knowledge. The third sphere expresses value judgments and prescribes rules: it contains the practical component of the human sciences (Dilthey 1988: 91). It is this 'practical constituent' that suggests an opening for the succession from human science to practical science.

As we have seen, natural science emerged first, followed by human science that seceded from it. Practical science appears in that zone where the natural and human sciences converge.

The significance of 'imagination' in practical science

I place Miki Kiyoshi's concept of imagination in technological study at the core of the methodology of practical science. Departing from neo-Kantianism and absorbing diverse influences, from Dilthey and Heidegger to Marxism, Miki founded the so-called Nishida school of thought and developed a unique philosophy of technology. Not stopping with the so-called 'logic of technology', Miki reached further and sought both the 'logic of history' (the logic that has shaped the historical evolution of the world) and the logic of the organic, natural world in imagination (Miki 1967 [7–8]).

According to Miki, 'science is neutral when it comes to questions of value; it merely strives to *describe* and *explain*, therefore value judgments lie beyond its boundaries' (1967 [7]: 52). Furthermore, although science was originally born out of practical and technological necessity, it seeks 'knowledge of general facts and laws' by setting itself strictly apart from the world of action. Hence, Miki explains, technology develops upon the foundations laid down by science. For Miki, applied science means the study of technology. Technology here refers simply to 'making things' and can be described as a 'model for action' that is both creative and formative. It thus relies on imagination. However, for Miki, the imagination extends to all spheres of action to the extent that 'all human action is technical'. Hence Miki's concept of technology transcends the boundaries of practical science as I have defined it, having a much broader meaning (1967 [7]: 221). As we have seen, there are subtle differences between practical science and Miki's 'technological study' regarding how science and value, the relationship between science and technology, and the breadth of technological concepts are described. Regardless, the concept of 'imagination' is of great significance to the fields of engineering, agriculture, medicine, pharmacology, pedagogy, management and economics because it lies at the very root of cognition and productive action.

According to Miki (1967 [7]: 250–6), 'technology gives concrete form using the general laws of science and produces goods for the external world'. He points out that technology is 'not merely science but rather creativity that flows from intuition and formative imagination'. 'By creating new forms, meanings and values that previously did not exist in nature, it creates what is referred to as culture…The formative imagination is the force behind this process.' Miki further argues that science 'is not purely objective, but rather a synthesis of objective natural laws and human goals'. Therefore, while on the one hand science encompasses a 'cognition of natural laws', on the other hand it embraces 'goal-setting' in order to realize specific objectives. Technology and human imagination bring these two together and formulate some sort of creative response. Through these three factors – cognition of natural laws, goal-setting and synthesis – science produces specific models (1967 [8]: 242).

'Cognition of natural laws' refers to objective and logos-like matters – that is, theoretical and general matters that have to do with natural science. 'Goal-setting' concerns subjective and pathos-like matters – i.e. matters of concern to the human sciences and with concrete, specific practices or actions. Through imagination these two sets of factors are mediated and integrated, the result of which is technology capable of producing things.

Practical science assumes this role of mediating and integrating natural science and human science, causation and teleology, humans and nature, subject and object, and mind and matter. In practical science, human imagination plays a crucial role. Accordingly, practical science also mediates and integrates the *explanation* of natural science and the *understanding* of human science. The reason of imagination is at the same time also a concrete, practical and goal-oriented 'logic of action'. It is furthermore a 'logic of creation' and by definition, a 'logic of modeling' that makes it possible to realize human values in the material world.

In short, *imagination* is the core concept behind productive action. An advanced mental and practical human function, it enables the invention and production of hitherto nonexistent goods and the creation of models.

Research methodology in practical science

When considering imagination as the basis for the methodology of practical science, I focus on the 'logic of inquiry' as formulated by Dewey and Peirce.

The philosophy that Dewey and Peirce established is known as *pragmatism* or *instrumentalism*. These two authors lived in the newly-created and rapidly growing United States of America, which, despite having become independent of Europe and launching its own industrial revolution, was still experiencing tremendous change, and thus remained practically in a state of chaos. It was under these circumstances that pragmatism – a highly practical stream of philosophy – sought to chart its own course on a trial-and-error basis. Tsurumi Shunsuke asserts that pragmatism was a 'movement that attempted to mediate the rivalry of religious principles of everyday life (in other words, Puritanism) and the technological way of thinking' (1950–51: 98). Ueyama Shunpei points out that in contrast to Marxism, which dismissed religion as 'false consciousness', and existentialism, which in many respects rejected science, pragmatism proposed that religion and science should be viewed as complimentary – occupying a position that was neutral and eclectic (1968: 13). Meguri Masatami suggests that by not falling into nihilism but constantly striving to discover new avenues of human thought and action, pragmatism is a 'philosophy of change and development' (1963: 4–7).

Bollnow writes, with particular reference to Dewey, that 'pragmatism at times bears striking similarity to the philosophy of life in terms of its methodology' (1958: 8, 66). He makes this observation based on pragmatism's fight against traditional forms of philosophy; its prioritization of the everyday phenomena of human life; its marked fluidity that is suggestive of relativism; and its tendency to reduce the meaning of life to the pursuit of truth. Where the German philosophy of life becomes overly abstract, Bollnow advocates consideration of Dewey's more sensible school of thought. Miki criticizes pragmatism for its lack of insight into the historicity of human societies, yet values its comprehensive practicality based on experience, and for its attempt to grasp the truth in a dynamic and process-based way (1967 [3]: 78–97). He thus endorses Dewey's concept of an *imaginative forecast of the future* that subsumes empiricism, futurism and creativity, likening it to his own *logic of imagination* (1967 [7]: 275).

Using Miki's theory of technology as the lens, we have seen that practical science occupies a position in which, in almost all respects, it mediates and synthesizes the polarized natural and human sciences. Similarly, pragmatism presents a theory of science that attaches great importance to the unique experience and goal-based practice of applied science. Dewey argues that 'surely there is no more significant question facing humanity than this question of the possibility and method of reconciling the attitudes of practical science and contemplative aesthetic appreciation. Without the former, man will be the sport and victim of natural forces which he cannot use or control. Without the latter, mankind might become a race of economic monsters, restlessly driving hard bargains with nature and one another, bored with leisure or capable of putting it to use only in ostentatious display and extravagant dissipation' (Dewey 1948: 127). In short, from its inception, pragmatism has warned against humans turning into economic animals, positing the 'logic of inquiry' as a methodology for practical science that supports both ethical and productive practice.

Dewey's philosophy – as was perhaps required by the domestic situation in America at the time – introduced and synthesized a range of European philosophies from various schools of thought that could be useful in the clarification of practical, real-life problems. Among these were aspects of biology, which made Dewey's theory more approachable from the point of view of agricultural science.

I will now focus on the relevance of the pragmatists' logic of inquiry from the perspective of agriculture. Beyond the terminology of H. Weber, I consider agricultural science, engineering and other technological fields to be 'practical science' on the basis of Miki's philosophy of technology and Dewey's realist pragmatism.

Dewey's concept of 'inquiry' refers to transforming 'indeterminate' or problematic situations into 'determinately unified situations' in which problems have been resolved. To summarize this transformational process, we first recognize that (1) human reality is always constituted by various problematic situations (e.g. contradiction, confusion, confrontation, disparity and constraints). Secondly, we note that (2) under these circumstances, it is up to the researcher to determine what ought to be defined as a problem requiring a solution. This 'location of the problem' is the point of departure for inquiry. Next, an (3) idea (hypothesis) considered conducive to solving a specific problem is conceived and formulated. Then, a (4) certain conclusion is reached by way of logical inferences based on the ideas embodied in the hypothesis. A symbolic proposition (of characters and/or numbers) is subsequently formulated, taking into account the premises of the hypothesis and conclusions. Next, (5) the conclusions drawn by inference are tested and verified. This testing ascertains the truth of the ideas by comparing them against facts. If verified, the (6) hypothesis becomes a truth, 'as a declaration with a guarantee'. This is not, however, an absolute truth, but a *relative* truth that is subject to revision at any time based on new data from further inquiries (Dewey 1938: 107–19; Eames 1977: 67–9).

Dewey's logic of inquiry is typically referred to as the 'experimental method'. Yet, the word 'experimental' here does not refer merely to procedures carried out within a laboratory: it encompasses, as terms such as 'verification' and 'testing' imply, methods that include the application of mental reasoning capabilities as well. In this sense, Dewey's logic of inquiry is consistent with Peirce's theories. Peirce identified three modes of inference; 'abduction', 'deduction' and 'induction', and saw inquiry as a three-stage process (Hartshorne 1960: vol.6, 94–131). The first stage, abduction, consists of the discovery of a new theory and formulating a corresponding hypotheses that lead to new ideas. This process progresses from observing a phenomenon to formulating a hypothesis that might explain it. In the second stage, deduction, the hypothesis is further clarified through a logical examination and the conclusions that emerge are tested by abstract principles of logical thought. The third stage, induction, is the process by which the theoretical findings reached by deduction are compared against facts to ascertain the accuracy and veracity of the hypothesis. This process consists of 'classification', 'testing' and 'evaluation'.

Having passed through all three stages, inquiry completes one full cycle. By repeating such cycles, we are able to advance knowledge from the superficial to substantial; from incomplete to complete; from biased and fragmentary to comprehensive and holistic, and move as close as possible to 'the truth'.

Figure 10.1 illustrates the dynamic and continuous processes that form the basis of the pragmatists' methods of inquiry. This figure is based on Peirce's assertion that the constituents of all concepts must first pass the gate of perception to enter the world of logical thought (or the world of signs), and subsequently they must exit this realm through the gate of goal-based action. Those unable to present their passports at either of these gates must be arrested as constituents that have not received the approval of reason (Hartshorne 1960: vol.5, 212). From this perspective, thought is not something that can be limited to the world of concepts, but rather must constantly interact with the practical, real world. In other words, in essence, pragmatism emphasizes the substantial links between cognition and action. Moreover, Peirce invites reflection on the practicability of science itself.

Figure 10.1: Method and process of inquiry (problem solving)

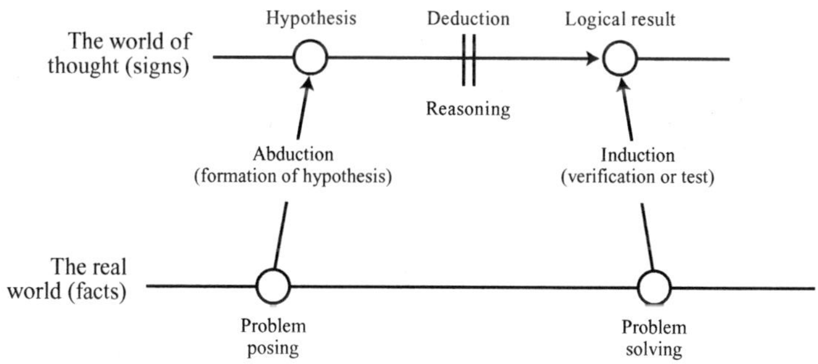

Ueyama Shunpei showed us this figure in one of his philosophy lectures in 1966. I was inspired by this model (Miyanaga and Soda 1976), and developed it in figure 10.4. Kawakita (1967) and Kikuchi (2000) also present similar ideas.

Following this line of thinking, we might say that natural science focuses on hypothetical 'laws', human science on hypothetical 'webs of meanings', and practical science on hypothetical 'models'. Furthermore, the methods employed to check these hypotheses against facts and verify their conclusions consist of 'experiment' in the natural sciences, 'field survey, classification and comparison' in the human sciences, and 'testing' in the practical sciences. For the natural sciences, the ultimate goal is to 'establish universal laws', for the human sciences it is the 'resolution of particular webs of meaning', whereas practical science aims ultimately to create 'practical models'. The central concepts underlying the processes of inquiry in these sciences and supporting their comparative uniqueness as well as giving their methodologies their special characteristics are 'explanation' in the case of natural science (below, 'analogical inference' will be added especially when referring to living organisms), 'understanding and interpretation' in human science, and 'imagination' in practical science. As I pointed out earlier, Miki's imagination is a dialectical concept that integrates logos with pathos. However, it is not simply abstract and formal but indicative of comprehensive human capability to produce goods and form society creatively. Imagination is in tune with the essence of practical science that plays the productive and formative roles in the relationship between human beings and nature.

We have now illuminated the main features of the main types of science. However, if we continue to focus on the differences in their methods, our comparisons will of necessity remain merely relative. Although each has different characteristics, all sciences follow the three stage process of 'abduction – deduction – induction' to clarify and solve problems.

Through this investigation of the qualities and methods of the main fields of science, we have reached a stage where we can more precisely discuss the qualities and methods of agricultural science – a practical science in its own right.

Characteristics of agricultural science

The characteristics and methods of agricultural science correspond almost entirely with the field of practical science defined above. However, with its focus on specific local environments and with biological production as the primary object of its research, agricultural science differs significantly from other fields of practical science – to the extent that it should appropriately be called 'the comprehensive science of living systems'. From this perspective, I now detail the characteristics and research methods of agricultural science.

Agricultural science as the pursuit of value

As stated above, practical science is value-driven, adopting specific value objectives prior to commencing any research. Agricultural science is identical to practical science in this respect. Miki claims that how the goals are determined is of foremost importance in agricultural science and therefore high ethical standards are required. I outlined the value objectives that contemporary agriculture and agricultural science should strive for in Chapter 3. To briefly revisit the argument there, I demonstrated how the value objectives of agricultural science have evolved through several stages: (1) the 'agricultural science of *production*', where economic value is the main pursuit; (2) the 'agricultural science of *life and environment*' where ecological value begins to be pursued; (3) the 'agricultural science of *daily life*' or '*social* agricultural science' where the (social and cultural) values of everyday life begin to be pursued; (4) the stage where (2) and (3) are combined into an 'agricultural science of *life*'; (5) the stage where total value (the harmonization of the previous stages) is pursued in a given locality, or the 'agricultural science of *ba*'. It is imperative to embrace the challenge of resolving difficult issues in such a way as to realize economic, ecological and life values – three fields that tend to conflict with one another when pursued separately – in a harmonious and comprehensive way.

Agricultural science as the science of bio-life

As is well-known, agricultural science emerged to address problems concerning raising and utilizing plants and animals (biological life forms) for human benefit. Those species most directly useful for such ends are referred to as crops (cultivated plants) and livestock (farm-raised animals). Species that are not useful as crops or livestock (e.g., weeds) may also be objects of agricultural science research, but only in so far as the research contributes to realizing the production objectives of raising and utilizing crops and livestock. I discussed the evolution of crops and livestock in Chapter 2. Here I need to differentiate between practical science fields that primarily take non-living/non-organic nature as their object of research (e.g. engineering), and those such as agriculture that focus on the cultivation of organic nature. This is the single most significant characteristic of agricultural science.

I have discussed how following the birth of modern science, first natural science and soon thereafter human science asserted their originality and established their scientificity. Yet, we must note that within the natural sciences, the organic or biological sciences must be granted special positions. Dilthey argues that the 'three major substantial orders of reality' consist of (1) mechanical nature, (2) organic nature and (3) human social and historical reality, and notes that 'organic (life) shows the limit to the mechanical explanation of nature'. Early modern biology attempted to explain living organisms as mere exquisite machines. However, due to the spectacular diversity of living organisms and the fact that observers (scientists) tend to focus on the characteristics of specific organisms, formulating natural laws in biology is frequently problematic. Kant observes that just as it is delusional to ignore the mechanical side of nature in favor of a wholly teleological explanation, so too, it is illusory to reject all purposiveness of nature in favor of an exclusively mechanical theory. It is particularly important in the case of biology to heed the lessons learned from reductive elementary investigations that offer only partial cognition, remembering that 'what may be valid regarding an isolated part may be completely invalid when applied to the whole' (Uzawa/Teramoto 1989: 258).

On this point, Bertalanffy lists (1) openness, (2) wholeness, (3) stratification and (4) activeness, as the properties of organic life forms. Whitehead lists (1) correlational nature, (2) creation of environment,

(3) self-forming process and (4) creativeness. Although these listings differ in terminology, they point to pretty much the same properties of organic life forms (Murata 1994: 256–91), namely: (1) Life forms have an open, flexible system to cope with unforeseeable disturbances. This is different from a machine, which is a closed, fixed system that does not work when faced with the unforeseen. (2) In the mechanical system, each part exists as such and the whole system merely represents the sum total of individual parts, whereas in life forms, each part is characterized in terms of the whole, exists only in relation to the whole, and is governed by the whole. (3) The natural world has a stratified structure, consisting of different levels, from elementary particles, to atoms, molecules, high molecular compounds, cells, organisms, and finally supra-individual organizations. This organized stratification of the whole is essential to life forms. (4) Life forms are active and creative, forming themselves to adjust to their environment but also change the environment to some extent to suit themselves. As described earlier, natural life has technological and creative powers that rupture the uniformity found in physical and chemical nature, changing both its environment and itself. Is this creativity a biological instinct, a law, or technology? This question raises essential issues about scientific methodology. If organic life forms have these various characteristics, they require research methods that differ from those applied to inorganic, mechanical nature.

We should therefore consider the perception method called 'analogical inference'. It has long been thought that analogical inference is 'unfounded, and thus of little logical value.' Analogy, however, occupies a very important position in philosophy in antiquity and the Middle Ages. Its value has diminished in modern times because analogical inference has been detached from its ontological foundation and become formalistic, a result of the mechanical view of nature that has formed around physics and mathematics. Inference by analogy is 'bottom-up logos', reasoning inductively from facts and seeking to find one general truth out of many phenomena. Importantly, however, inference by analogy 'seeks to find a general truth not by obliterating and parting from many phenomena but by accepting many phenomena as they are.' It is 'nothing like incomplete induction; it is in fact a more concrete logic than induction' (Miki 1967[8]: 455–8). This form of logic is well-suited to contemplating organic life forms and is perhaps indispensable to understanding the truth about them. Imanishi Kinji has revived the use of analogical inference in biology.

Imanishi argues that traditional biology has been 'dead science' that tends to focus on morphology and classification of collected organisms. He changed direction toward 'living science' that directly observes organisms living in their ecosystem (1972: 18–26). Grasshoppers live in fields and mountains by eating leaves of grass, mating, reproducing and enduring rain and wind. This is the true, living grasshopper. Imanishi also asserts that analogy based on 'affinity relationships' is important in understanding the truth about life forms in their *ba* of life as an 'extension of life forms themselves'. The term affinity means 'blood relationship and relationship through land…It means not only historical, but also social, intimate or close relationship surrounding the formation of things.' Imanishi states that the basis for analogy exists because 'humans, animals and plants are similar beings that have affinity relationships with each other in that they are all organic life forms.'

The anthropologist E. M. Thomas studied dogs, the most familiar animals to us, and wrote about their 'inner life' in *The Hidden Life of Dogs* (1993). Thomas accepts the possibility that dogs may, like humans, have awareness, an inner life, and act based on thought. She attempts to infer analogically to understand dogs' inner mind. Similarly, in *Inu ga uttaeru shiawasena seikatsu* (Happy lives wanted by dogs) (1995), Hayashi Yoshihiro pursued the question of what dogs want from humans. In the process Hayashi raises doubts about human-centered views of nature. This perception is particularly significant to agriculture, forestry and fishing industries and agricultural science, each of which are concerned with biological production.

Supplementary note

Hinoue Heiichirō, an exemplary farmer in Shiga Prefecture, wrote that he would stand on a raised footpath between rice fields and 'talk to Mr. Rice' to his heart's content (1966: 33; cf. Kondō 1953: 78). Hinoue listens to the voices of growing rice by assuming the viewpoint of the rice crop. This approach is more apparent in raising livestock such as cattle and horses; farmers watch them and make analogical inferences through certain kinds of emotional exchanges with the animals. A farmer who lived in the Tōhoku district at a time when they kept cattle indoors, says, 'I often came home late at night after a neighborhood association meeting, to find that my wife was already in bed, but my cattle always greeted me by stamping their hooves.' Similarly, as a child, I used to saddle up my family's plow horse in the evening and take it to the Oki-no-kawa River to let it drink water. I

> loved to be refreshed by running the horse through farm roads, while the horse enjoyed going into the shallows and leisurely drinking water. Both the horse and I would grow restless as evening approached, and I could really feel the intimacy growing between us on every occasion.

Bollnow states that the 'understanding' of human science 'extends beyond the domain of humans to include that of irrational animals. The domain of communication that is solid and can be prescribed in more detail also exists in such relationship with animals (e.g. horses and dogs)', and argues that it is possible to understand organic life forms from the perspective of human science (1949: 100). Traditional agricultural management and animal science regard domestic animals as objects of study, and their research is directed toward the efficient management of raising livestock. References to the psychology of domestic animals tend to only go as far as to say that when you move around an animal such as a horse, it is advisable to provide a sense of security by touching part of its body (Mimura 1980). Although this aspect of agriculture is being lost through an over-emphasis on rationality and economics, such intimate exchanges between domestic animals and humans should have been a source of pleasure in agricultural production activities. The responsibility for this issue lies not only with human science but also animal science.

In Japan, the 21st century is often regarded as an age of spiritual richness, with increasing numbers of people leaving towns and joining the agricultural workforce. It therefore seems to have become necessary for agricultural science to begin to contemplate the relationship between domestic animals and humans. Worthington (1993) proposes a revision of human-animal relations from the perspective of animal welfare. The Society for the Study of Human Animal Relations has recently been established in Japan, for which we can have high hopes. Under these circumstances it appears that agricultural science must expand its use of methodological analogy to enhance understanding. The effectiveness of analogy as a method of perception has become relatively well-established in biology, but physicists such as Ichikawa Kikuya and Yukawa Hideki also proclaim its relevance in the processes of scientific invention and discovery (cited in Nakamura and Furuta 1968: 198).

Modern biology, in fact, was born from concerns about issues in agriculture and agricultural science. This can be seen in Mendel's laws of heredity, Pasteur's fermentation microbiology in brewing, and Liebig's clarification of the correlation between crops and inorganic substances. Other important contributions include Thaer's and Thünen's respective clarifications of the managerial economics of biological production (cited in Haushofer 1963).

Recently, agricultural science has taken a turn toward expansion and integration to incorporate ecological perspectives. But at the same time, it has become more fragmented. Over a long period, agricultural science studies have progressed from the level of individual organisms to the level of the cells that constitute such organisms and then to the level of molecules such as genes. Stratifying into studies of groups, individuals, cells and molecules, research has progressed at each level. These studies are developing a strong tendency toward molecular biology, a microscopic field that physically elucidates and manipulates biological phenomena. This has increased the potential for understanding biological phenomena at the deepest levels and for modifying biological organisms through rapid advances in physical and chemical control methods, such as cell fusion, genetic engineering and tissue culture. In agricultural science, this biotechnology has taken the form of molecular agriculture and molecular breeding, particularly in the US but also in Japan (Murakami 1984; Mori 1987; Nishio 1996).

Traditional breeding science improves animals and plants to create breeds useful for human lives, maintaining and propagating such breeds in accordance with nature's providence. In contrast, biotechnology designs and creates new life forms by engineering such as genetic manipulations that would not otherwise have existed in a state of nature (Okada 1994: 38). These new life forms have been manipulated to display characteristics such as high yield, disease resistance, cold resistance, drug resistance, intense flavors, uniformity, or enhanced beauty. Genetically engineered agricultural products such as soybean, corn and rapeseed are already common in the marketplace. Although we expect much from biotechnology, its greatest potential remains at the level of possibility.

But many technological and ethical issues remain unresolved. Biotechnology must make its products socially acceptable by carefully considering potential effects on individuals, groups, ecology, food safety, productivity, etc. There is a growing demand for agricultural science to adopt a comprehensive perspective that can take all of these concerns and more into account.

At the same time, the social sciences, particularly economics, have rapidly developed a growing interest in life itself out of reflection on environmental issues. The integration of economics and ecology, and the establishment of an economics of the living system and a social science of the living system are becoming urgent tasks, as argued by Polanyi, Elkins, Alier, Tamanoi Yoshirō and Kamisato Isao (Tamanoi 1978). In many of his works, Tamanoi has pointed out limitations in the modern economic system, which has been constructed on the basis of industrial production and market mechanisms. He argues that the logic of industrial production should be reconstructed towards an 'economy of life systems', epitomized by agricultural production and agricultural economics, with the concept of entropy as the pivot of conversion. However, as discussed in previous chapters, agriculture itself has become increasingly industrialized, that is, moving in the opposite direction to the ecosystem advocated by Tamanoi. Nevertheless, the logic of agricultural production seems more likely than industrial production to provide the key to a new production system.

In short, agricultural production and agricultural science are directly concerned with biological production and thus offer the strongest position for developing new sciences of the life system.

Agricultural science as the regional science

Agriculture is heavily influenced by natural conditions, namely, climate (temperature, humidity, rainfall, etc.), topography, vegetation, soil, and others. Various natural patterns, such as the tropical, temperate and arctic regions, deserts, savannas, and forests, significantly shape diverse styles of agriculture and human lifestyles.

Watsuji Tetsurō (1935) developed a theory of *fūdo*, which sees the Earth's climate as a significant determinant of culture. The term *fūdo* is not directly translatable, but refers to this conjunction of climate and culture. Watsuji divides *fūdo* into three categories: monsoon (coastal regions in East Asia), desert (Arabia and Africa), and pasture (Europe), later adding the steppes in America, Siberia and Mongolia as a fourth category. Watsuji explains how these natural features characterize the style of agriculture and the state of human existence in each region. In the process, he explores both the spatiality and temporality of human existence, arguing that a *fūdo* pattern is also a pattern of human self-understanding. Although Watsuji's theory has been subjected to various criticisms, his contributions should be recognized in, among other things, demonstrating that differences in the natural environment have a great influence on the formation of human societies and understanding cultural differences in terms of interactions between nature and humans.

Misawa Katsue (1941) extended the conception of *fūdo*, defining it as a contact surface between the surface of the earth and the bottom of the atmosphere. For Misawa, *fūdo* is formed by an inseparable mixture of organisms, including humans, and other matter that have been created by such contact. *Fūdo* is cooperative work by human and regional nature. Based on this idea, Misawa started a very intriguing field of region studies for agriculture, forestry and fishing industries – the development of *fūdo* industries.

Inspired by the French climatologist de Martonne's aridity index, among others, Īnuma Jirō explains the history of world agriculture in terms of the flow of two styles of agriculture, namely, fallow and intertillage, whose respective technological differences arise from climatic conditions (Īnuma 1985; cf. Parry 1990). In other words, Watsuji, Misawa and Īnuma each explain differences in agriculture and society in terms of different natural environments and climatic conditions. Similarly, agricultural geography has developed as an academic discipline in Europe, attempting to describe and explain regional differences in agricultural activities (Johnston et al. 1994; Sick 1993). Although Japan's regions vary widely in climate and soil conditions, agricultural geography has not developed very far.

From a different perspective, Thünen's (1886) typology of regional farm management identifies differences that correspond to the distance from town (farm product market). Kurihara Tōshichirō contrasts Oriental and European agriculture in terms of a comparison of their staple crops, rice and wheat, based on the idea that regional aptitudes and differences in cultivated crops result in different styles of agriculture (Kurihara 1964; Tsukuba 1969).

From a modern industrial perspective, we could classify regions according to the scale of production and economic efficiency, resulting in a classification such as: regions where agriculture started from subsistence farming and is practiced mainly by small farmers, such as Japan, Korea, Taiwan and Switzerland; regions in vast new settler societies where agriculture started and developed as large-scale farming to produce export products, such as the US, Canada and Australia; regions that are intermediate between these two groups, such as Europe; and developing countries where wages are low and agriculture is practiced mainly by small farmers.

As discussed in Chapter 4, these regional differences are clearly seen in the comparison of the two most contrasting farming styles in advanced countries, namely, Japan and the US. Japanese rice farming is practiced mainly by small-scale farmers cultivating an average area of 1.4 hectares per farm. Farm work is conducted mainly by family labor using medium- or small-size machines, and tends to require high levels of investment that raise production costs. In contrast, new continent farming such as that found in the US uses vast, flat and fertile land and is practiced on an average scale of 189 hectares per farm. It started and has developed as farming to produce export products, where farm work relies on both family members and employees, using large-size machines in large farmland sections. These differences in historical and natural conditions result in great disparities, especially in economic efficiency. In this era of globalization, such differences have resulted in economic friction and distress for agricultural producers.

In short, agriculture is subject to numerous factors, such as climatic conditions, geographical locations, differences in principal crops, land conditions and historical conditions, all of which intertwine, making the agriculture of each region distinctive and complicated. Such factors are almost irrelevant to industry, because industrial production is based on inorganic nature while agricultural production is based on organic nature.

The regionality or placeness (following Relph's terminology described in Chapter 7) I have been discussing is a very broad global-scale classification. The regional issues that arise in actual instances of agricultural production, however, are much more particular. There are countless examples, including: (1) north and south facing fields differ in conditions, affecting the selection of crops and methods of cultivation management. (2) Fields along a river require different farm tools, including differently shaped hoes, depending on whether they are located along the upper, middle or lower reaches of the river. (3) Different fields have different soils, which affects the selection of crop varieties and their management. (4) Over use of land without consideration for the ecosystem and topography may result in soil erosion or flooding, leading to destruction of the regional environment. (5) Modern farming practices, especially specialization and monoculturization result in pollution of the regional environment unless complementary crops are cultivated in the region. (6) Viable crops are totally different in regions with snow than those without, and vary depending on the amount of snow. 'Regional limitations' are unavoidable factors in developing agricultural technologies. These are very acute practical issues for individual farmers and for agricultural science.

As I have argued throughout this book, to ensure that such 'issues' are not 'problems', we must encourage the formation of autonomous 'local regions', whose basic unit is a lifeworld in which individual persons interact with each other and have direct contacts with others. The optimal rationalization of agricultural production requires cooperative measures by community organizations, etc, such as consideration of topography and climatic conditions, maintenance of water supply, and control of noxious animals. The village community must be expanded into a regional community as the appropriate place to cope with agriculture. The local region is the place of ecology, of production, and of life. The integration of these places forms *ba* (see Chapter 1 for details) as regional community. I have given a full account of this in Chapter 6. The harmonious, comprehensive realization of economic, ecological, and life values requires the formation of a local region where a fully human style of life is assured. Today, these three values should be self-consciously adopted as the only value objectives to be addressed by agricultural science. Studies of regional agriculture must be promoted vigorously in the fields of technology, economic management, regional ecology, and rural village community.

Since agriculture is inseparable from issues of regionality, agricultural science must be framed as the science of local region.

Agricultural science as comprehensive science

Comprehensiveness as a characteristic of agricultural science has already been discussed. Thus part of what follows will be repetitive; but my aim here is to clarify this issue in three points.

Integration of scientific knowledge and human objectives

Agricultural science is one field of practical science. I have described natural science and human science in a contrasting manner – in an attempt to reflect the historical process of their formation and to describe their uniqueness. However, although claims to their *relative* uniqueness can be substantiated, neither is *absolutely* unique – neither can claim exclusive providence to any particular methods or fields of inquiry, or to knowledge and 'truth'. In fact, recent scientific discussions tend to emphasize their similarity and complementarity.

Natural science attempts to discover and explain the natural laws at work in natural phenomena. Human science attempts to understand and interpret the meanings of, and relations between, mental and social phenomena which contains the results of purposive human actions. Practical science deals with the interaction between humans and nature, and attempts to invent and establish models that can appropriately achieve human objectives. Thus, it is in practical science where the relationship, or interactivity, between humans and nature is revealed. Practical science exists where human science and natural science meet. In the production of various machines or space satellites, one observes the collection of laws of nature as well as that of human desires, values and histories. Technology connects the laws of nature with the purposes of human beings and transforms them into concrete shapes. In this sense, practical science integrates two other sciences. In practical science, the characteristics and methods of human science and natural science meet and are integrated and fused. In the process, the relative uniqueness of practical science also appears. Agricultural science as a field of practical science is thus comprehensive in the sense of integrating the human and natural sciences.

Basic, applied and developmental research

Natural science and human science, or physical science and humanities, are often called pure science, basic science, or basic research in Japan. Practical science is regarded as their applied research, or even developmental research. Richards demonstrates, however, that agricultural science contains both basic research and applied/developmental research, and attempts to develop technologies and realize value objectives (cited in Sakamoto 1986: 1). However, in practical science, any distinction between basic research and applied research is only for convenience.

Revisiting Krzymowski's discussion of research on termites as an example, agricultural science, aiming to protect houses from termites, first elucidates the ecology of termites and then looks for methods to exterminate them. The ecological research here is no different from the basic research that may be conducted in any other field of natural science. Yet, the research is directed at finding practical methods of exterminating termites. That is, it is 'directed basic research' that aims to realize an objective. All studies in agricultural science are conducted with the aim of realizing particular human objectives. Such studies are supported by an enormous amount of basic scientific research, part of which is conducted by agricultural science itself.

In agricultural science, basic research and practical applied research work in harmony and are inseparable. Similarly, medicine can be divided into basic medical sciences and clinical medicine – or basic research and practical applied research – the former aiming to obtain knowledge about humans as life forms and the latter to protect human health and treat disease. Yet, although distinct, they coexist in an inseparable relationship. Basic medical sciences aim to elucidate the basics of the human body, but are directed at maintaining human health. Similarly, clinical medicine which directly treats individual humans must, in order to develop treatment techniques, refer to basic medical sciences (Omodaka 1960: 1–4).[3] Medicine cannot exist without intricate knowledge of the human body or disease prevention and treatment.

Thus, although agricultural science is directed at specific objectives, it entails basic research, which elucidates facts unrelated to these objectives, and practical applied research, which applies this basic research to realizing specific objectives. In agricultural science, both basic research and practical applied research are subsumed under the logic of conceptual powers, whereby individual parts are directed harmoniously, toward the creation of models, fitted together into a whole that fulfils the objectives. Thus, agricultural science is comprehensive in the sense that it integrates basic research and intentional applied research; and in the sense that it draws on all other fields of science to do so.

Comprehensive science of bio-life and local regions

Agricultural science aims at producing organic life forms for human use. Organic life forms can be elucidated by analyzing the particular functions of their individual organs, but a life form can only exist through the comprehensive and harmonious functioning of the totality of its organs. Agriculture, forestry, fishing and related industries focus on the raising and rearing, as well as the processing and utilization, of life forms. As agricultural science deals with these industries, it can be said that there have been demands, from the beginning, for agricultural science to have multifarious areas and characteristics.

The success of activities in these industries requires research on the life forms themselves, but also research on environmental conditions such as ecosystems, soil, the atmosphere, water, geography and weather as well as various production materials, such as fertilizer, agrochemicals, machines and facilities. This requires the biological and life sciences, ecology, physics, chemistry, astronomy, meteorology, hydrology, geology and geography, and civil engineering. Veterinary medicine, pharmacology and food science are among the other necessary sciences in these fields. Most fields of natural science must thus be mobilized for agricultural science to be comprehensive in its approach.

Agricultural science also contains various fields of human science. It is true that some humanities or social science subjects are offered at natural science schools. However, only agricultural schools offer subjects of humanities and social science that have a system linked inseparably to technology, though their names may vary such as natural resources economics, agricultural and forestry economics, or regional economics. Agricultural science includes various fields, such as agricultural economics and management, agricultural accounting, agricultural information science, regional economics, environmental economics, rural sociology, agricultural geography, rural planning methodology, agricultural policy, agricultural history, forest economics and fisheries economics.

Why have only agricultural schools, among all other schools of practical science, been formed with fields of humanities and social science systematically incorporated in them? Because: (1) agriculture has strong regionality in relation to biological production; and (2) many farms are family-run, where the places of life and production are the same. These points were discussed in Chapter 4.

Integration of experiential and scientific knowledge

Science appears in its true form after resolution and synthesis. The purpose of science, particularly natural science, lies in breaking a certain phenomenon down into its constituent elements and elucidating how they behave. In this sense, then, science is elementary, reductive and analytical. This reductive tendency has become increasingly dominant, resulting in extreme specialization and diversification of science. This tendency appears in many respects to be inevitable as science becomes ever more advanced and sophisticated; science is fated to reductivism, so to speak. It might be argued that this reductive specialization and diversification has enhanced science's ability to produce knowledge.

Extreme specialization is not, however, sufficient for the development of science. Science becomes true science by synthesis. The purpose of resolving a phenomenon into its constituent elements for analysis is to explain the phenomenon in question as consistently as possible. Such explanation requires that the findings of the elementary analysis be put together, reconstructing or synthesizing the results, and thereby the phenomenon itself. Resolution is therefore premised on and aims at synthesis. Science is deepened by resolution and completed by synthesis. According to Murakami, 'As long as science employs resolution as its means, specialization is inevitable and need not be opposed. The core of this issue is to free ourselves from the dogma that being scientific is equal to being analytical, and to allow science to have a more flexible framework' (1986: 116). This is necessary to overcome the negative effects of science's disciplinary fragmentation, enabling science to emerge from its partiality and move toward totality. From this perspective, agricultural science must be analytical and synthetic to be comprehensive.

Even if science achieves such comprehensiveness, another problem still remains. The development of science and technology and the industries based on this development have produced numerous environmental problems, including pollution of our immediate surroundings such as rivers, drinking water, air and food. On a global scale, such problems now threaten human existence *per se*. Under these circumstances, scientists' social responsibility and restructuring scientific practice have become pressing issues.

Maruyama Takashi focuses on experiential knowledge in the everyday lifeworld and, in these terms, attempts to reconsider the scientific world, more specifically, the world of scientific research. Maruyama (1987) argues that the lifeworld provides a clue to finding the right approach toward resolving many issues and must be regarded as the 'ultimate basis of science'. Returning to simple questions and intuition arising from our experience as ordinary citizens in the everyday lifeworld serves as a starting point for reorganizing the scientific world. As shown in Figure 10.2, Maruyama calls the link between the lifeworld and the scientific world the 'hermeneutic circle' and hopes that science will develop soundly through this circular process. This perspective is similar to the 'clinical knowledge'

Figure 10.2: Integration of the lifeworld and the scientific world

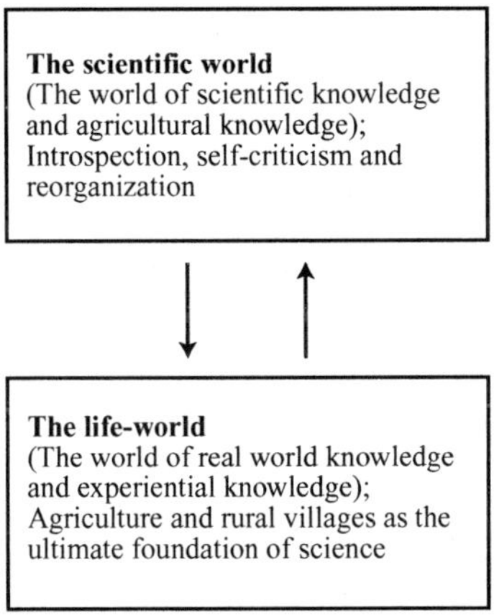

proposed by Nakamura Yūjirō. For agricultural science this means constructing scientific knowledge using, as a guide, both the experiential knowledge of the consumers of agricultural products and of farmers as producers. By doing so, agricultural science can bridge the gap between producers and consumers and meet both sets of demands. Aspects of this link between scientific and experiential knowledge were discussed in Chapter 9.

On this basis, agricultural science might generally be called the 'comprehensive science of life systems'. In what follows, I discuss the methodology appropriate to agricultural science if it is to fulfill this promise.[4]

Multidimensional and comprehensive methodology

It is clear that agricultural science is a field of practical science characterized by its deep relationship with biological production and regionality. As a practical science, it synthesizes the findings and methods of the natural and human sciences (Figure 10.3).

It should also be clear, then, that the methodology of agricultural science is not a single method but a combination or synthesis of three broad sets of methods, namely, methods based on explanation, understanding and imagination, which are, respectively, typical of natural science, human science, and the practical sciences. I say 'typical of' here because none of the sciences is reducible to a single

Figure 10.3: Position of agricultural science as the comprehensive science of living-system

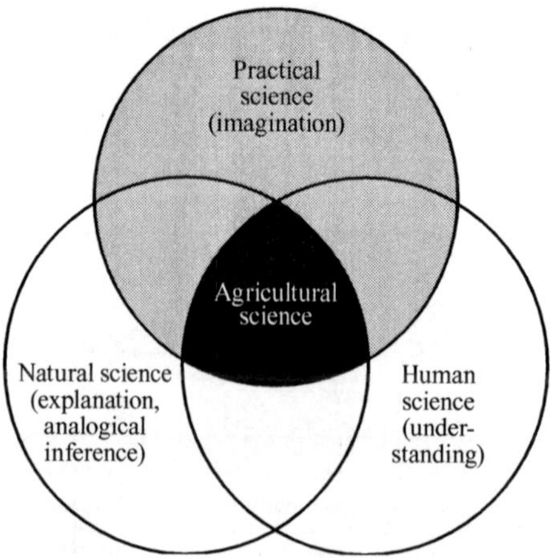

method: human science entails both understanding and explanation, natural science entails both explanation and analogical-understanding, and practical science subsumes all of these. Thus, although the agricultural science methodology is, as discussed shortly, ultimately characterized by 'element-integrating methods' using imagination and testing, it is not restricted to such methods. Rather, the research objectives of agricultural science can only be achieved by treating all three sets of methods as complementary. Since agricultural science must take into consideration the diversity, complexity, regionality and creativity of nature, it requires a higher degree of comprehensiveness than other fields of practical science. This is why I regard agricultural science as the 'comprehensive science of life systems'.

Supplementary note

A brief detour to consider the complexities of treating cancer patients may shed some light on the type of comprehensive synthesis I have in mind here. The human being is a unity of body and soul. For the cancer patient, there are two distinct dimensions to consider: (1) his/her physical conditions must be explained based on knowledge of physical and chemical processes. But, (2) his/her psychological response to this knowledge – to being confronted with death – must be understood and interpreted as meaningful thoughts and actions. In conventional terms, the former is mainly a consideration for natural science while the latter is primarily a matter for the human sciences. Furthermore, there are two dimensions to be considered by the human sciences. One is to deal with a general, step-by-step tendency that almost all patients display: after a patient is informed that s/he has cancer, initially s/he is typically stunned and thrown into despair, then develops an unstable psychological condition and loses interest in life, and then gradually restores him/herself and comes to terms with life once again. The other is to have thorough knowledge of the particular patient's history and personality. The first dimension cannot precisely be called a law, but it displays statistical regularity, and may thus be regarded as a quasi-law. Finally, deciding the appropriate type of care for the particular patient in question is a matter of practical science; the three methods are employed, as necessary, to explain, interpret and understand the relevant matters, and optimal care is conceived by high-level, synthetic penetration. In this way, practical science, as the field faced with concrete realities, shows its real strength through synthesis. This applies equally to caring for plants and animals in farm management.

The particular characteristics and methods of the three fields of science were outlined in the first half of this chapter. However, that discussion was based on the conventional classification of characteristics that are most typical of each field. As noted, rather than focusing on the distinguishing characteristics of each field, it has become necessary to identify 'continuity and similarity' (as per Bollnow and Ricoeur), or complementarity (as per Habermas) (Maruyama 1987: 234), between the different fields. As Bollnow argues, 'explanation…can only be made where the fundamental understanding of life has been lost', and, 'understanding is a more fundamental function, on which basis alone can explanation develop under certain circumstances' (1958: 140). Bollnow regards understanding in the broad sense as the most fundamental of all methods.

In what follows, I summarize the three methodological frameworks – explanation (and analogy), understanding, and imagination – as they apply in agricultural science. I believe that integrating or synthesizing these frameworks will deepen the capacity for further development in agricultural science.

Natural scientific method: explanation and analogical inference

Agricultural science makes practical use of the laws of nature discovered by other fields of science, as well as conducting its own basic research in accordance with natural scientific method.

The termite research described above is one example of agricultural science conducting its own basic research. From another perspective, if your intent is to invent a tomato harvester, you must first have full knowledge of the physiology and ecology of tomatoes as a plant, plus the physical and chemical characteristics of the fruit, a basic theory of a tomato ripeness sensor, and the various physical laws necessary to produce the composite parts. In short, such a project requires a combination of basic knowledge from other fields plus original research – both basic and applied – in agricultural science. Meanwhile, new research in the life sciences that produces knowledge about the molecular constitution of living organisms can be developed to provide a foundation for improving crops.

Natural science aims at discovering and formalizing natural laws. It attempts to elucidate and explain, by observation and experimentation, the underlying causative relations between and among natural phenomena. For this purpose, it reduces a complex phenomenon to its constituent elements, makes these elements describe themselves through artificial experiments, and discovers the laws at work.

These activities typically go through a three-stage process: hypothesis formulation by induction; drawing a conclusion by deduction; and finally, verifying the conclusion by observation and experimentation. This is the typical process through which the natural sciences explain a phenomenon.

However, even in the realm of natural science, life forms, as described above, not only transform themselves to suit the environment, but can only exist as comprehensive life forms. Here, an aspect different from the reduction and synthesis of physics or chemistry arises. Life forms can be understood by analogical inference. This understanding has resulted in the emphasis of comparative methods in biology by Ewald (1931), among others. Indeed, given that agricultural science is the science of biological production, analogical methods are arguably the key methods in agricultural science. But as we have seen, the vast array of phenomena that concern the agricultural scientist demand that different methods must be employed to suit the particular circumstances. Agricultural science therefore must be methodologically multidimensional in its search for synthesis.

Human scientific method: understanding and interpretation

Agriculture utilizes various regional natural resources. In order to make the best use of a region's natural resources, farms – as producers and members of the community – must cooperate in their agricultural activities. Cooperation in village communities may have undergone changes over time – both in nature and intensity – but it remains fundamentally different from the cooperation of industrial production and the urban communities that undertake such production. Rural community is *ba* of totality, place of production, place of ecology, and place of daily life. It is therefore *ba* of total human life.

It is therefore essential that agricultural science consider the human dimensions of agricultural production. What social organizations are required to secure and manage water and other natural resources? What activities are required in the village community to enhance the identity of villagers and maintain stable organizations? How should personal relations, such as disputes or trouble among villagers, be managed?

The essential premise for conducting practical, conceptual research towards resolving these socio-economic issues is to consider social and historical aspects of human beings, such as: What is the history of the village community and what are its current structures?

What are the normative values of the community that regulate the behavior of villagers? What factors are responsible for regional and individual differences between different communities? What historical and contemporary factors underlie issues that arise in the village community?

Human science attempts to elucidate the experiential world rather than the natural world, social structures rather than natural laws, functions rather than causes, and intentional connections rather than mechanical linkages. Through self-reflection, the human sciences aim to re-interpret, and thus re-create the human/social world itself. Agricultural science draws on the reflections and interpretations of practitioners in the various fields of the human sciences as well as conducting research into these questions for itself. It then proceeds to intentional and imaginative research into the concrete interactions between humans and nature in the course of agricultural production.

Practical scientific method: imagination

The essence of agricultural science is practical research aimed at the development of technology. Although it relies heavily on the basic research of both the natural and human sciences, its own research is primarily directed toward practical problem solving. The basic research, nevertheless, constitutes a prerequisite for practical or goal-directed research.

According to Miki, imagination is the 'ability to represent an object in intuition even in the absence of the object' (1967[7]: 348). In practical science, we must first expect that we know the laws of nature, because we cannot put any technology into practice against the laws of nature. Next, we must set human value objectives. Imagination arises from purposeful, intentional activities. Tools and machines are never produced by nature itself; they are only produced through the intentional, creative activities of human beings. Practical science, while synthesizing these laws and goals, achieves its goals by creating what we might call technological or exemplary models, in actual changes of matter finally through imagination (Miki 1967[7]: 241–2, 422).

In this sense, agricultural science: conceptualizes specific mechanisms and models to solve technological, economic, and social issues surrounding agricultural production; verifies the practical efficacy of such mechanisms and models by experimentation, testing and investigation; and presents the results of this process as exemplary models. Based on the combined methods for natural scientific and human scientific research, agricultural science constitutes an applied, practical and creative field of science.

In the following section, I will more specifically consider the methods of agricultural science.

Research as a dynamic process

Complexity and element-integrating method

The practical scientific methodology of Peirce and Dewey is very instructive in that it not only compares various sciences in relative terms but also attempts to find methods for science in a dynamic cognitive process. From this point of view, scientific cognition in agricultural science includes all methods, but it is still characterized by an element-integrative process which, as shown in Figure 10.4 (see also Figure 10.1), is completed through three stages (abduction, deduction and induction) that involve moving back and forth between the factual world and the symbolic world.

The cycle of cognitive processes consisting of these three stages will have, when repeated, two additional dimensions of 'synthesis'. First, this is a trial-and-error process, seeking a better solution to a problem. In this process, researchers' cognition develops from superficial to substantial cognition, incomplete to complete cognition, and partial (fragmentary) to overall (comprehensive) cognition.

Second, this process starts from the stage of element-reductive experimentation, followed by gradual addition and integration until

Figure 10.4: Method of agricultural research as a comprehensive science

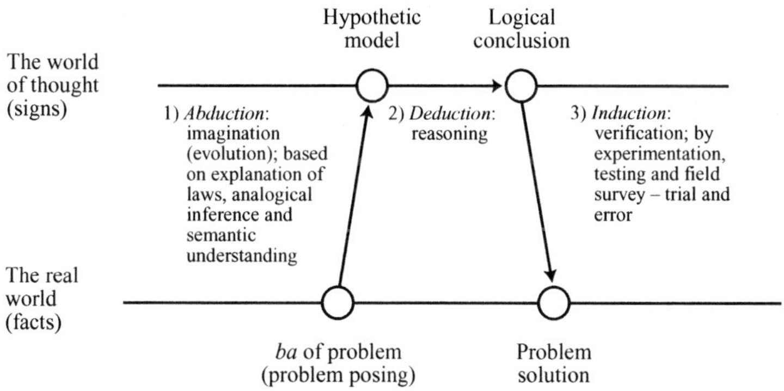

finally reaching the stage of element-integrative testing. This cycle from experimentation to testing, to deeper testing, is the essence of practical science, particularly agricultural science. This is also the case in social scientific field surveys conducted in relation to agriculture. Agricultural science researchers should renew and deepen their understanding of the meaning of testing and field surveys. Kashiwa Sukekata's great achievement lay in clearly distinguishing between experimentation and testing, and identifying the latter as the characteristic methodology of agricultural science (1962: 406–14).

Testing has various stages. In the end, testing is conducted under the same conditions as the production sites, where all elements must work together, leading to the production of a technology that can be extensively employed. Production sites are complex places that form complex, divergent systems, but can be broken down into four dimensions: (1) those elements that have already been elucidated; (2) those elements that have not been elucidated; (3) those elements that are unforeseeable; and (4) complex effects generated by the interaction among different elements. Research in agricultural science is conducted by adding influential elements together through the process of repetition and deeper testing, which finally results in approaching the sites at the level where all elements work together to establish a stable technology. Thus, agricultural science is characterized by 'element-integrating methods', in contrast to the element-reductive approach of natural science. In fact, agricultural science is already characterized by methodological perspectives that have recently been advocated, such as the synthesis of sciences, the necessity of 'science from a bird's eye view' (*fukanteki kagaku*) proposed by Yoshikawa Hiroyuki at the Science Council of Japan, and the science of complexity (Santa Fe Institute) (Mainzer 1994; Waldrop 1992; Ueda et al. 1999).

Agricultural science can perhaps only achieve its goals by applying these methods for dynamic development to its specific value objectives. In the following, I discuss the stages of this process in detail from the perspective of agricultural science.

Stages of research and their meanings

'Problems in *ba*' as the starting point of research

Agricultural science is a field of practical science that stands by agriculture, forestry and fishing industries and related industries – such as production materials (machines, fertilizers, feeds, pipes, etc.), food processing, agricultural products, and food distribution industries – and contributes to their development. Agricultural science is thus a collection of activities that attempt to solve practical problems. We face ever more varied problems as our attempts to make life better make it more complicated. In this book, I have argued that the value objective of modern society, and more specifically of modern agricultural science, is the harmonious realization of economic, ecological, and everyday life values – three, so to speak, comprehensive values that are often conflictual and demand trade-offs. The 'place of life', where we live and act, is rife with conflicts and problems surrounding individuals and social conditions. It is thus also a place of questioning and posing problems in our efforts to resolve these problems. Our existence is always in question; and life is formed in the cognition and practice aimed at resolving these questions. In this sense, questioning is the ultimate source of creation and world-formulation.

Problems exist before us in strata

A fundamental issue for agricultural science is how it can best contribute to creating a desirable development of agriculture, forestry and fishing and related industries in order to secure a healthy, stable environment for humans. More broadly, agricultural science is enmeshed in an incredible array of modern problems, such as: international and regional trade frictions; deteriorating rates of national food self-sufficiency and the corresponding security issues; adverse environmental effects from activities that overemphasize economic efficiency and the international division of labor; new technologies and management practices aimed at realizing environmental- and natural resources-protective industries; food safety and associated agricultural distribution and processing methods; the deterioration of rural communities due to industrialization and urbanization and the corresponding difficulty in finding successors for (family-based) agriculture, forestry and fishing industries; a shortage of marriageable women in rural villages; improving family relationships, lives and culture in rural areas; reorganizing and rationalizing organizations related to agriculture; and global population growth. These issues were discussed in Chapter 3.

It is in the context of resolving these problems that I have proposed the 'pursuit of comprehensive values' – the harmonious realization of economic, ecological and everyday life values. Whether a problem is recognized and what methods are available to solve it depends on 'what is valued'. Problems in agriculture are therefore both epistemological concerns and very real practical issues affecting our lives.

Further, when we attempt to clarify and resolve technological problems in agricultural production, the problems are first reduced to their constituent elements. For example, in rice cultivation, simply pursuing higher-yielding varieties is no longer viable; consumers now demand inexpensive, palatable rice that is life-friendly. This consumer pressure adds yet another set of criteria to the almost infinite list of desirable qualities that producers seek in their rice varieties, including: varieties that are resistant to cold, drought, disease, insects, and wind, do not require excessive amounts of water or fertilizer, and are easy to handle and tolerate being handled by machines. Different researchers in agricultural science take up different objectives along these lines, depending on which of these features is in greatest demand, is more urgent, or best fits the particular researcher's qualifications and interests. For example, the pursuit of high-yielding varieties and highly palatable varieties typically occurs under different social conditions; high-yielding varieties tend to be sought under conditions in which there is not enough food to feed the populace, whereas increased palatability tends to only become an issue when food is plentiful. When palatable food is plentiful, we proceed to the pursuit of higher-level, more comprehensive values, such as varieties that are high-yielding, palatable and safe (i.e., life- and environmentally-protective).

In any of these processes, research into the basic physiology and ecology of rice is indispensable. The methods for such research have diversified and fragmented, from molecular biology to traditional breeding studies. Research in agricultural science thus develops and becomes more practical by moving back and forth between the processes of fragmentation and synthesis, and between the processes of basic research and practical/intentional research. This applies not only to research on crop varieties but also on agricultural machines, soil, fertilizers, chemicals, processing of agricultural products, etc.

This also applies in fields typically associated with human science, such as attempting to create a humane, rich living place. In these fields, too, agricultural science begins with actual problems and resolves them by moving from fragmentation to synthesis, and from basic research to intentional research. All of these activities aim at improving human life.

As discussed, agricultural science begins with a problematic situation and attempts to resolve it. In practical science, identifying a problem is an important duty. Identifying a problem as such and logically developing it are essential prerequisites to resolving it. In short, research should not be simply driven by scientific concerns, but must begin by the incorporation of issues that arise in actual production sites and the lifeworlds of agricultural producers. Following this approach, the value objectives for resolving the problems will be set in due course. The lifeworld where real plants, animals and humans live with all of their numerous contradictions is the starting point and the 'ultimate basis' of science (Maruyama et al. 1987).

Imagining hypothetical models: the abduction stage

Identifying a problem of course requires that the researcher recognize certain values that are somehow hindered by the 'problem' or will be enhanced by its resolution. A model for solving the problem posed is creatively formulated using the researcher's imagination – including examining relevant theories and considering social trends and community demands. As previously mentioned, Miki Kiyoshi and Kashiwa Sukekata consider this modeling approach as the ultimate goal of technology. Yet I argue that this is only one dimension of a much broader and multifaceted methodological approach. The creation of a theory and model for problem solving is the most laborious, yet also the most romantic and creative, aspects of a scientist's work. Such imagination requires the researcher's passion for solving the problem, combined with dispassionate reasoning, extensive knowledge and, above all, a capacity for abduction.

But the model thus obtained, for the time being at least, remains only a possibility. Since the agricultural scientists are not searching for abstract theories or general laws, but for real, practical solutions to real, intractable problems, the formation of a possible model is only the beginning of the process.

The next step is based on the logic of invention, which applies laws to problems in which both nature and humans are involved, and attempts thereby to create specific means to realize specific value objectives. The imagination required here should be distinguished from and go beyond the supposedly pure sciences to synthesize explanation and analogy, understanding and interpretation, by being prospective, normative, systematic, formative, creative and constructive.

This type of imaginative activity should not be confined to 'mere' technical creation, or invention. It can be likewise applied to problem solving such as improving the profitability of agriculture, farmers' well-being, and the public peace in rural communities. We must stress that the ultimate task of agricultural science is not merely to offer insightful analyses of current issues in agriculture and rural villages, but also to conceive and formulate measures for improving economic, managerial, social, and policy issues, for example, that arise through such analyses.

Creation is typically regarded as making something from nothing – Einstein's theory of relativity is frequently cited as a typical example of creation. Yet, the theory of relativity builds upon Newton's physics and Maxwell's electromagnetics. Hence, although creation obviously entails a leap, it is not making something from nothing. Instead, creation is 'accomplished, often in a very awkward manner, based on an old framework and old tools' (Karakida 1995: 107). 'A theoretical development is more often a result of the reorganization of conventional "facts" using a different conceptual framework than a result of the rapid acquisition of new "facts"' (Murakami 1986: 32). In other words, creation is establishing new connections between existing elements, often elements that conflict and seem to be in a trade-off relationship with each other, differing from and conflicting with each other when placed in the same '*ba*'. By putting two conflicting elements in the same place and considering them from a new perspective, the conflict may be sublated and synthesized, opening the world of creation. The German term '*aufheben*', the equivalent of 'sublate', includes both 'to eliminate, to deny' as well as 'to preserve' and 'to enhance', which are seemingly opposite meanings. In a sense, then, 'to sublate' means 'to preserve while denying'. While conflicting elements oppose and fight against each other and are exposed to criticism, they internally permeate each other and are thus synthesized. Criticism here means to position, and give meaning to, an object in the context of new developments, rather than breaking up and eliminating the object (cf. Bollnow 1958/1975: 50–1). To solve a problem is thus to overcome conflicts, or to create new *ba* as logical space and an exemplar model based on such space.

The importance of 'intuition' or 'inspiration' has often been pointed out in the context of scientists' creative activities. Dewey says that a clue to an 'idea for potential solution...pops into one's head' (1938/1968: 497). Itō Kunitake describes it as 'something like an irrational flash that is left to researchers' discretion' (1985: 197). However, as per creation, intuition does not mean making something from nothing. Intuition may occur suddenly, in a lightning flash, only if the scientist has extensive knowledge about existing theories and goes through the struggle and labor of creative conception, attempting to overcome the limitations of existing theories. Intuition is a passive–active mental activity. We can see the light only at the end of the struggle.

This imagination of hypothetic models is what Peirce and Dewey call 'abduction'.

Logical validity of models: the stage of deduction

Once a solution for the problem has been imagined the researcher must logically assess whether it can truly synthesize/integrate natural laws and human objectives. That is, whether it constitutes an appropriate framework for formulating an adequate model.

Questions that must now be asked include: 'What would be the consequences of employing this model?' This mode of thinking is also an exercise of the imagination, but now under the sort of logical conditions of trial and error as used in a laboratory or field testing. Through this thought process, we may arrive at a single conclusion that is suitable to verification. This conclusion will determine whether the original idea is logically acceptable or not. In this way, an inference is reached in the form of a model that is more appropriate for resolving the problem in question than the original idea. This is still not a solution itself; we must now test this model against practical events.

This process of giving a hypothesis a logical basis is the process of deduction.

Experimentation, field testing and surveys: the induction stage

A logical conclusion drawn by inference must be verified through means such as experimentation, testing or field survey. Through this process, the proposed solution for a certain problem is evaluated, and the model solution can be presented to the specific production site as a valid model. This model is not, however, perfectly generalizable. For example, a particular machine (model) may be ideal under certain conditions, but ineffective under different geographic, climatic or soil conditions. Agricultural technology is fated to be location-bound. It also varies according to the managerial culture and economic conditions of the region. This is where the diverse regionalism of agriculture differs from industry.

Let us now discuss the methods that are used in verification (or proof) in this stage of induction, in particular: experimentation, testing, and field surveys.

Firstly, in agricultural science, particularly in research other than social scientific, a model is verified by experimentation or field testing. The terms 'experimentation' and 'examination' are typically used interchangeably in most research laboratories. Togari Yoshitsugu (1957) systematically describe the research methods used for studying crops, but the distinction between testing and experimentation is never mentioned. Indeed the two terms are used almost synonymously. Are experimentation and testing the same thing, or is there some essential difference between them? Behind this question lies an essential characteristic of the methodology of agricultural science.

Crop research intends to determine methods to breed or grow varieties of domesticated plants that have been improved in some way or another. To satisfy this intention, we must first conduct basic research, such as morphological, physiological and ecological studies of the plants per se. Morphological studies require anatomical methods to study both external and internal morphology. This, however, provides only a static view of lifeless objects. Fully comprehending plants as life forms requires an understanding of their physiology and ecology, that is, their ability to survive in nature – their living, dynamic aspects. Based on this basic research, we need to develop practical research that can establish models that can, in turn, be employed in places of production.

These different activities can be classified into basic research and practical, applied research. The objective of agricultural science is not yet achieved merely because the researchers have had success in creating a desirable plant variety in their laboratory. They must then test the physiology and ecology of the plant in the field; the mission of agricultural science is achieved only when the new crop variety can be fruitfully employed in places of production. In other words, agricultural science begins with basic research, including that conducted in other fields of science, and then proceeds to practical, applied research aimed at realizing specific value objectives. These research results go through (1) experimental studies in laboratories, and then (2) testing in test fields that more closely emulate the actual place of production, before being employed in general production. Togari identifies two methods of agricultural testing: research institute-type and local experiment station-type (1957: 16). This indicates a staged process of agricultural research, which starts from the creation of a general model and proceeds to establish regional models that take regional differences, such as land conditions, into consideration.

Hence, we can say that experimentation is a verification method associated with basic research under simplified or general conditions, whereas testing is a verification method associated with more practical, applied research under complicated and particular 'real-world' conditions. In basic research, it is important to simplify a hypothesis, to control as many variables as possible. However, as we proceed to practical research we must gradually subject our hypothetical model to real-world conditions in which all variables are, indeed, variable. The mission of agricultural science is achieved when it makes contributions at the practical level of satisfying regional and economic needs. This holds true not only in crop research but also in research on livestock, forests and forestry, and oceans and fisheries.

Secondly, although there has been discussion about the use of field surveys in rural villages, their importance for agricultural research is not well recognized. Among the many areas of agricultural science, field surveys are critically important in those associated with the human sciences, such as agricultural economics, rural sociology, rural planning methodology, agricultural management, and agricultural policy.

According to Miki Kiyoshi, in those areas of social science concerned with agriculture and rural communities, the concept of economic technology is as well established as mechanical technology, etc, and there is significant reliance on imagination. In particular, agricultural policy and rural planning methodology are among the fields that have been directed toward model formation by imagination. However, it is difficult in these fields to freely use natural scientific methods such as experimentation and testing. Artificial manipulation of human events for research purposes is associated with extremely severe restrictions. Although experimentation is not totally impossible, its size and extent in these fields are definitely different from those in natural science. In addition, such experimentation may have social consequences, which may cause ethical problems.

In this context, field surveys hold an important position in the human science fields of agricultural science. We have learned methods that can elucidate human/social events by understanding. However, beyond simple understanding of what *has happened*, both Dilthey and Bollnow focus on the creativity of understanding and interpretation, or the 'dynamism of life'. Scientific studies from the philosophy of life perspective aim at increasing the 'wealth of human life' by creative interpretation that changes human/social events, beyond simple historical explanation. Interpretation creates awareness of what lies behind an event, a reflection leading to self-awareness for both the scientist and others. Thus, human science is 'not only a "mirror" that reflects past and present human affairs as they are, but also a "creative mirror" that creates new human beings' through 'reflection on humans by themselves' (Maruyama 1985: 235). From this perspective, human life and society are products of their own creation. This is an important understanding for practical sciences that aim to resolve the problems inherent in human life and society through the use of imagination.

As mentioned, this is not creation of something from nothing; imagination gives insight into the true state of affairs in human society, including social forces and trends, and can lead to 'endogenous development' (Tsurumi and Kawata 1989) that extends such insight (Brokensha et al. 1980). This brings us back to H. Weber's observation that practical science is a method for integrating being (what is) and oughtness (what ought to be) (1919: 38). Through the methodological process described, human/socially oriented fields of agricultural science start from understanding history, proceed to identifying problems of the present and future, and attempt to present models that might resolve these problems. The field survey is the methodology that provides concrete supports to these activities.

Chapin's *Field Work and Social Research* (1920) appears to have been the first publication on this methodology. Toda Teizō's *The Method of Social Research* (1949) is the first comparable work in Japan. Various works on field survey have appeared since then, including some high-quality large-scale investigations that can serve as models. In agricultural science, field surveys (including statistical analyses) are conducted in fields such as agricultural management, farm economics, regional patterns of agriculture, farmers' lives, rural village communities, and rural village culture. These fields of study are supported by field surveys to such an extent that we can say that they begin and end with field surveys (Sakamoto 1994: 90).[5]

In identifying a problem and conceiving and formulating a model for solving it, we often use general observation, information obtained through the media, and a preliminary survey. To learn 'what issues exist in agricultural management or rural village communities?', and 'what is the desired resolution, and what measures must be taken towards resolution?', we begin with interviews, questionnaires and data collection. The subjects for these field surveys are farmers or rural communities, agricultural cooperatives, local chambers of commerce and industry, and local governments, as the case may be.

Based on our analysis of these observations and interviews, a problem is identified, a model conceptualized and formulated to solve the problem, the logical consequences of this model are examined, and the validity of the model is checked via re-survey. Since human science cannot experiment on its subjects via manipulations in a laboratory, it has no choice but to verify conclusions by repeating field surveys. Therefore, what is intended as a field survey for verification is destined to continually return to the preliminary survey to re-constitute its model until a valid conclusion is obtained.

Regional investigations in agricultural science must therefore be supplemented and completed by basic 'area studies' which elucidate the natural, social, cultural and historical events of the region, and by practical 'regional science' which presents measures to improve agriculture and rural villages by using problem-oriented, imaginative methods.

Importantly, field surveys in agricultural science are not limited to human/social fields but are also an important research method in natural scientific fields. Agricultural science deals with organic life forms such as crops, wood, livestock and fish, and, furthermore, must elucidate the true state of organisms not only in laboratories but also in their *ba* of life, or the field where they live.

Supplementary note

Let us consider the relationship between large wild animals, and agriculture and forestry for example (Furubayashi 1996). The damage to Japanese forests and agriculture brought about by the Japanese deer (*Cervus nippon*) that inhabit the Tanzawa Mountains is a serious problem affecting the lives of people who are reliant on agriculture and forestry there. However, as concerns about environmental issues have grown, so too has awareness of the 'value of injurious animals' as people's awareness of natural ecology and wildlife has been enhanced. While productive agriculture and forestry is essential to human survival, the protection of natural ecology and wildlife is increasingly understood to hold important meaning for humans. Achieving coexistence between nature and humans is now important. At present, however, this involves issues and values that are often in opposition or conflict with each other. In order to resolve problems such as the damage that Japanese deer inflict on productive industries, then, we must develop a precise and comprehensive understanding of the living conditions of the deer, including their feeding habits and geographical distribution – including changes in distribution brought about by changes in their living environment wrought by human intervention for productive industries – and conceptualize measures that might both preserve the species and prevent further damage to agriculture and forestry. Analyses of deer feces and field surveys of their feeding areas, etc, have revealed that (1) the Japanese deer feed on 106 different species of plants in 53 families, and that increased habitat size and deer population density have

> led to increased damage to artificial forests. Based on these findings, (2) the establishment of proper wildlife reserves and hunting zones, revisions to hunting regulations and standards, and protection of forested areas by barrier fences will be proposed following further theoretical reviews, field surveys and demonstration of the specific effects of such measures.

In the example described in the supplementary note, process (1) belongs to basic field survey studies and (2) to practical field survey studies. Hence, through cooperative interaction between basic research and goal-oriented practical research, imaginative agricultural science research is developed to the stage of model formulation. Field surveys in agricultural science have methodological significance only through the complementary interaction between these two types of studies.

Methods used in field survey include literature reviews, observation, interviews, and questionnaires. In addition, there are specific sampling methods, methods for identifying cases for field survey, and methods for summarizing. These are, however, involved in the 'techniques' of survey which must be discussed separately (cf. Kobayashi and Funabiki 1994; Itō 1992; Trusted 1979).

The system of agricultural science today

Since the birth of agricultural science, its organization has grown and fragmented, and in response to various demands in different periods, has undergone major changes. A full historical account of the development of the structures of agricultural science is beyond the scope of this work, but I would like to outline the organizational structure of agricultural science in Japan today. The structure depicted in Table 10.1 is derived from materials collected in 1998 from approximately 80 different organizations across Japan, including undergraduate and graduate schools of agricultural science in universities and agricultural colleges, details of the composition of departments and courses at that time, and by classifying fields into several categories based on my own observations.[6]

Faculties (schools) of agricultural science, that is, of agriculture, fishery science, dairy farming, veterinary medicine, textile science, etc., were gradually reorganized from the 1980s to the 1990s. The fields of study that were emphasized in the reorganization include environmental science, bioscience, biotechnology, ecology, regional science, international relations, and information science. The keywords involved include environment, life, biotechnology, ecology, globalization, regions, information, preservation, living, population and food, coexistence and cycle, related industries, developing countries, interdisciplinary, and synthesis.

These fields and keywords indicate that social trends were reflected in the reorganizations. My study also revealed that while, on the one hand, fields of study were further fragmented and specialized, on the other hand, synthetic and interdisciplinary approaches were being sought at the same time. My attempt to classify the fields of agricultural science after this reorganization indicates seven broad fields of study:

1. Philosophy of agricultural science: studies regarding the history of, the essence of and issues in, the value objectives, research methodology, and system of agriculture, forestry, fishing industries, rural villages and agricultural science.
2. Bioproduction environmental studies: studies regarding various environmental conditions surrounding bioproduction. Examples include meteorology, space information science, hydrology, soil science, microbiology, and regional environment studies. In particular, with the progress of ecological studies, development of geographic information systems (GIS), which make use of information from an artificial satellite that is capable of sending detailed geographic information in one-meter squares (for example, soil condition and the degree of some fertilizers), is desired at present.
3. Agricultural biosciences: studies for elucidation and utilization of biological functions at molecular and cellular levels.
4. Bioproduction sciences: studies regarding raising, managing and utilizing plants, animals, marine life and microorganisms. This is the most traditional field of agricultural science and includes crop production science, silviculture, animal production science, marine bioresources, and microbial production science, among others.
5. Related industrial sciences: studies regarding materials industries supporting bioproduction, distribution and processing of biomaterials. Examples include agricultural chemistry such as fertilizer science and pesticide science, development of agricultural machinery and facilities, food engineering, wood processing, aquatic product processing, and textile engineering.
6. Bioproduction social and economic sciences: studies regarding management, economics and societies surrounding bioproduction and utilization of products. This field includes all human scientific fields.

Table 10.1: The system of agricultural science in Japan (2005)

Philosophy of agricultural science

Studies regarding the history of agricultural science, and the essence of and issues in, and the values and aims, research methodology, and system of, agriculture, forestry, fishing industries, rural villages and agricultural science.

Bioproduction environmental sciences

Studies regarding various environmental conditions surrounding bioproduction.

Meteorology; space information science; hydrology; soil science; weed science; wildlife damage management; pestology (entomology); farmland ecology; forest ecology; marine ecology; microbial environment studies; crop evolutionary history; global environment studies; geographic information system; etc.

Agricultural biosciences

Studies for elucidation and utilization of biological functions at molecular and cellular levels.

Applied biochemistry; molecular cell breeding; bioinformatics; microbiological science; biofunctional science; food life sciences; etc.

Bioproduction sciences

Studies regarding raising, management and utilization of plants, animals, oceanic life and microorganisms.

Crop production science
Crop science; breeding; fruit trees, vegetable, flower and horticultural sciences; cultivation science; tropical agricultural science; arid land agricultural science; etc.

Silviculture
Forestry; silviculture; etc.

Animal production science
Genetics and breeding science; reproduction physiology; animal nutritional science; biomechanism; grassland management; animal production management; veterinary medicine; etc.

Marine bioresources
Marine life resources; applied science of marine resources; aquatic resources management; etc.

Microbial production science
Enzyme chemistry; microbial breeding; fermentation chemistry, zymurgy; etc.

Related industrial sciences

Studies regarding materials industries supporting bioproduction, and distribution and processing of biomaterials.

Agricultural chemistry
Fertilizer science; pesticide science; chemical materials science; etc.

Agricultural machinery
Production and processing machinery; facilities engineering; work system; robotics for agricultural work; production energy science; etc.

Biological materials processing
Food processing; food chemistry; food functional technology science; food and nutritional sciences; wood processing; aquatic product processing; biological medicinal products; pharmacy; natural textile engineering; biomass conversion; etc.

Bioproduction social and economic sciences

Studies regarding management, economics and societies surrounding bioproduction and utilization of products

Demography; food economics; agricultural and forestry environmental economics; agriculture, forestry and fisheries management accounting; agricultural finance; natural resources economics; management and information science; regional economics; rural sociology; rural-urban relations; international agricultural, forest and fisheries economics; agricultural history; agricultural geography; technology diffusion; bioproduction ethics; home economics (home management); food culture studies; health science; etc.

Environmental science and technology

Studies regarding environment creation that integrate bioproduction, living and ecological environments.

Environment improvement
Farmland improvement; water environment engineering; global environment science and technology; forest environment studies; landscape architecture; environmental design; landscape studies; rural planning methodology; etc.

Environment preservation and restoration
Water quality management; natural environment conservation; forest conservation; regional environment conservation; desertification prevention; arid land afforestation; rural area rehabilitation; water purification technologies; air purification technologies; recycling studies; etc.

---

7. Environmental science and technology: studies regarding the environment that integrate the fields of bioproduction, living and ecological environments. This includes the fields of environment improvement, preservation and restoration that aim at creating sustainable agriculture and sustainable rural village areas.[7]

Modern agricultural science covers these seven extensive fields. Collectively they aim at the harmonious realization of the comprehensive value that synthesizes the three values – economic, ecological and life values – in production activities carried out in each region.

Conclusion

Agricultural science is a science of: practical pursuit of values, life, regions, and synthesis. Although it is a practical science, it can be called the 'comprehensive science of life systems'. The practicable methods for agricultural science are based on natural science's experimental explanatory methods, particularly biology's analogical methods of observation, and human science's understanding-based methods and interpretations, all centered around practical science's imaginative methods and testing. In this regard, my argument has been built upon Miki Kiyoshi's theory of imagination, Kant's judgment, Dewey's imaginative interpretation and practical cognition, Imanishi Kinji's concept of analogical inference, and Kashiwa Sukekata's analysis of testing, among others.

Although the significance of agricultural science in the establishment and practical implementation of technological models that are useful in production is unambiguous, many aspects of the relationships between agricultural science and other issues await clarification, such as: improving lives in rural villages; activities for the formation of sustainable regions; consumer awareness and behavior; and related processing and distribution industries. The value of agricultural science is greatest, however, where it is closely involved in consumer demand or the farmers' problems and behavior. Agricultural science must be based on demands and problems in the lifeworld. As long as this foundation is maintained, it is not possible for agricultural science to flourish while agriculture declines.

Finally, I must mention the current state and ethical demands of science. Through practical science and corresponding technological development, humans have dramatically adapted to and modified nature, thereby, in many respects, securing a more stable survival. In this sense, science has significantly contributed to a material civilization of enormous reach and affluence.

However, the world is today feeling increasingly uneasy about the burden of this material civilization. Through science and technology, human beings have developed a devilish power that is capable of destroying all of the organisms on Earth, including themselves – nuclear weapons. At the same time we have created a vast array of environmental problems by industrial development and the exploitation of nature. Natural ecosystems have been disturbed by soil and water pollution, resulting in the death of numerous organisms. Forests have been destroyed by a combination of unrestrained logging, air pollution and acid rain. The ozone layer of the atmosphere has been depleted by the emission of so-called 'greenhouse gases', which has contributed to global warming and accompanying abnormal weather patterns and increased desertification. Dependence on agricultural chemicals has contributed to further contaminating food stocks already at risk from the aforementioned air, water and soil pollution. Furthermore, a culture that prioritizes economic growth above all other considerations has generated management organizations and institutions that have often and in many ways oppressed and exploited human 'life'. This is but a small sample of a litany of modern complaints that appear before us as consequences of modern science and economic development.

Aristotle regarded human technology as an imitation and completion of nature. Heydebreck claimed that nature's creativity provides a role model for technology. Miki argues that 'human technology creates new forms, meanings and values that do not exist in nature' – i.e., culture and history (1967[7]: 256). Human technology thus exceeds natural technology, and will eventually conflict with nature. On this point, Bacon famously argued, nature cannot be conquered but by obeying her. Miki states that technology is based on the inherent, transcendental nature of humans. Human technology is in conflict with nature by the combination of modern technology and economy-oriented ideologies; it may even be heading for self-destruction in some cases.

As though Miki had expected this situation, he demands the 'morality of technology'. For Miki, technology 'both develops and restricts human lives' (1967[7]: 270). Machines alienate and degrade human beings by depriving them of the freedom to exercise their creative capacities and subjecting them to the relentless rhythm of the mechanized workplace. In short, technology gradually 'controls humans by its self-governing mechanism'. Miki notes, however, that 'neither simple pessimism nor simple optimism' is warranted.

Today it is beyond doubt that moral self-restraint is necessary in the development and deployment of technology, regardless of social and economic circumstances. While the issues surrounding modern technology, such as the increasing systematization of society and the corresponding loss of humanity, and environmental issues, undoubtedly indicate market failures, the consequences of socialism in the Soviet Union and China at the end of the 20th century are also obvious failures of systems of government.

Although practical science is inventive, it 'invents not only means but also objectives. Particularly in the invention of objectives', it is necessary for technology to be moral (Miki 1967[7]: 285). Agricultural science must, like all other sciences, adopt new, truly human value objectives and set its course with self-awareness in order to resolve the various problems we face today.

Notes

Chapter 1

1 Morinaga's book incorporated a path breaking viewpoint that emphasized the roles of ecology and biology within agricultural science.
2 I have learnt much from Sakamoto's concept of 'human life' which has drawn heavily on the *lebens* philosophy tradition.
3 I have examined the nature of the local place from a number of standpoints in, for instance, Soda 1980; 1984/1997 and with Ōhara and Kako 1996.

Chapter 2

1 In England, mutton was mixed into animal feed until around 1986 when one cow after another began collapsing. This was a result of the so-called mad cow disease that attacks the cerebral nerves and transforms them into a sponge-like state. As it became apparent that the disease could be contracted by humans, export embargoes were placed on English beef, massive numbers of cows were slaughtered and their carcasses incinerated. Furthermore, as fears grew that the growth-inducing drugs given to cows in America cause abnormalities in the reproductive systems of young women, there was an uproar demanding a ban on all beef imports from America. Among other problems that have been pointed out, there have been concerns that antibiotics may lose their effectiveness for people who have eaten the meat of cows that have been given antibiotics when being bred.
2 Rachel Carson's *Silent Spring: Warnings Against an Ecological Tragedy* (1962) and Theo Colborn's *Our Stolen Future: Are We Threatening Our Fertility, Intelligence, and Survival?* (1996) shocked people by exposing the extent and depth of the influence of pollutants and environmental hormones on living beings.
3 In recent years, research in environmental ethics and economical ethics has made great progress. I have included an example of this in note 32 of Chapter 5.
4 For more on the concept of 'formative equilibrium', see Soda Osamu, 1999. In addition, for sources relevant to this chapter, see, among others, Kokuritsu Rekishi Minzoku Hakubutsukan (ed.), 1997; Fukui Katsuyoshi, 1995; Kawai Masao and Hanihara Kazurō (eds), 1995.

Chapter 3

1 Yoshida Tamito added 'liberal, reasonable discussion of values' to these two freedoms (1999: 12)

2 See, for instance, Soda Osamu (1984), and Soda Osamu, Ōhara Kōtarō and Kako Toshiyuki (eds, 1996). The literature attempting to understand trends in modern society, agriculture, etc. from the aspect of three values include: Stratmann-Martens (1991) and Thiede (1988). Grigg (1984, 1986) conducts similar analyses from a geographical perspective.

3 European countries tend to provide financial support for the 'environment-preserving function' of agriculture. This tendency is most obvious in Germany. See, for instance, Yokokawa Hiroshi (1996), Alois Heissenhuber (1994).

Chapter 4

1 The second chapter of Eduard David's *Sozialismus und Landwirtschaft* (Socialism and Agriculture, 1922), entitled 'The Fundamental Difference between Organic and Mechanical Production', laid the foundations for the theory of the distinctiveness of agriculture. In the Japanese literature on the issue, works, such as Zenji Yagisawa's *Nōgyō no tokushitsu ni kansuru ichi kōsatsu* (An examination of the distinctive features of agriculture) (1925), Seiichi Tōbata's *Nōgyō mondai no shosō* (Some aspects of the agrarian question) (1938), and Masao Ōtsuki's *Nōgyō rōdō ron* (Theory of agricultural labor) analyzed the issue. Ōtsuki's work is outstanding in this regard. Sukekata Kashiwa in *Nōgaku genron* (Philosophy of agricultural science) (1962) and *Shihon shugi no mekanizumu* (The mechanisms of capitalism) (1957) extended Schumpeter's dynamic theory and discussed the meaning of land ownership and the specifics of this ownership. The study of agriculture and agricultural products has continued both in Japan and other countries. One new contribution to this discourse is the work of Makoto Hoshi, *Nōgyō to kōgyou no sai ni tsuite* (On the difference between agriculture and manufacturing). In what follows, I will summarize these works and add my personal views.

2 This categorization was proposed by Akinori Enoki of Kinki University.

3 Nei Masahiro (1999) predicts that in the future, economic theory will consider methods for correcting market mechanisms.

Chapter 5

1 Calculations were made by the Ministry of Agriculture, Forestry and Fisheries.

2 We also suffer much anxiety about many other environmental problems, including the destruction of the ozone hole, acid rain, environmental hormones and dioxin problems.

Chapter 6

1 A situation where children lock themselves in a room and do not want to communicate with the outside world.

Chapter 7

1 The concept of '*chiiki sangyō fukugōtai* (regional industrial complex)' was constructed by combining, through research and study of the Tango area, Sakamoto's concept of '*chiiki fukugōtai* (regional complex)', which focuses on agriculture (Sakamoto and Takayama 1983) and Soda's (1975) concept of '*takakuteki sangyō kōsei no mura* (village with multiple industries)', which incorporated other industries.

2 My book *Jizokuteki nōson no keisei* (The formation of the sustainable farming village) describes the farming village region that I have in mind here in more detail, and concretely analyzes it in terms of the three aspects outlined in this chapter.

Chapter 8

1 Urban-rural combination, especially the combination of small-to-medium cities and country areas, has been central to my research and publishing in Japan since 1973, when I conducted research on local policy in Germany. For more details, refer to my (1997) *Toshi to nōson no ketsugō* (The combination of city and country), an expanded edition of the 1984 *Nishi Doitsu no chiiki keikaku – toshi to nōson no ketsugō* (Regional planning in West Germany: the combination of city and country).

2 This conceptual map was introduced to Germany from the US by O. Boustedt, and then interpreted and applied in a specifically German way (Gildemeister 1973: 76).

Chapter 9

1 The framework of this chapter was devised in 1968 based on an investigation into the actual conditions of a farming village in Nara Prefecture, Japan.

2 See, for instance, Nishio Toshihiko (1998). For a general explanation of postwar agricultural technology, see Editing Committee of Showa Period Agricultural Technology Development History (1998).

3 For a good description of people's behavior and spread of technology, see: Ōmura and Hōzuki, 1979; Y M. E. Bratman, 1987; E. M. Rogers, 1971. For a holistic discussion on the developmental process of concrete agricultural technologies, see Tōbata Seiichi and Morinaga Shuntarō (eds), 1953–59.

Chapter 10

1 In writing this chapter, I owe much to Kashiwa Sukekata's *Nōgaku genron* (The philosophy of agricultural science) and *Nōgaku no yukue* (The future of agricultural science) and Sakamoto Keiichi's studies on 'sei no nōgaku (agricultural science of life)'. I also learned much from Miki Kiyoshi's *Kōsōryoku no ronri* (The logic of imagination) and *Gijutsu tetsugaku* (The philosophy of technology), Bollnow's *Lebensphilosophie* (The philosophy of life), Dewey's *Logic: The Theory of Inquiry*, and Imanishi Kinji's *Seibutsu no sekai* (The biological world), among others.

2 Weber's 'value freedom' is an often-misunderstood concept. What Weber really referred to by this was his assertion that (1) individuals have the freedom to choose their own particular values, and that (2) they also possess a free and objective consciousness that is separate from their values. Sumiya 1970: 188.

3 On these counts, medical science has plenty in common with agricultural science. Medical science is put into practice by physicians. Physicians face their patients through the medium of medical science. While medical science includes the foundations of natural sciences such as the theory of life and biology, it is also well known that it subsumes elements of sociology, psychology, human studies and ethics, among other disciplines.

4 The concept of 'the comprehensive science of the life system' is explained in the 1993 publication 21*seiki ni mukete no nōgaku kyōiku kenkyū no tenbō* (The prospects of agricultural science education at the start of the start of the 21st Century) by the Faculty of Agriculture Kyoto University, the planning of which I took part. The origins of this position lie with the work of P. Elkins and his peers. Works that describe agriculture in a rather comprehensive fashion from the point of view of technology include, among others, R. S. Loomis and D. J. Connor, 1992; J. R. Raeburn, 1984.

5 For more regarding rural village research, see Fukutake Tadashi and Matsubara Haruo, 1967, *Shakai chōsahō* (Methods of social research), Tokyo: Yūhikaku. Other relevant sources include Ishikawa Kiyoshi, Hamatani Masaharu and Hashimoto Kazutaka, 1994, *Shakai chōsahō – rekishi to shiten* (Methods of social research – History and perspectives), Kyoto: Minerva Shobō; Hōgetsu Makoto, Nakamichi Minoru, Tanaka Shigeru and Nakano Masataka, 1989, *Shakai chōsa* (Methods of social research), Tokyo: Yūhikaku;

Peter H. Mann, 1968, *Methods of Sociological Enquiry*, Oxford: Blackwell; Nakamura Hisashi and Hirooka Hiroyuki, 2000, *Fīrudo wāku no shingihō* (New techniques for field work), Tokyo: Nihon Hyōronsha. Regarding the history of rural village research, see for instance Fukutake Tadashi and Shimazaki Minoru et al., 1977, *Sengo Nihon no nōson chōsa* (Rural village research in postwar Japan), Tokyo: Tokyo University Press.

6 Sources that described the underlying ideas and system of agricultural science at the time of the reorganization of agricultural faculties in Japan include Faculty of Agriculture at Tōhoku University, 1997, *Ningen to kankyō no komyunikēshon nōgaku* (Communicative agricultural science for humans and the environment), Tōkyō: Nōrintōkei Kyōkai; Tōkyō Daigaku Daigakuin Nōgaku Seimei Kagaku Kenkyūka (The Research Department at the Department of Life Sciences, Graduate School of the University of Tokyo) (ed.), 2000, *Nōgaku – 21-seiki he no chosen: Chikyū wo sukuu 50 no teian* (Agricultural science – Challenges at the turn of the century: 50 proposals to save the Earth), Tokyo: Sekai Bunkasha; Tōkyō Nōgyō Daigaku Nihon Nōgyō Wo Kangaeru Kai (Tokyo University of Agriculture, committee on Japanese agriculture) (ed.), 1992, *21-seiki nōgyō he no teigen* (Suggestions for agriculture in the 21st century), Tōkyō: Nōrin Tōkei Kyōkai; The Faculty of Agriculture of Kyoto University, 1993, *21-seiki he mukete no nōgaku kyōiku, kenkyū no tenbō* (The prospects of agricultural education and research in the early 21st century), Kyoto: Kyōto Daigaku Nōgakubu '21-seiki no nōgakuzō' Kentōkai; The Faculty of Agriculture of the University of Kyoto, 1996, *Asu no nōgaku wo kangaeru* (Thinking about agriculture's future), Kyoto: Kyōto Daigaku Nōgakubu '21-seiki no nōgakuzō' Kentōkai; Kokuritsu Daigaku Nōgakukei Gakubuchō Kaigi (The Conference of Deans of Agricultural Faculties of National Universities), 1997, *21-seiki no nōgaku no bijon* (A vision for agriculture of the 21st century); Nihon Gakujutsu Kaigi Dai 6 Bukai (The Science Council of Japan, Special Committee 6, Kitamura Teitarō) (ed.), 1998, *Atarashī nōgaku no tenkai* (The emergence of new agriculture), Tokyo: Nōrin Tōkei Kyōkai.

7 The following publications provide introductions to all the relevant fields: Shu Shinmin and Sai Hōin, (eds), 1991, *Nōgaku Gaku Gairon* (An introduction to agricultural science), Chūgoku Kagaku Gijutsu Daigaku Shuppansha (The China University of Science and Technology Press). Noguchi Yakichi, 1946, *Saibaigenron* (The principles of cultivation), Tokyo: Yōkendō. Ehara Kaoru, 1970, *Saibaigaku taiyō* (An introduction to agronomics), Tokyo: Yōkendō. Horie Takeshi et al., 1999, *Sakumotsugaku sōron* (A general introduction to agronomics), Tokyo: Asakura Shoten. Saitō Takashi, 1992, *Engeigaku gairon* (An introduction to horticulture), Tokyo: Buneidō Shuppan. Konohira Yūkichi, 1994, *Shinrin kagakuron* (A thesis on forest science), Tokyo: Asakura

Shoten. Tadaki Yoshiya, 1996, *Shinrin kankyō kagaku* (The science of forest environment), Tokyo: Asakura Shoten. Ōkubo Tadakatsu, Toyoda Yutaka and Aida Katsumi, 1996, *Dōbutsu seisangaku gairon* (An introduction to the study of farm animal production), Tokyo: Buneidō Shuppan. Ōgimoto Keiji, 1992, *Dōbutsu seisangaku gairon* (An introduction to the study of farm animal production), Tokyo: Kawashima Shoten. Mizuma Yutaka et al., (ed.), 1998, *Saishin chikusangaku* (The latest developments in animal husbandry), Tokyo: Asakura Shoten. Chiiki Gyogyō Gakkai, 1998, *Gyogyō kōgengaku* (A study of the modern fishing industry), Tokyo: Nōrin Tōkei Kyōkai. Watanabe Hiroyuki et al., (ed.), 1996, *Nettai nōgaku* (Agricultural science of the tropics), Tokyo: Asakura Shoten. Kishimoto Osamu (ed.), 1984, *Nettai nōgyō nyūmon* (An introduction to tropical agriculture), Tokyo: Kokon Shoin. Tanaka Akira (ed.), 1997, *Nettai nōgyō gairon* (An introduction to tropical agriculture), Tokyo: Tsukiji Shokan. Lorentz C. Pearson, 1967, *Principles of Agronomy*, New York: Reinhold. Edward C. Hewitt and D. J. Brazier, 1968, *Agricultural Science for Schools and Colleges*, London: Lockwood. Gerar E. D'Souza (ed.), 1998, *Sustainability in Agricultural and Rural Development*, Aldershot: Ashgate. Colin Kirkpatrick and Norman Lee (eds), 1997, *Sustainable Development in a Developing World: Integrating Socio-economic Appraisal and Environmental Assessment*, Cheltenham: Edward Elgar.

Bibliography

Adorno, Theodor W., 1969, *Der Positivismusstreit in der deutschen Soziologie*, Berlin: Luchtehand Verlag.

Agricultural Research Institute of Ministry of Agriculture, Forestry and Fishery, 1998, *Nōgyō nōson no kōekiteki kinō no hyōka kentō ni yoru shisan* (A calculation based on the consideration of the public benefit functions of agriculture and farming villages).

Akademie für Raumforschung und Landesplanung, 1970, *Handwörterbuch der Raumforschung und Raumordnung*, Hannover: Gebrüder Jänecke.

Alfred Schütz, 1932, *Der sinnhafte Aufbau der sozialen Welt: eine Einleitung in die verstehende Soziologie*, Wien: J. Springer.

Alier, Juan Martinez, 1990, *Ecological Economics*, Oxford: Blackwell.

Amino Yoshihiko, 1994, *Nihonshakai saikō* (Rethinking Japanese society) Tokyo: Shogakukan.

Andō Shōeki, 1982–87, *Andō Shōeki zenshū* (The complete works of Andō Shōeki), Tokyo: Nōbunkyō.

Aoki, Masahiko, 1975, 'Rijun to keizai jikan' (Returns and economic time), *Shisō* (Thoughts), September and October Issues.

Asahi Shinbunsha, 1994, *Kankyougaku ga wakaru* (The basics of environmental science), Tokyo: Asahi Shinbunsha.

Asami Atsuyuki, 1989, *Nōgyō keiei sanchi hatten ron* (Theory of agricultural management and the development of regional production centers), Tokyo: Taimeidō.

Aust, Horst (ed.), 1985, *Raumplanung und Umweltschutz*, Stuttgart: Enke.

Barner, J., 1975, *Einführung in die Raumforschung und Landesplanung*, Stuttgart: Enke.

Bergmann, Klaus, 1970, *Agrarromantik und grosstadtfeindschaft*, Meisenheim am Glan: A. Hain.

Bernal, J. D., 1939, *The Social Function of Science*, London: George Routledge & Sons.

Bollnow, Otto Friedrich, 1949, *Das Verstehen: drei Aufsatze zur Theorie der Geisteswissenschaften*, Mainz/Rhein: Kirchheim Verlag.

Bollnow, Otto Friedrich, 1958, *Die Lebensphilosophie*, Berlin: Springer.

Bollnow, Otto Friedrich, 1959, edited and translated by Kojima Takehiko, 1960, *Kibō no tetsugaku* (The philosophy of hope), Tokyo: Shinkigensha.

Bollnow, Otto Friedrich, 1963, *Mensch und Raum* (Human and space), Stuttgart: W. Kohlhammer.

Bollnow, Otto Friedrich, 1970, *Philosophie der Erkenntnis: das Vorverstandnis und die Erhahrung des Neuen*, Stuttgart: W. Kohlhammer.

Bollnow, Otto Friedrich, 1986, *Dilthey and Husserl*, edited and translated by Takahashi Yoshito, *Dirutai to Fussāru: 20 seiki tetsugaku no genryuu* (Dilthey and Husserl: The origins of 20th century philosophy), Tokyo: Iwanami Shoten.

Bratman, M. E., 1987, *Intention, Plans and Practical Reason*, London: Harvard University Press.

Brokensha, David W., Dennis M. Warren and Oswald Werner, 1980, *Indigenous Knowledge Systems and Development*, Washington DC: University Press of America.

Bundesministerium für Raumordnung, Bauwesen und Städtebau, 1995, *Grundlagen einer Europäischen Raumentwicklungspolitik*, Bonn.

Carson, Rachel, 1962, *Silent Spring: Warnings Against an Ecological Tragedy,* Boston: Houghton Mifflin.

Chiiki Nourin Keizai Gakkai, 1999, *Chiiki nourin keizai kenkyū no kadai to houhou* (Topics and methodologies in the research of regional agriculture), Osaka: Fumin Kyoukai.

Childe, Vere Gordon, 1942, *What Happened in History*, Harmondsworth and New York: Penguin.

Chō Kenji, 1997, *Gendai Amerika kazoku nōgyō keieiron* (Modern American family farming), Fukuoka: Kyushu University Press.

Chō Yukio and Sumiya Kazuhiko, 1969, *Kindai Nihon keizai shisō shi* (A history of contemporary Japanese economic thought), Tokyo: Yūhikaku.

Christaller, W., 1933, *Die zentralen Orte in Süddeutschland: eine okonemisch-geographische Untersuchung über die Gesetzmässigkeit der Verbreitung und Entwicklung der Siedlungen mit stadtischen Funktionen*, Jena: Gustav Fischer.

Colborn, Theo, 1996, *Our Stolen Future: Are We Threatening Our Fertility, Intelligence and Survival?* New York: Dutton.

Council of Economic Advisors, 1988, 1992, *Annual Report of the CEA*, Washington DC.

Daly, H., 1977, 'The Steady-state Economy: What, Why and How?', in Dennis Pirages (ed.), *The Sustainable Society*, New York: Praeger.

Darwin, Charles, 1859, *On the Origin of Species by Means of Natural Selection.* London: Murray.

David, Eduard Heinrich, 1903, *Sozialismus und Landwirtschaft* (Socialism and agriculture), 2. Aufl. Kapitel.

Dethier, Vincent Gaston, 1976, *Man's plague?* Princeton, New Jersey: Darwin Press.

Dewey, J., 1938, *Logic: The Theory of Inquiry*, New York: H. Holt, Rinehart & Winston.

Dewey, John, 1948, *Reconstruction in Philosophy,* Boston: Beacon Press.

Dillard, Dudley D., 1950, *The Economics of John Maynard Keynes: The Theory of a Monetary Economy*, London: Crosby Lockwood & Son.

Dilthey, Wilhelm, 1961, *Die Philosophie des Lebens: eine Auswahl aus seinen Schriften, 1867–1910*, Stuttgart: Teubner.

Dilthey, Wilhelm, 1988, translated by Ramon J. Betangos, *Introduction to the Human Sciences*, Detroit: Wayne State University Press.

Dithey, Wilhelm, 1957, *Die Entstehung der Hermeneutik*, (Gesammelteschriften, Bd 5.6.7), Stuttgart: Teubner.

Dittrich, Erich, 1996, *Report: Gutachten des Sachverstündigenausschusses für Raumforschung*, Bonn.

Eames, Morris, 1977, *Pragmatic Naturalism: An Introduction*, Carbondale and Edwardsville: Southern Illinois University Press.

Economic Planning Agency of Japan, 1961, *The Economic White Paper 1960*, Tokyo.

Editing Committee of Showa Period Agricultural Technology Development History (ed.), 1998, *Shōwa nōgyō gijutsu hattatsushi* (The history of Showa period agricultural technology development), 7 volumes, Tokyo: Nōrin Suisan Gijitsu Jōhō Kyōkai.

Ehrlich, Paul R. and Ehrlich, Anne H., 1990, *The Population Explosion*, London: Hutchinson.

Elkins, Paul, 1986, *The Living Economy: A New Economics in the Making*, London: Routledge & Kegan Paul.

Ewald, Peter Paul, 1931, *Das Weltbild der Naturwissenschaften*, Stuttgart. Translated by Nagano Kōji, 1940, *Gijutsu to shizenkagaku teki sekaizō* (Technology and world image of natural science), Tokyo: Hakuya sha.

Federal Ministry for Regional Planning, Building and Urban Development, 1994, Report-*Synopsis of Spatial Planning Objectives of European Countries.*

Ferry, Luc, 1995, *The New Ecological Order*, Chicago: University of Chicago Press. Translated by Katō Hiroyuki, 1994, *Ecorojī no shin chitsujo*, Hōsel University Press.

Fichte, Johann Gottlieb, 1846, *Fichtes Werke*, Bd. 2, Berlin: Walter de Gruyter.

Fritsch, Theodor, 1896, *Stadt der zukunft*, Leipzig: Fritsch.

Froriep, S., 1970, 'Der siedlungsverband und ruhrkohlenbezirk-sein werden und wirken 1920–1980', *Raumforschung und raumordnung,* no. 2, Hannover.

Fukuda Minoru, Yori Taira and Mese Morio, 1971, *Shokuryō keizaigaku* (Economics of food products), Tokyo: Meibun Shobō.

Fukui Katsuyoshi, 1995, *Kōza chikyū ni ikiru 4: Shizen to ningen no kyōsei*

(Lecture series Living on Earth lecture 4: Symbiosis of nature and humans), Tokyo: Yūzankaku.

Fukuoka Katsuya, 1987, *Mori to mizu no keizaigaku* (Economics of forests and water), Tokyo: Touyou Keizai Shinpōsha.

Fukuoka Masanobu, 2004, *Shizen nōhō* (Natural farming), Tokyo: Syunjūsya.

Fukutake Tadashi, 1964, *Nihon nōson shakairon* (Social theory of Japanese farming villages), Tokyo: Tokyo Daigaku Shuppankai.

Furubayashi Kenkō, 1996, *Tanzawa sanchi no nihonjika no hogo ni kan suru kenkyū* (Researching the conservation of the Japanese deer in the mountainous Tanzawa region). PhD thesis, Agricultural Research Department of Kyoto University.

Furuta Hikaru and Nakamura Hidekichi, 1968, *Kagaku no hōhō* (Methods of science), Tokyo: Iwanami Shoten.

Galpin, Charles Josiah, 1918, *Rural Life*, New York: Century.

Galpin, Charles Josiah, 1924, *Rural Social Problems*, New York: Century.

Galpin, Charles Josiah, 1980, *The Social Anatomy of an Agricultural Community*, Ann Arbor: Michigan University Microfilm.

Gerth, H. H., and C. Wright Mills (eds and trans), 1948, *From Max Weber*, London: Routledge and Kegan Paul.

Gibson, James Jerome, 1979, *The Ecological Approach to Visual Perception*, Boston: Houghton Mifflin.

Gildemeister, Reinhard, 1973, *Landesplanung*, Braunschweig: Westermann.

Goethe, Johann Wolfgang, 1794–96 *Wilhelm Meisters Lehrjahre,* München: Hanser.

Gotō Shigeru and Sakamoto Keiichi, 1975, *Ōwen San-Shimon Fūrie* (Owen, Saint-Simon, Fourier), Tokyo: Chūō Kōronsha.

Grigg, D., 1995, *An Introduction to Agricultural Geography*, London: Routledge.

Habermas, Jurgen, 1969, *Technik und Wissenschaft als 'Ideologie'*, Frankfurt am Main: Suhrkamp.

Harlan, Jack R., 1984, *Sakumotsu no shinka to nōgyō shokuryō* (Crop evolution and agriculture and foods), Nagoya; Japan: Gakkai Syuppan Centre.

Hartshorne, C. and P. Weiss, 1960, *The Collected Papers of Charles Sanders Peirce*, vol. 6, Cambridge: Balknap.

Hartshorne, Richard, 1939, *The Nature of Geography*, Washington, DC: The Association of American Geography.

Hasebe Tadashi (ed.), 1999, *Report: Nōgyō Keizai Rinrigaku no Kōchiku* (Establishing Environmental Economic Ethics), Sendai: Tōhoku University Press.

Hasegawa Akihiko, 1986, *Nōson no kazoku to chūiki shakai* (Farm families and community), Tokyo: Ochanomizu Shobō.

Haushofer, Heinz, 1963, *Die deutsche Landwirtschaft im technischen Zeitalter*, Stuttgart: E. Ulmer.

Hayashi Tatsuo and Shimonaka Hiroshi (eds), 1971, *Tetsugaku jiten* (A dictionary of philosophy), Tokyo: Heibonsha, revised edition.

Hayashi Yoshihiro, 1995, *Inu ga uttaeru shiawase na seikatsu: Wakattekudasai! Nani wo kangae, nani wo nozondeiru ka* (The happy life that dogs long for: Please understand us! What dogs think and what they wish), Tokyo: Kōbunsha.

Heissenhuber, Alois (ed.), 1994, *Landwirtschaft und Umwelt*, Bonn: Economica.

Hinoue Hei'ichirō, 1966, *Mujinzō* (Inexhaustible), self-published.

Hirasawa Hisao, 1996, *Dōbutsu ni naniga ohite iruha* (What happened to animals), Tokyo: Sanichi Shobō.

Hishinuma Tatsuya, 1973, *Watashi no nōgaku gairon – Nōgaku ni okeru ningen no kenkyū* (My own introduction to agricultural science: Researching human beings in agriculture), Tokyo: Nōsan Gyoson Bunka Kyōkai.

Honda Yukio, 1982, *Suiden wa chikyū o sukuu* (Paddy fields to the rescue), Tokyo: Ienohikari Kyoukai.

Hoshi Makoto, 1998, 'Nōgyou to kōgyou no sai ni tsuite (On the difference between agriculture and manufacturing)', *Nōson kenkyū* (Rural studies), No. 88, March.

Hoshikawa Kiyochika, 1987, *Saibaishokubutsu no kigen to denpa* (The origins and propagation of cultivated plants), revised and enlarged edition, Tokyo: Ninomiya Shoten.

Howard, Ebenezer, 1965, *Garden Cities of To-Morrow*, London: Faber & Faber.

Hübler, Karl-Heimann, 1987, *Raumordnungspolitik und Wertwandel, Beiträge-Akademie für Raumforschung und Landesplanung*, Band 103, Hannover: Vincentz.

Ikegami Kōichi, 1991, *Nihon no mizu to nōgyō* (Water and agriculture in Japan), Tokyo: Gakuyō Shobō.

Ikegami Kōichi, 1996, 'Nōgyō kaihatsu purojekuto to ODA wo shiraberu (Examining the agricultural development project and ODA)', in Satō Makoto (ed.), *Chiiki kenkyū chōsahō wo manabu* (Studying the methods of local research), Kyoto: Sekaishisōsha.

Imai Kenichi, 1984, *Jōhō nettowāku shakai* (Information network society), Tokyo: Iwanami Shinsho.

Imanishi Kinji, 1972, *Seibutsu no sekai* (The biological world), Tokyo: Kōdansha.

Imanishi Kinji, 1994, *Seibutsushakai no ronri* (The logic of biological societies), Tokyo: Heibonsha.

International Egg Commission, 1999, Report: *Welfare of Living Hens in the European Union*, London: IEC.

International Farm Management Association, 1980, *Report: The Role of Agriculture in Society.*

Īnuma Jirō, 1985, *Nōgyō kakumei no kenkyū: Kindai nōgaku no seiritsu to hatan* (The research of the agricultural revolution: The development and failure of modern agricultural science), Tokyo: Nōbunkyō.

Isbary, Gerhard, 1965, *Zentrale Orte und Versorgungsbereiche* (Central Place and Service Areas), Bad Godesberg: BfLR.

Isobe Toshihiko, 1993, *Kiki ni okeru kazokukeie I* (Family farming in crisis), Tokyo: Nippon Keizai Hyōronsha.

Itō Kunitake, 1985, *Pāsu no puragumatizumu* (The pragmatism of Peirce), Tokyo: Keisōshobō.

Itō Mitsuharu, 1962, *Keinzu* (Keynes), Tokyo: Iwanami Shoten.

Itō Motoshige and Ōyama Michihiro, 1985, *Kokusai bōeki* (International trade), Tokyo: Iwanami Shoten.

Itō Shuntarō and Kaneko Tsutomu (eds), 1992, *Kagaku to wa nanika: 12 no fōkasu* (On science: twelve points of focus), Tokyo: Hokuju Shuppan.

Iwakata Isoo, 1977, *Nōrinsuisan genron* (The basic principles of agriculture, forestry and fisheries), Tokyo: Yōkendō.

Iwatsuki Kunio, 1993, *Tayōsei no seibutsugaku* (Biology of diversity), Tokyo: Iwanami Shoten.

Izumi Mari, 1989, *Eikoku no nōgyō kankyō seisaku* (Agricultural environment policies in the UK), Osaka Fumin Kyōkai.

Jaspers, Karl, 1976, *Was ist Philosophie? Ein Lesebuch*, München: Piper.

Johnston, Ronald, Gregory Derek and David Marshall Smith (eds), 1994, *The Dictionary of Human Geography*, Oxford: Blackwell.

Kamisato Isao, Kiyonari Tadao and Masamura Kimihiro (eds), 1978, *Keizaigaku no furontia* (The frontier of economics), Tokyo: Tōyō Keizai Shinpō Sha.

Kamiya Keiji and Tōbata Seiichi (eds), *Gendai Nihon no nōgyō to nōmin* (The agriculture and farmers of modern Japan), Tokyo: Iwanami Shoten.

Kanemitsu Hideo, 1991, *Kokusai keizai seisaku: Samitto, hogoshugi, gatto no sekai* (International economic policy: The world of summits, protectionism, and GATT), Tokyo: Tōyō Keizai Shinpōsha.

Karakida Kenichi, 1995, *Riron no sōzō to sōzō no riron* (The creation of theory and the theory of creation), Tokyo: Asakura Shoten.

Kashiwa Sukekata, 1957, *Shihon shugi no mekanizumu* (The mechanism of capitalism), Tokyo: Sekishoin.

Kashiwa Sukekata, 1962, *Nōgaku genron* (The philosophy of agricultural science), Tokyo: Yōkendō.

Kashiwa Sukekata, 1980, *Nōgaku no yukue* (The future of agricultural science), Tokyo: Fumin Kyōkai.

Kashiwa Sukekata, 1968, *Kiki no rekishi kan* (Historic views on crises), Tokyo: Miraisha.

Katō Hisatake, 1991, *Kankyō rinrigaku no susume* (Recommended readings for environmental ethics), Tokyo: Maruzen.

Katō Ichirō, 1986, *Kyōiku to nōson* (Education and farming villages), Tokyo: Chikyūsha.

Katō Kunioki (ed.), 1991, *Shizenkagaku gairon* (An introduction to the natural sciences), Tokyo: Aoki Shoten.

Katō Masashi (ed.), 1994, *Kisei kanwa no keizaigaku* (Economics of de-regulation), Tokyo: Tōyō Keizai Shinpōsha.

Katō Hisatake (ed.), 1998, *Kankyō to rinri* (Environment and ethics), Tokyo: Yūhikaku.

Kawaguchi Akira, 1995, *Ie to mura* (Family and village society), Tokyo: Nōbunkyō.

Kawaguchi Hiroshi, 1977, *Keinzu ippan riron no kiso* (Foundations of the general theory of J. M. Keynes), Tokyo: Yūhikaku.

Kawai Kazuyuki, 1988, *Kokusaika jidai no shokuryō jukyū* (Supply and demand of food products in the age of internationalization), Tokyo: Nōbunkyō.

Kawai Masao and Hanihara Kazurō (eds), 1995, *Dōbutsu to bunmei* (Animals and civilization), Tokyo: Asakura Shoten.

Kawakita Jirō, 1969, *Hassōhō* (The way of ideas), Tokyo: Chūō Kōronsha.

Kawamoto Akira, 1972, *Nihon nōson no ronri* (Logic of Japanese farming villages), Tokyo: Ryūkeishōsha.

Kawano Kenji, 1967, *Keizaigaku nyūmon* (Introduction to economics), Kyoto: Minerva Shobō.

Keynes, John Maynard, 1927, *The End of Laissez-fair*, London: L & Virginia Woolf.

Keynes, John Maynard, 1936, *The General Theory of Employment, Interest, and Money*, London: Macmillan.

Kikuchi Takurō, 2000, *Nōgaku no yagaikagaku teki hōhō* (Method of agricultural science as a field of science), Tokyo: Nōbunkyō.

Kiley-Worthington, Marthe, 1993, *Eco-agriculture: Food First Farming: Theory and Practice*, London: Souvenir Press.

Kitahara Atsushi, 1996, *Kyōdōtai no shisō* (Idea of community), Kyoto: Sekaishisōsha.

Kiuchi Shinzō, 1968, *Chiiki gairon* (An introduction to regional study), Tokyo: Tokyo Daigaku Shuppankai.

Kobayashi Sumie, 1991, *Rōsaku kyōiku shisōshi* (The history of thought on work education), Tokyo: Tamagawa University Press.

Kobayashi Yasuo and Funabiki Takeo, 1994, *Chi no gihō: Tōkyō daigaku kyōyō gakubu kisoenshū tekisuto* (Academic skills for the human sciences: Guide for first-year students, University of Tokyo), Tokyo: Tokyo University Press.

Kokuritsu Rekishi Minzoku Hakubutsukan (ed.), 1997, *Dōbutsu to ningen no bunkashi* (Cultural history of animals and humans), Tokyo: Yoshikawa Kōbunkan.

Kondō Tadashi, 1953, *Shinnen no hyakushō – Hinoue, Hei'ichirō-shi no nōgyō keiei to gijutsu* (Farming with conviction: Hinoue Hei'ichirō's agricultural management and technique), Ōsaka: Fuminsha.

Kroner, G., 1964, *Die zentralen Arte in Wissenschaft und Raumforschungspolitik*, edited by Institut für Raumforschung, Inforamtionen, No.13.

Kropotkin, Peter, 1968, *Fields, Factories and Workshops*, New York: Benjamin Blom.

Krzymowski, Richard, 1919, *Philosophie der Landwirtschaftslehre*, Stuttgart: Eugen Ulmer.

Kuhn, Thomas S., 1962, *The Structure of Scientific Revolutions,* Chicago: University of Chicago Press.

Kuki Kazuto, 1989, *Shin Kanto gakuha no kachitetsugaku – taikei to sei no hazama* (The value philosophy of the new Kantian school: Between systems and life), Tokyo: Kōbundō.

Kunigō Shigeru, 1960, *Nishi Doitsu no nōgyō keizai* (The agricultural economy of West Germany), Tokyo: Council for the Enhancement of Productivity in the Agriculture, Forestry and Fishing Industries.

Kurihara Toshichirō, 1964, *Tōyō no kome, seiyō no komugi* (The East's rice, the West's wheat), Tokyo: Tōyō Keizai Shinpōsha.

Kurita Kenzō and Ueyama Shunpei (eds), 1968, *Tetsugaku 9: Kachi* (Philosophy vol. 9: Values), Tokyo: Iwanami Shoten.

Kurosaka Hiroyuki, 1990, *Seikatsu no naka no kankyō mondai* (Environmental problems in daily life), Tokyo: Azusa.

Kyūma Kazutake and Soda Osamu (eds), 1995, *Nōgyō to kankyō* (Agriculture and environment), Ōsaka: Fumin Kyōkai.

Kyūma Kazutake, 1997, *Shokuryou Seisan to Kankyō* (Food Production and Environment), Kyoto: Kagaku Doujin.

La Mettrie, Julien Offray de, 1747, *L'homme-machine*, Paris: Denoel/Gonthier. Translated by Sugi Toshio 1957, *Ningen kikairon* (The theory of the human machine), Tokyo: Iwanami Shoten.

Leiss, William, 1976, *Limits to Satisfaction*, Tront: University of Tront Press.

List, Friedrich, 1917, *On the Principles of Political Economy and Taxation,* 1st ed., London.

Loomis, R. S. & Connor, D. J., 1992, *Crop Ecology*, Cambridge: Cambridge University Press.

Lovelock, Ephraim James, 1979, *Gaia: A New Look at Life on Earth*, Oxford: Oxford University Press.
Mainzer, Klaus, 1994, *Thinking in Complexity: The Complex Dynamics of Matter, Mind, and Mankind*, Berlin: Springer.
Malthus, Thomas Robert, 1798, *An Essay on the Principle of Population as it Affects the Future Improvement of Society* and *Principles of Political Economy*, London, 1st ed. Johnson.
Marshall, Alfred, 1898, *Principles of Economics*, London: Macmillan.
Maruyama Takashi, 1985, *Ningenkagaku no hōhōronsō* (The debate on the methods of human science), Tokyo: Keisōshobō.
Maruyama Takashi, Ogawa Tadashi and Noe Kei'ichi (eds), 1987, *Chi no riron no genzai* (The current theories of knowledge), Kyōto: Sekai Shisōsha.
Matsumoto Tadao, 1993, *Seitai to kankyō* (Ecology and environment), Tokyo: Iwanami Shoten.
Matsuo Takane, 1974, *Kankyō nōgaku gairon* (An introduction to environmental agricultural science), Tokyo: Nōsan Gyoson Bunka Kyōkai.
Meadows, Donella, 1972, *The Limits To Growth: A Report For The Club Of Rome's Project On The Predicament Of Mankind*, London: Potomac Assoc.
Meadows, Donella, Dennis L. Meadows, Jorgen Randers, 1992, *Beyond the Limits: Global Collapse or A Sustainable Future*, London: Earthscan.
Meguri Masatami, 1963, *Dyūi kenkyū: Kasetsu to tankyuu no shisō taikei* (Studying Dewey: The philosophical system of hypothesis and investigation), Tokyo: Shunjusha.
Meguri Masatami, 1968, *Ba no tetsugaku to gendai shakaigaku – Tetsugaku kyōikugaku kara shakaigaku e* (The philosophy of place (circumstances) and modern sociology: From philosophy and education to sociology), Tokyo: Shunjūsha.
Miki Kiyoshi, 1966–86, *Miki Kiyoshi zenshū* (The complete works of Miki Kiyoshi), vols. 1–20, Tokyo: Iwanami Shoten.
Mimura Kō, 1980, *Kachiku kanrigaku* (The study of livestock control), Tokyo: Yōkendō.
Ministry of Agriculture, Forestry and Fisheries, Japan, 1991, *An Outline of the Results of the 1990 Agricultural Census*, Tokyo.
Ministry of Agriculture, Forestry and Fisheries, Japan, 1999, *A Report by the Committee Contemplating the Direct-payment Scheme for the Mountainous Areas*, Tokyo.
Ministry of Environment, 1996, *Nettairin no genshō* (Decrease of tropical forest), Tokyo: Chūō Hōki.
Misawa Katsue, 1941, *Fūdo sangyō* (The climate industry), Nagano: Shinano Mainichi Shinbunsha.

Misch, Georg, 1947, *Die Idee der Lebensphilosophie in der Geistwissenschaften*, Frankfurt: Von Lebens-und Gedankentrei.

Miyanaga Masao and Soda Osamu, 1981, *Keizai seisaku* (Economic policy), revised edition, Kyoto: Keibunsha.

Mizuta Hiroshi and Tamanoi Yoshirō, 1978, *Keizai shisōshi dokuhon* (A guidebook to the history of economic thought), Tokyo: Tōyō Keizai Shinpō Sha.

Mori Wataru (ed.), 1987, *Baiotekunorojī to shakai* (Biotechnology and society), Tokyo: Tokyo University Press.

Morinaga Toshitarō, 1951, *Nōgakukō* (On agricultural science), Tokyo: Yōkendō.

Morita Shirō, 1971, *Nōgyō wa nōgyō de aru: Kindaika ron no sakuryaku* (Agriculture is agriculture: The tactics of modernization theory), Tokyo: Nōsan Gyoson Bunka Kyōkai.

Murakami Kazuo, 1984, *Baiotekunorojī: Idenshikōgaku ga hiraku sentangijutsu* (Biotechnology: The possibilities in high technology that genetic engineering creates), Tokyo: Kōdansha.

Murakami Yōichirō, 1986, *Kindaikagaku wo koete* (Transcending modern science), Tokyo: Kōdansha.

Murata Haruo, 1994, 'Ippan shisutemu riron ni okeru yūkitai no shisō (The organismic philosophy in general systems theory)', *Gendai shisō 12 – Seimei to shisutemu no shisō* (Contemporary philosophy 12: The philosophy of life and systems), Tokyo: Iwanami Shoten.

Murdock, G. P., 1971, 'How Culture Changes?' in H.L. Shapiro (ed.), 1960, *Man, Culture and Society*, New York: Oxford University Press.

Mutai Risaku and Kozai Yoshishige, 1967, *Tetsugaku no kadai* (Issues in philosophy), Tokyo: Iwanami Shoten.

Nagai Hiroshi, 1987, *Kagaku gairon: Kagaku no tetsugaku* (An introduction to science: The philosophy of science), Tokyo: Sōbunsha.

Nakamura Hisashi, 1993, *Chiiki jiritsu no keizaigaku* (Economics for independ-ent communities), Tokyo: Nippon Hyōron Sha.

Nakao Sasuke, 1966, *Saibaishokubutsu to nōkō no kigen* (The origins of cultivated plants and agriculture), Tokyo: Iwanami Shoten.

Nakayama Ichirō, 1939, *Hatten katei no kinkō bunseki* (The equilibrium analysis of developmental processes), Tokyo: Iwanami Shoten.

National Land Agency of Japan, 1972, 1993, *Kasotaisaku no genjō* (The present state of counter-depopulation policy), Tokyo.

National Research Council (ed.), 1989, *Alternative Agriculture*, Washington, DC: National Academy of Sciences.

Negishi Takeo, 1993, 'Chiiki nōgyō shinkō to toshika no sōkankankei' (The relationship between the promotion of local agriculture and urbanization), in *Kinkiken ni okeru toshi, nōson kankei no atarashii tenkai* (The new developments in rural-urban relations in the Kinki district), No. 2, Nōgakugenron Research Laboratory.

Nei Masahiro, 1999, *21 seiki no keizaigaku: Shijo shugi o koete* (Economics of the 21st century: After laissez-faire), Kōdansha.

Nihon Keizai Gakkai Rengō, 1975, *Keizaigaku no dōkō* (Trends in economics), Tokyo: Tōyō Keizai Shinpōsha.

Nishida Kitarō, 1953, *Tetsugaku gairon* (An introduction to philosophy), Tokyo: Iwanami Shoten.

Nishio Toshihiko, 1996, *Baiotekunorojī no nōgyō tetsugaku: Chiiki no kosei wo ikasu* (Philosophy of biotechnology in agriculture: Utilizing the special characteristics of the local society), Tokyo: Nōbunkyō.

Nishio Toshihiko, 1998, *Nōgyō gijutsu o tsukutta hitotachi* (The people who created agricultural technology), Tokyo: Ie no Hikarikyōkai.

Nitobe Inazō, 1905, *Nōgyō honron* (The discourse on agriculture), Tokyo: Shokashobō.

Noda Matao, 1963, *Tetsugaku nyūmon* (An introduction to philosophy), Kyoto: Mineruva Shobō.

Noguchi Yakichi, 1950, *Nōgaku gairon* (An introduction to agricultural science), Tokyo: Yōkendō.

Noguchi Yakichi (ed.), 1987, *Nōgaku daijiten* (Comprehensive dictionary of agricultural science) Tokyo: Yōkendō.

Nojiri Taketoshi, 1995, 'Shakaiteki shijō keizai' (Social market economy), in *Ōsaka gakuin daigaku keizairon shū* (Osaka Gakuin University compilation on economics), vol. 9, number 3, pp. 109–55.

Nurkse, Ragner, 1953, *Problems of Capital Formation in Underdeveloped Countries*, Oxford: Blackwell.

Oguni Hiroshi, 1978, *Kaseigaku genron shinkō* (New principles of domestic science), Tokyo: Meibun Syobō.

Okada Tokindo, 1994, *Seimeitai no kagaku: Tekunorojī to bunka* (Organic science: Technology and culture), Kyōto: Jinbunshoin.

Omodaka Hisayuki, 1945, *Igaku gairon* (An introduction to medical science), parts 1 and 2, Tokyo: Seishin Shobō.

Omodaka Hisayuki, 1959, *Igaku gairon* (An introduction to medical science), Tokyo: Seishin Shobō, part 3.

Omodaka Hisayuki, 1960, *Igaku gairon* (An introduction to medical science), Tokyo: Sōgen Shinsha.

Omodaka Hisayuki, 1987, *Igaku gairon to wa* (An introduction to medical science), Tokyo: Seishin Shobō.

Omodaka Hisayuki, 1989, *Igaku no tetsugaku* (The philosophy of medical science), Tokyo: Seishin Shobō, revised edition.

Ōmura Hideaki and Hōzuki Makoto, 1979, *Itsudatsu no shakaigaku* (The sociology of deviancy), Tokyo: Shinyōsha.

Ōnishi Takeo, 1992, *Doitsu no keizai: Shakaiteki shijō keizai no kōzō* (Economy of Germany: Structure of the social market economy), Tokyo: Waseda Daigaku Shuppanbu.

Ōtsuka Hisao, 1955, *Kyōdōtai no kiso riron* (Basic theory of communities), Tokyo: Iwanami Shoten.

Ōtsuki Masao, 1941, *Nōgyō rōdō ron* (Theory of agricultural labor), Tokyo: Nishigahara Kankōkai.

Owen, Robert, 1972, *A New View of Society*, New York: AMS Press.

Parry, Martin, 1990, *Climate Change and World Agriculture*, London: Earth-scan.

Parsons, T., 1977, *The Social System*, London: Routledge and Kegan Paul.

Pestalozzi, Johann Heinrich, 1801, *Wie Gertrud ihre Kinderlehrt*, Zürich: Roscher.

Phannschmidt, M., 1970, 'Robert Schmidt', *Raumforschung und raumordnung*, no.2, Hannover.

Pigou, Arthur Cecil, 1920, *The Economics of Welfare*, New Brunswick: Transaction.

Pilz, F., 1974, *Das System der sozialen Marktwirtschaft*, München: Reinhardt.

Polanyi, Karl, edited by Tamanoi Yoshirō, 1975, *Keizai no Bunnmeishi* (The economic aspects of the history of civilization), Tokyo: Nihonkeizai Shinbunsha.

Polanyi, Karl, 2002, *Economy and Society*, Montreal: Black Rose Books.

Priebe, Hermann, 1970, *Landwirtschaft in der sozialen Marktwirthschaft*, Düsseldorf: Ecom Verlag.

Raeburn, J. R., 1984, *Agriculture: Foundations, Principles and Development*, Chickester, NY: J. Wiley.

Relph, Edward, 1976, *Place and Placelessness*, London: Pion.

Ricardo, David, 1815, *An essay on the influence of a low price of corn on the profits of stock, shewing the expediency of restrictions on importation: with remarks on Mr Malthus' last two publications*, London: John Murray.

Rickert, Heinrich, 1898, *Kulturwissenschaft und Naturwissenschaft: ein Vortrag*, Freiburg: Mohr.

Ricoeur, Paul, 1977, *Expliquer et Comprendre*, translated by Kume Hiroshi, 1985, *Kaishaku no kakushin* (The reform of interpretation), Tokyo: Hakusuisha.

Robinson, Joan Violet, 1971, *Economic Heresies: Some old-fashioned questions in economic theory*, New York: Basic Books.

Rogers, E. M., 1962, *Diffusion of Innovations*, New York: Free Press of Glencoe.

Rogers, E. M., 1971, *Communication of Innovations*, New York: Free Press of Glencoe.

Röpke, Wilhelm, 1996, *The Moral Foundation of Civil Society* (*Civitas humana*), New Brunswick: Transaction.

Rousseau, Jean-Jacques, 1972, *Emile*, Braunschweig: Schulbuchhandlung.

Rümker, Kurt Heinrich Theodor von, 1905, *Landwirtschaft und Wissenschaft: Ein offenes Wort zur Klarung der Lag*, Berlin: P. Parey.

Sakai Junichi, 1981, *Chiiki fukugō nōgyō no riron to jissen* (Theories and practice of regional composite agriculture), Tokyo: Ienohikari Kyōkai.

Sakamoto Keiichi (ed.), 1989, *Ningen ni totte nōgyō towa* (What agriculture is to man), Tokyo: Gakuyō Shobō.

Sakamoto Keiichi and Takayama Toshihiro, 1983, *Chiiki nōgyō no kakushin* (Reform of regional agriculture), Tokyo: Meibun Shobō.

Sakamoto Keiichi, 1975, 'Ōyō kagaku no hōhō 3 (Methods of applied science, vol. 3)', *Fukui kenritsu daigaku ronshū* (Fukui Prefectural University thesis collection), no. 5.

Sakamoto Keiichi, 1981, 'Nōgyō ni okeru kachi no mondai (The problem of values in agriculture)', *Nōringyō mondai kenkyū* (Journal of rural problems), vol. 15, Issue 3, September.

Sakamoto Keiichi, 1989, *Ningen ni totte nōgyō to wa* (What is agricultural science for human beings?), Tokyo: Gakuyō Shobō.

Sakamoto Keiichi, Takayama Toshihiro and Soda Osamu (eds), 1986, *Chiiki sangyō fukugōtai no tenkai* (The development of regional industrial complex), Tokyo: Meibun Shobō.

Sakamoto Keiichi, 1986, 'Nōgaku no kenkyū hōhō (Research methods in agricultural science)', *Nōgaku genron kenkyū nō to*, vol. 3, Kyoto: Nōgaku Genron Kenkyūshitsu.

Sakamoto Kenichi, 1994, 'Ōyō kagaku no hōhō 2 (Methods of applied science 2)', *Fukui Kenritsu Daigaku ronshū* (Fukui Prefectural University's collection of theses), vol. 3, Fukui: Fukui Prefectural Universtiy.

Sakuta Keiichi, 1963, *Bunka to kōdō* (Culture and behavior), Tokyo: Bai-fūkan.

Samuelson, Paul A., 1995, *Economics: An Introductory Analysis*, New York: McGraw-Hill.

Sauer, Carl Ortwin, 1952, *Agricultural Origins and Dispersals*, New York: The American Geographical Society. Translated by Takeuchi Tsuneyuki and Saitō Kōkichi, 1960, *Nōgyō no kigen* (The origin of agriculture), Tokyo: Kokon Shoin.

Sawa Takamitsu, 1991, *Kore kara no keizaigaku* (Economics from now on), Tokyo: Iwanami Shinsho.

Schmidt, Robert, 1912, *Denkschrift betreffend grundsäze zur aufstellung eines general-siedlungsplanes für den regierungsbezirk Düsseldorf*, Düsseldorf.

Schteinhauer, Gerhard, 1967, *Robert Schmidt*, Köln: Westdeutscher Verlag.

Schumacher, E. F., 1974, *Small is Beautiful*, London: Blond and Briggs.

Schumpeter, J. A., 1939, *Business Cycles*, New York: McGraw-Hill.

Schumpeter, J. A., 1989, *The Theory of Economic Development*, New Brunswick/ London: Transaction.

Science Council of Japan, 1999, *Aratanaru kenkyū rinen wo motomete* (In search of new principles of research), Tokyo: Science Council of Japan.

Sekiya Shinichirō, 1992, *Suiden no hataraki* (The role of paddy fields), Tokyo: Ienohikari Kyōkai.

Shichinohe Chōsei, 1990, *Nōgyō no kyōikuryoku* (The educational power of agriculture), Tokyo: Nōbunkyō.

Shimizu Hiroshi, 1992, *Seimei to basho: Imi o sōshutsu suru kankeikagaku* (Life and place: The meaning-creating science of relationships), Tokyo: NHK Shuppan.

Shimizu Hiroshi, 1996, *Seimeichi toshite no ba no ronri* (Logic of *ba* as living knowledge), Tokyo: Chūō Kōronsha.

Shisō no Kagaku Kenkyūkai, 1950–1951, *Amerika shisōshi* (The history of American thought), Tokyo: Nihon Hyōronsha.

Sick, W-D., 1993, *Agrargeographie*, Köln: Westerman.

Singer, Charles Joseph (ed.), 1954, *A History of Technology*, vol. 1–2, Oxford: Clarendon Press.

Smith, Adam, 1776, *An Inquiry into the Nature and Causes of the Wealth of Nations*, Glasgow.

Smith, Philip E. L., 1976, *Food Production and its Consequences*, Menlo Park, California: Cummings.

Soda Osamu and Ōhara Kōtarō (eds), 1994, *Gendai Nihon no nōgyōkan: Sono genjitsu to tenbō* (Contemporary Japan's views on agriculture: Realities and trends), Ōsaka: Fumin Kyōkai.

Soda Osamu, 1975, *Soni, Mitsue chiiki no jittai to hatten hōsaku* (Actual situation of the Soni and Mitsue areas and measures for their development), Kyoto: Ryūkoku University.

Soda Osamu, 1980, *Chihōsangyō no shisō to undō* (Philosophy and movements of local industries), Kyoto: Minerva Shobō.

Soda Osamu, 1984, *Nishi Doitsu no chiiki keikaku – Toshi to nōson no ketsugō* (Regional planning in Western Germany: The combination of city and country), Tokyo: Taimeidō.

Soda Osamu, 1985, *Seiyōshokoku no kokudoseibi to nōsansonseibi* (The improvement of national land and agricultural and mountain villages in West European countries), Tokyo: Nōson Kaihatsu Kikaku Iinkai.

Soda Osamu, 1990, 'Nōringyō ni totte chiiki towa nani ka – Ba no nōgaku josetsu (What is community to agriculture and forestry? An introduction to the agricultural science of 'ba')', *Nōringyō mondai kenkyū* (Journal of rural problems), vol. 26, Issues 4 and 5.

Soda Osamu, 1991, 'Nōgyō no shakaiteki yakuwari to gaibu kouka (Social functions and external effects of agriculture)', *Nōgyō to keizai* (Agriculture and economics), vol. 57, Issue 13, December.

Soda Osamu, 1992, *Shimin nōen no susume* (The advancement of small citizen gardens), Iwanami Booklet.

Soda Osamu, 1993, *Nōson seikatsu no yutori no sōzō* (For the creation of relaxed country life), Tokyo: Nōson Seikatsu Sōgō Kenkyū Sentā.

Soda Osamu, 1994, 'Nōgaku no kachimokuhyō (The value objectives of agricultural science)', *Nōgakugenron kenkyū* (Researching the philosophy of agriculture), vol. 2, Nōgakugenron Kenkyūshitsu.

Soda Osamu, 1996, *Tanzania, Kironbero bonchi no inasaku nōson* (Rice-farming villages of Kironbero basin in Tanzania), Research series, No. 14, Tokyo: Japan Association for International Collaboration of Agriculture and Forestry.

Soda Osamu, 1997, *Toshi to nōson no ketsugō* (The combination of city and country), Tokyo: Taimeidō.

Soda Osamu, 1999, 'Kyōsei to wa nanika (What is symbiosis?)', *Gakujutsu no dōkō* (Trends in science), vol. 4, no. 7, Tokyo: Nihon Gakujutsu Kaigi (The Science Council of Japan), July.

Soda Osamu, 1999, 'Nōson chiiki seisaku no kadai (Issues in rural area policy)', *Nōson keikaku gakkaishi* (Journal of Rural Planning Association), vol. 17, no. 4, March.

Soda Osamu, Horiguchi Kenji and Yamaguchi Mitoshi, 1993, *Kokusai nōgyō funsō: Hogo to jiyū no hazamade* (International agricultural disputes: Between protectionism and freedom), Tokyo: Kōdansha.

Soda Osamu, Ōhara Kōtarō and Kako Toshiyuki, 1996, *Jizokuteki nōson no keisei – sono rinen to kanōsei* (The formation of a sustainable rural village: principles and possibilities), Osaka: Fumin Kyōkai.

Soda Osamu (ed.), 1998, *Daichi to ningen* (Humans and the earth), Tokyo: Hōsō Daigaku Kyōiku Shinkōkai.

Sorokin, Pitirim and Carle C. Zimmerman, 1929, *Principles of Rural-Urban Sociology*, New York: Henry Holt.

Souri Norio and Hattori Yoshitaka (eds), 1999, *21 Seiki no Keizai Seisaku* (Economic Policy in the 21st Century), Tokyo: Nihon Hyōronsha.

Stern, Alfred, 1962, *Philosophy of History and the Problem of Values*, Hague: Mouton.

Stratmann-Mertens, Eckhard (ed.), 1991, *Wachstum*, Frankfurt: Fischer.

Strecker, Reinhard, 1938, *Die philosophischen Probleme in der Forstwirtschaftslehre*, Berlin: Neumann.

Sugawara Tomota, 1993, *Nouringyou ga chikyuu wo sukuu kore dake no riyou* (The only reason that agriculture should save earth), Tokyo: Nōbunkyō.

Sumiya Kazuhiko, 1970, *Makkusu Vēbā: Gendai e no shisōteki shiza* (Max Weber: Modernity as seen from a philosophical perspective), Tokyo: Nihon Hōsō Shuppan Kyōkai.

Tajima Yoshihiro, 1994, *Kisei kanwa: Ryūtsū no kaikaku bijon* (Deregulation: A vision with regard to reform of distribution), Tokyo: NHK Bukkusu.

Takagi Jinzaburō, 1989, 'Ekorojī no kangaekata (Thinking ecologically)', in Uzawa Hirofumi (ed.), *Tenkanki ni ōkeru ningen* (Mankind in transition), *Vol.2: Shizen to ha* (What is nature?), Tokyo: Iwanami Shoten.

Takechi Akihiro (ed.), 1976, *Tetsugaku to gendai* (Philosophy and the modern era), Kyoto: Sanwa Shobō

Takemura Takuji, 1981, *Yao zoku no rekishi to bunka* (The history and culture of Yao), Tokyo: Kōbundō.

Takeuchi Kazuhiro and Tanaka Manabu (eds), 1998, *Seibutsu shigen no jizokuteki riyou* (The sustainable utilization of biological resources), Tokyo: Iwanami Shoten, SAFO Report.

Tamai Torao, 1980, *Nōgyō hogo no riron to seisaku* (Agricultural protection: Theory and policy), Tokyo: Rakuyō Shobō.

Tamanoi Yoshihirō, 1974, 'Doitsu keizaigaku no dentō (Traditions of German economics)', in *Shisō* (Thought), December issue, pp. 1–19.

Tamanoi Yoshirō (ed.), 1978, *Chiiki shugi* (Regionalism), Tokyo: Gakuyō Shobō.

Tamanoi Yoshirō and Hayasaka Tadashi (eds), 1978, *Keizaigakushi* (History of economic theory), Tokyo: Seirin Shoin Shinsha.

Tamanoi Yoshirō, 1977, *Chiiki bunken no shisō* (Thought of local decentralization), Tokyo: Tōyō Keizai Shinpō Sha.

Tamanoi Yoshirō, 1978, *Ekonomī to ekorojī: Kōgi no keizaigaku he no michi* (Economy and ecology: The road to more comprehensive economics), Tokyo: Misuzu Shobō.

Tamanoi Yoshirō, 1982, *Seimeikei no ekonomī: Keizaigaku, butsurigaku, tetsugaku eno toikake* (Living economics: Questioning economical and physical sciences and philosophy), Tokyo: Shinhyōron.

Tamura Shinpachirou, 1997, Report in: *Shizen to ningen wo musubu* (Connecting mankind with nature), Tokyo: Nōbunkyō.

Tanabe Hajime, 1918, *Kagaku gairon* (An introduction to science), Tokyo: Iwanami Shoten.

Tarde, G., 1903, *The Laws of Imitation* (*Les lois de l'imitation*). Translated by E.C. Parsons, New York: Henry Holt.

Teramoto Ei, 1989, 'Seibutsugaku – Seimei no shikumi to shinka (Biology: The mechanisms and evolution of life)', in Uzawa Hirofumi (ed.), *Iwanami kōza*

'Tenkanki ni okeru ningen' hen 2: Shizen to wa (Iwanami lecture series 'Humans in transition', volume 2: On nature), Tokyo: Iwanami Shoten.

Thiede, Günther, 1988, *Landwirt im Jahr 2000*, Frankfurt: DLG.

Thomas, Elizabeth Marshall, 1993, *The Hidden Life of Dogs*, Boston: Houghton Mifflin Co.

Thompson, P. B. (ed.), 1994, *Ethics, Public Policy, and Agriculture*, New York: Macmillan.

Thünen, Johann Heinrich von, 1910, *Der isolierte Staat in Beziehung auf Landwirtschaft und Nationalökonomie*, Jena: G. Fischer.

Tōbata Seiichi and Morinaga Shuntarou (eds), 1953–59, *Nihon nōgyō hattatsushi, zen 12 kan* (The history of development of Japanese agriculture, 12 volumes), Tokyo: Chūō Kōronsha

Tōbata Seiichi, 1936, *Nihon nōgyō no tenkai katei* (The development process of Japanese agriculture), Tokyo: Iwanami Shoten.

Tōbata Seiichi, 1938, *Nōgyo mondai no shosō* (Some aspects of the agrarian question), Tokyo: Iwanami Shoten.

Togari Yoshiji, 1957, *Sakumotsu shikenhō* (The crop testing method), Tokyo: Nōgyō Gijutsu Kyōkai.

Trusted, Jennifer, 1979, *The Logic of Scientific Inference: An Introduction*, London: Macmillan.

Tsukuba Hisaharu, 1969, *Beishoku nikushoku no bunmei* (The rice-eating and meat-eating civilizations), Tokyo: Nippon Hōsō Shuppan Kyōkai.

Tsuno Yukindo, 1975, *Nōgaku no shisō – Gijutsuron no genten o tou* (The thought of agricultural science: Questioning the original position of technology), Tokyo: Nōsan Gyoson Bunka Kyōkai.

Tsurumi Kazuko and Kawata Tadashi, 1989, *Naihatsuteki hattenron* (The theory of spontaneous development), Tokyo: Tokyo University Press.

Ueda Yoshisuke, Nishimura Kazuo and Inagaki Kūsaku, 1999, *Fukuzatsukei wo koete: kaosu hakken kara mirai he* (Transcending complexity: from the discovery of chaos towards the future), Tokyo: Chikumashobō.

Uemura Kenji (ed.), 1997, *Seitai kankyou kagaku gairon* (A brief treatise on environmental science), Tokyo: Kōdansha.

Ueyama Shunpei, 1963, *Benshōhō no keifu: Marukusushugi to puragumatizumu* (The genealogy of dialectics: Marxism and pragmatism), Tokyo: Miraisha.

Ueyama Shunpei, 1968, *Pāsu, Jeimuzu, Dyūi* (Peirce, James, Dewey), Tokyo: Chūō Kōronsha.

Uzawa Hirofumi (ed.), 1989, *Tenkanki ni okeru ningen 2: Shizen towa* (People in a transitional period, vol. 2: What nature is), Tokyo: Iwanami Shoten.

Uzawa Hirofumi (ed.), 1990, *Iwanami kōza 'Tenkanki ni okeru ningen' hen 6: kagaku to wa* (Iwanami lecture series 'Humans in transition', volume 6: On science), Tokyo: Iwanami Shoten.

Vidal de La Blache, Paul, 1922, *Principes de Géographie Humane*, Paris: Armoud Colin.

Wagner, Adolph, 1892 1893, *Lehr- und Handbuch der politischen Ökonomie*, Leipzig: Winter.

Waldrop, Mitchell, 1992, *Complexity: The Emerging Science at the Edge of Order and Chaos*, New York: Simon & Schuster.

Watabe Tadayo (ed.), 2000, *Gendai no nōkō jyōkyō wo tou* (Questioning the current state of farming), Tokyo: Taimeidō.

Watanabe Yukihiro, 1977, 'Bunka ni okeru shizen, doushokubutsu no shiiku, saibai to ningen (Nature, breeding of plants and animals, cultivation and humans in culture)', in Yajima Sadao (ed.), *Gendai no tetsugaku 4: Shizen to hanshizen* (Modern philosophy 4: Nature and anti-nature), Tokyo: Kōbundō.

Watsuji Tetsurō, 1935, *Fūdo: ningengakuteki kōsatsu* (Climate and culture: a humanistic investigation), Tokyo: Iwanami Shoten.

Weber, Heinrich Wilhelm von, 1919, *Grundlinien einer neuen Forstwirtschafts philosophie*, Tübingen: Laupp.

Weber, Max, 1904/1922, 'Die Objektivität sozialwissenschaftlicher und sozialpolitischer Erkenntnis', in *Gesammelte Aufsätze zur Wissenschaftslehre*, Tübingen: J. C. B. Mohr.

Weddigen, Walter, 1966 *Grundzüge der Allgemeinen Volkswirtschaftspolitik*. Translated by Yajima Kisoji and Watanabe Shigeru, 1979, *Gendai no keizai seisaku genri* (Contemporary principles of economic policy), Shunjusha.

Wehling, Hans-Werner, 1982, 'Probleme der Raumordnung am Beispiel des Verstädterungsprozesses ländlicher Gemeinden', *Geographische Zeitfragen*, Hannover: Schroedel Schulbuch Verlag, no.7.

Windelband, Wilhelm, 1911, *Geschichte und Naturwissenschaft*, Strassburg: Univeristy of Strassburg Press.

Wirth, Louis, 1938, 'Urbanism as a Way of Life', *American Journal of Sociology*, 44 (1): 1–24.

Worster, Donald, 1985, *Nature's Economy: A History of Ecological Ideas*, New York: Cambrige University Press.

Yagisawa Zenji, 1925, 'Nōgyō no tokushitsu ni kansuru ichi kōsatsu (An examination of the distinctive features of agriculture)', *Hōgaku Shinpō*, February Issue.

Yaguchi Yoshio, 1995, *Shokuryō to kankyō no seisaku kōsō* (Policy conceptions for food and environment), Tokyo: Nōrin Tōkei Kyōkai.

Yamamoto Ichirō, 1972, *Kūkyo to jikken: kindaikagaku no seishin to sono ronri* (Void and experiment: the spirit of modern science and its logic), Kyōto: Hōritsu Bunkasha.

Yanai Hiroyuki, 1992, *Kankyō tetsugaku: Yokubō no yasurakana nemuri*

(Environmental philosophy: Our desires rest in peace), Tokyo: Nikkan Kōgyō Shimbunsha.

Yokoi Tokiyoshi, 1925, *Yokoi Hakase zenshū* (The complete works of Dr. Yokoi), vol. 3, Tokyo: Yokoi Zenshū Kankōkai.

Yokokawa Hiroshi, 1996, 'Nihon Nōgyō Keizai Gakkai taikai hōkoku (A report for the Agricultural Economics Society of Japan Meeting)', *Nōgyō Keizai Kenkyū* (The Japanese journal of rural economics), vol. 67, Issue 2.

Yonemoto Shōhei, 1994, *Chikyū kankyō mondai to wa nani ka* (What are the earth's environmental problems?), Tokyo: Iwanami Shinsho.

Yoshida Kiichirō and Okabe Mamoru (eds), 1985, *Chiiki shakai nōgyō* (Community agriculture), Research Center of Nōrinchūkin Bank, Tokyo: Ie-no-hikari Kyōkai.

Yoshida Tamito, 1998, '21 seiki no kagaku (Science for the 21st century)', *Soshiki kagaku* (Organizational science), vol. 32, Issue 3.

Zenkoku Nōgyō Shinbun Newspaper, 9 December 1977, *The National Agricultural Newspaper.*

Index